Biliary Tract and Gallbladder Biomechanical Modelling with Physiological and Clinical Elements

Biliary Tract and Gallbladder Biomechanical Modelling with Physiological and Clinical Elements

Wenguang Li

CRC Press is an imprint of the
Taylor & Francis Group, an **informa** business

First edition published 2021
by CRC Press
6000 Broken Sound Parkway NW, Suite 300, Boca Raton, FL 33487-2742

and by CRC Press
2 Park Square, Milton Park, Abingdon, Oxon, OX14 4RN

CRC Press is an imprint of Taylor & Francis Group, an Informa business

Visit the Taylor & Francis Web site at
http://www.taylorandfrancis.com

and the CRC Press Web site at
http://www.crcpress.com

Library of Congress Cataloging-in-Publication Data
Names: Li, Wenguang, 1964– author.
Title: Biliary tract and gallbladder biomechanical modelling with physiological and clinical elements / Wenguang Li.
Description: First edition. | Boca Raton : CRC Press, 2021. | Includes bibliographical references and index. |
Summary: "Gallstone and other diseases of the biliary tract affect more than around twenty percent of the adult population. The complications of gallstones, acute pancreatitis and obstructive jaundice can be lethal. This is the first book to systematically treat biliary tract and gallbladder modelling with physiological and clinical information in a biomechanical context. The book provides readers with detailed biomechanical modelling procedures for the biliary tract and gallbladder based on physiological information, clinical observations and experimental data and with the results properly interpreted in terms of clinical diagnosis and with biomechanical mechanisms for biliary diseases. The book can be used as reference book for university undergraduates, postgraduates, professional researchers in applied mathematics, biomechanics, biomechanical engineering and biomedical engineering as well as related surgery doctors"—Provided by publisher.
Identifiers: LCCN 2020046935 (print) | LCCN 2020046936 (ebook) | ISBN 9780367722296 (hardcover) | ISBN 9781003153986 (ebook)
Subjects: MESH: Biliary Tract—metabolism | Gallbladder—metabolism | Models, Biological
Classification: LCC RC845 (print) | LCC RC845 (ebook) | NLM WI 750 | DDC 616.3/65—dc23
LC record available at https://lccn.loc.gov/2020046935
LC ebook record available at https://lccn.loc.gov/2020046936

ISBN: 978-0-367-72229-6 (hbk)
ISBN: 978-0-367-72233-3 (pbk)

Typeset in Minion Pro
by Apex CoVantage, LLC

To my parents, wife Yuan, son Zaiyang
and daughter Zifan

Contents

Preface

GALLSTONES AND OTHER DISEASES OF THE BILIARY TRACT AFFECT MORE THAN 10%–20% of the adult population. The complications of gallstones, acute pancreatitis and obstructive jaundice can be lethal, and patients with acalculous gallbladder pain often pose diagnostic difficulties. Moreover, surgery to remove the gallbladder in these patients gives variable results. Extensive research has been made to understand the physiological and pathological function of the biliary systems, but the mechanism of the pathogenesis of gallstones, pain production and gallbladder wall biomechanical property still remain poorly understood; especially, mathematical/biomechanical modelling of these aspects is lacking in the literature. There has not been a book to elaborate mathematical modelling of the human biliary system in terms of biomechanical context so far.

The author has paid attention to biomechanical modelling of the human biliary system since 2004. The tackled problems include one-dimensional rigid and elastic wall models of the human biliary system for Newtonian and non-Newtonian bile, three-dimensional fluid-structure interaction of cystic duct, biomechanical model for the pain experienced in patients based on *in vivo* gallbladder volume measured in response to a standard cholecystokinin stimulus, gallbladder wall anisotropic and heterogeneous properties identification, viscoelasticity modelling of intact gallbladder pressure-volume curve, constitutive law of gallbladder wall with damage effect, generation of three-dimensional geometrical models based on two ultrasound static images of human gallbladder as well as the corresponding finite element analysis, and bile flow through percutaneous transhepatic biliary drainage catheters.

This is the first book dealing with biliary tract and gallbladder from a biomechanical point of view in terms of analytical mathematical methods. There are comprehensive literature review and contemporary content as well as detailed biomechanical modelling procedures in each chapter. A broad content, such as bio-fluid mechanics and bio-solid mechanics, image processing, geometrical modelling and diagnosis-orientated interpretation, is embedded in the book to bridge the gap between applied mathematics and clinical practices. In the book, analytical methods are dominant and can be easily applicable in clinical practice.

The book can serve as a reference for university undergraduates, postgraduates, professional researchers in applied mathematics, biomechanics, biomechanical engineering and biomedical engineering as well as related surgery doctors.

Dr Wenguang Li

Nomenclature

a, a_1, a_2, a_3	Model constants	
A	Cross-sectional area of collapsed duct	m^2
A_1	Cross-sectional area of flow at point 1 in Figure 2.4	m^2
A_2	Cross-sectional area of the flow at point 2 in Figure 2.4	m^2
b, b_1, b_2, b_3	Model constants	
B	Constant to be determined	ml
c	Property constant of matrix in soft tissue	kPa
c_0, c_1, c_2, c_3, c_4	Model or empirical constants	
C	Compliance of gallbladder	ml/mmHg
$\mathbf{C}$	Cauchy-Green deformation tensor	
C_{DF}	Fluid friction loss coefficient in elastic cystic ducts	
C_{DR}	Fluid friction loss coefficient in rigid cystic ducts	
C_p	Inlet pressure coefficient of cystic duct	
D	Scalar damage variable	
D	Inner diameter of duct material	mm
D	Axis length of ellipsoid	mm
e	Velocity deformation rate or shear strain rate	1/s
E	Young's modulus or incremental Young's modulus of material	Pa or kPa
$\mathbf{E}$	Green-Lagrange strain tensor	
E_{cb}	Spring constant of cross bridges	Pa or kPa
EF	Bile ejection fraction	%
En	Elastic energy stored	J
f	Darcy friction factor	
f_0	Darcy friction factor for duct without baffles	
f_{app}	Apparent attached rate constant of two-state cross bridges	1/s
fun	Objective function	
F	Function of a variable	
$\mathbf{F}$	Deformation tensor	
g_{app}	Apparent detached rate constant of two-state cross bridges	1/s
G	Shear modulus	

h	Thickness of wall or baffle	mm
h_0	Artery wall thickness at zero pressure	mm
H	Baffle height	mm
I_1	Stretch invariant of matrix	
I_4, I_6	Stretch invariants of two families of collagen fibres	
$\mathbf{I}$	3 × 3 Identity tensor	
j	Number of node	
J	Total number of nodes	
k_1, k_2, k_3, k_4	Property constants of two families of fibres	
k, g	Gallbladder refilling rate constant	min^{-1}
$K_1, K_2, \ldots, K_7$	Rate constants of four-state cross bridge of smooth muscle contraction	
l_0	Artery length at zero pressure	mm
L	Length of duct or length of gallbladder muscle strip	mm
L_0	Length of gallbladder muscle strip in which a peak active stress is developed	mm
L_{12}	Arc length of the ellipse in the plane *x*-*y*	mm
L_{13}	Arc length of the ellipse in the plane *x*-*z*	mm
L_{23}	Arc length of the ellipse in the plane *y*-*z*	mm
L_k	Equivalent length due to minor pressure loss	m
m_1, m_2, m_3	Model parameters	
m_{1ADINA}, m_{2ADINA}	Parameters in the Carreau's model in ADINA	
m_d	Parameter for damage in matrix	
M	Fluid-structure interaction coefficient	
n	Number of baffles	
n_{4i}	Direction cosine of the first family of fibres, $i = 1, 2, 3$	
n_{6i}	Direction cosine of the second family of fibres, $i = 1, 2, 3$	
n_c	Critical number of baffles	
n_d	Parameter for damage in fibres	
N	Number of samples or total number of time steps	
p	Internal duct pressure	mmHg
p_e	External duct pressure	mmHg
P	Success rate of trials	
q	Auxiliary variable or constant	
Q	Bile flow rate	ml/min
r	Inner radius of duct, $r = \sqrt{A/\pi}$ or radial coordinate in a polar-coordinate system ($\theta - r$) or in a spherical system (φ, θ, r) or artery inner radius	mm
r_0	Artery inner radius at zero pressure	mm
r_1	Radius based on cross-sectional area of baffle clearance, $r_1 = \sqrt{A_1/\pi}$	m

R	Flow resistance	mmHg/ml/min
R^2	R-squared value in curve fitting	
Re	Reynolds number	
S	Ratio of the first principal stress in three-dimensional gallbladder to the stress in gallbladder ellipsoid model	
t	Time	min
$t_{1/2}$	Time at half gallbladder initial volume in emptying	min
t_1, t_2, t_3	Time moments at the start of infusion, the end of infusion and the end of withdrawal	min
t_e	Time for emptying in gallbladder pain test	min
t_f	Time for refilling in gallbladder pain test	min
t_i	Time for isometric contraction in gallbladder pain test	min
t^*	Time of chemical reaction of two-state cross bridges	s
T	Tension in muscle	g/cm^3
T_0	Constant in Hill's equation	g/cm^3
T_1	First principal tension in gallbladder wall	N/mm
U	Bile velocity in cystic duct, $u = Q/A$	m/s
V	Muscle contracting velocity or dimensionless gallbladder bile volume	
V	Gallbladder volume	ml
V_0	Gallbladder initial volume at emptying	ml
V_{3D}	Volume of gallbladder in three-dimensional model	ml
V_e	Gallbladder volume at emptying	ml
$V_{election}$	Gallbladder volume ejected in the emptying phase	ml
V_{el}	Ellipsoid model volume of gallbladder	ml
V_g	Bile volume discharged per gallbladder contraction	ml/min
V_{sc}	Volume of gallbladder based on sum-of-cylinders method	ml
V_t	Measured gallbladder volume at time moment t in emptying	ml
V_{total}	Total bile volume secreted by the liver	ml
w	Concentration of attached two-state cross bridges	
W	Work done the viscous response by loading	J
x, y, z	Three coordinates in Cartesian coordinate system (x, y, z)	mm
x_c, y_c	Two coordinates of a gallbladder cross section	mm
y_{neck}, z_{neck}	Two coordinates of the middle point of gallbladder neck in the y-z plane	mm
y_{bmax}, z_{bmax}	Two coordinates of the point at gallbladder bottom	mm
x_{surf}, y_{surf}, z_{surf}	Three coordinates of a gallbladder surface	mm
z_{rot}	z-coordinate of gallbladder cross section after rotation	mm
z_{rot1}, z_{rot2}	z-coordinates of the first and second points of gallbladder cross section from the neck after rotation	mm

Greek symbols

α	Area ratio, $\alpha = A / A_{eq}$	
β	Mean fibre angle measured from the circumferential direction of tube	deg
γ	Shear rate	1/s
Γ	Ratio of the work done on the viscous response in the hysteresis loop to the elastic energy	
$\delta(x)$	Distribution function of total number of attachments along cross bridges	
δ_0	Uniform distribution function of $\delta(x)$	
$\Delta D_1, \Delta D_2, \Delta D_3$	Peak displacements of three principal axes of ellipsoid gallbladder	mm
$\Delta D_r, \Delta D_\varphi, \Delta D_\theta$	Radial, circumferential and longitudinal displacements	mm
ΔL	Distance between two successive baffles in cystic duct	
Δp	Pressure drop or gallbladder pressure change in different time	Pa or mmHg
Δp_k	Minor pressure drop in cystic duct	Pa
Δp_{te}	Minor pressure drop in T-junction during emptying	Pa
Δp_{th}	Minor pressure drop in T-junction during refilling	Pa
Δr	Increase of artery radius due to increasing internal pressure	mm
Δx	Interval between two nodes	m
ΔV	Gallbladder volume change	
$\Delta \varepsilon$	Incremental strain	
$\Delta \sigma$	Incremental stress	kPa
$\Delta \chi_\sigma$	Change in χ_σ	%
$\varepsilon_1, \varepsilon_2$	first and second principal strains	
ε_{12}	Mean strain of arc length of the ellipse in the plane x-y	
ε_{13}	Mean strain of arc length of the ellipse in the plane x-z	
ε_{23}	Mean strain of arc length of the ellipse in the plane y-z	
ε_{ij}	Normal strain ($i = j$) or shear strain($i \neq j$), $i, j = 1, 2, 3$	
ζ	Parameter for damage in matrix	
η	Parameter for damage in fibres	
θ	Longitudinal coordinate in a spherical system (φ, θ, r)	rad
θ_b	Half central angle of baffle edge	rad
κ	Ratio of two principal axis lengths of ellipsoid gallbladder	
λ	Stretch ratio	s
μ	Bile dynamic viscosity	mPa·s
μ_0	Dynamic viscosity at zero shear rate	mPa·s
μ_c	Non-Newtonian coefficient of viscosity	mPa·s
μ_∞	Dynamic viscosity at infinite shear rate	mPa·s
ν	Bile kinematic viscosity, $\nu = \mu / \rho$	mm²/s
ξ	Dimensionless baffle height, $\xi = H / d_{CD}$	

ρ	Density of bile	kg/m^3
ρ_s	Density of solid	kg/m^3
σ	Normal stress	mmHg
σ_1, σ_2	first and second principal stresses	Pa or kPa or mmHg
$\sigma_{xx}, \sigma_{yy}, \sigma_{zz}$	Normal stresses in the x-, y- and z-axis directions	Pa
σ_φ	Hoop stress in common bile duct wall	kPa
τ	Shear stress	Pa or mmHg
υ	Poisson's ratio of elastic cystic duct or gallbladder wall	
φ	Circumferential coordinate in a spherical system (φ, θ, r)	rad
χ	Error in gallbladder volume between prediction and measurement	%
χ_1, χ_2	Errors in gallbladder volumes	%
χ_{seg}	Error in image segmentation	
χ_σ	Error in stress between model prediction and experiment	%
Ψ	Strain energy density function	kPa
ω	Gallbladder contraction frequency	min^{-1}
Ω	Constant in gallbladder p-V equation	ml

Superscripts

a	Active state of gallbladder smooth muscle
dam	Damage state
exp	Experimental
mod	Model prediction
p	Passive state of gallbladder wall tissue
vir	Virgin state
′	Dimensionless

Subscripts

1	Major axis of ellipsoid
1f	Refusion process in passive state
1w	Withdrawal process in passive state
2	Minor axis of ellipsoid
2f	Refusion process in active state of gallbladder
2w	Withdrawal process in active state of gallbladder
3	Shortest minor axis of ellipsoid
a	Active stress
A	Sample A
e	Elastic
eq	Equivalent
exp	Experimental
f	Fibre
b	Baffle

B	Sample B
c	Circumferential
d	Duodenum
dc	CCK decaying
EM	Emptying for 1D model
eq	Equivalent
F	Failure
fc	Circumferential direction of fibres
fl	Longitudinal direction of fibres
H	High value
i	Coordinate index, j = 1, 2, 3
in	Inlet of duct
j	Coordinate index, j = 1, 2, 3
l	Longitudinal
max	Maximum value
mc	Circumferential direction in matrix
min	Minimum value
ml	Longitudinal direction in matrix
n	Normal direction
out	Outlet of duct
r	Radial direction
ref	Reference
RF	Refilling
S	Success
w	At wall
φ	Circumferential direction
θ	Longitudinal direction

Abbreviations

1D	One-dimensional
2D	Two-dimensional
3D	Three-dimensional
^{99m}Tc-HIDA	^{99m}Technetium-labelled hepato imino diacetic acid
A	Actin (thin filament)
ADINA	Automatic dynamic incremental nonlinear analysis software
AM	Attached dephosphorylated cross bridge (latch bridge)
CBD	Common bile duct
CCK	Cholecystokinin
CD	Cystic duct
CFD	Computational fluid dynamics
CHD	Common hepatic duct
DAG	Diacylglycerol

EF	Ejection fraction
Exp	Experimental
FE	Finite element
FEA	Finite element analysis
FEM	Finite element method
FSI	Fluid-structure interaction
GB	Gallbladder
M	Detached dephosphorylated cross bridge
MLCK	Myosin light-chain kinase
MLCP	Myosin light-chain phosphatase
Mp	Detached phosphorylated cross bridge
PKC	Protein kinase C
VAS	Visual analogue scale

CHAPTER 1

Physiology of the Human Biliary System

THE HUMAN BILIARY SYSTEM

The human biliary system consists of the gallbladder (GB), cystic duct (CD) and common duct and sphincter of Oddi (Figure 1.1). The common duct includes both the common bile duct (CBD) and common hepatic duct (CHD). The GB is a thin-walled, pear-shaped sac and generally measures 7–10 cm in length and ~3 cm in width as well as 2.7-mm-thick wall (Oluseyi 2018). This muscular sac is located in a fossa in the posterior of the liver's right lobe. The GB is divided into three regions: (1) the fundus, (2) the body and (3) the neck. Its average storage capacity is about 20–30 ml (Dodds et al. 1989), depending on sex and age; for healthy men, mean GB volume is 18.7 ± 0.3 (Palasciano et al. 1992) and 33.3 ± 20.5 ml (Caroli-Bosc et al. 1999); for healthy women, 17 ± 0.3 (Palasciano et al. 1992), 27.1 ± 16.9 (Caroli-Bosc et al. 1999). At ages less than 50 years, the average GB volume is 25.6 ± 16.2 ml, otherwise 31.8 ± 19.6 ml (Caroli-Bosc et al. 1999).

The CD, about 3.5 cm long and 3 mm wide (Dodds et al. 1989), merges with the CHD. The mucosa of the proximal CD is arranged into 3–7 crescentic mucous membrane folds known as the spiral valves of Heister. For CDs without gallstones, their mean diameter is 2.63 ± 0.67 mm (Castelain et al. 1993). For men with healthy CDs, the CD lengths are 24.3 ± 9.0, 26.9 ± 5.5 and 29.1 ± 4.6 mm in age groups such as 10–20, 21–40 and

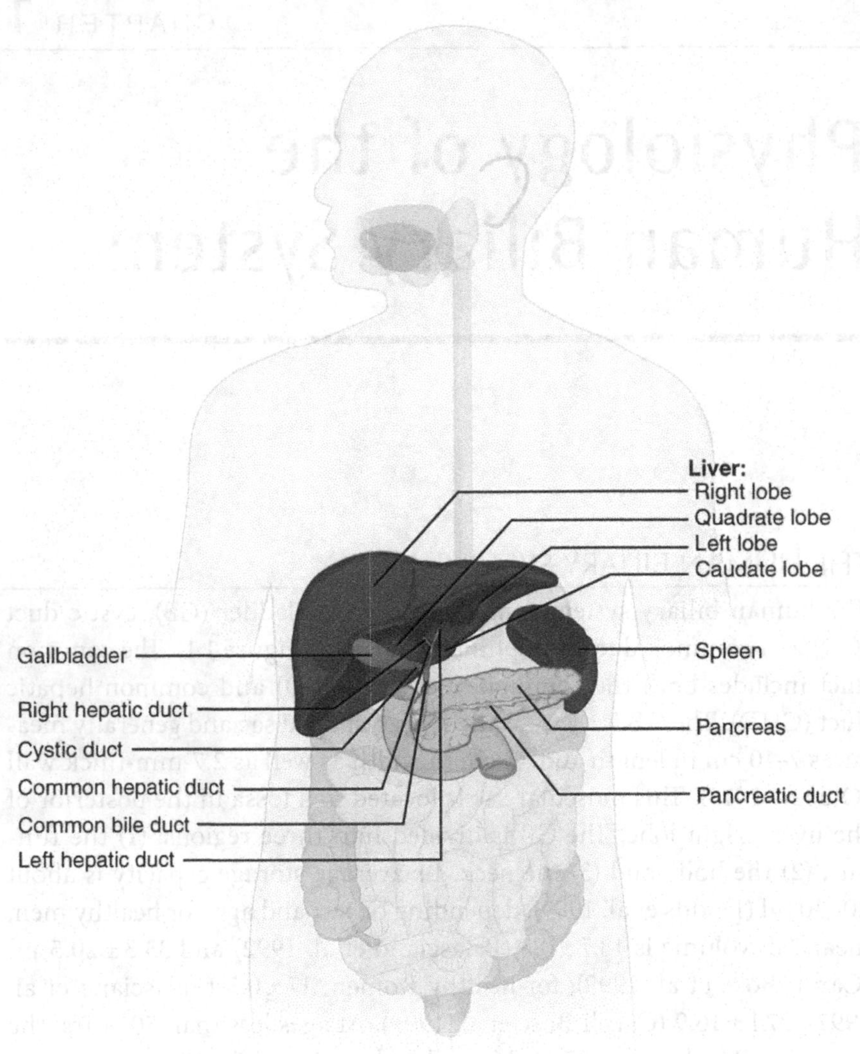

FIGURE 1.1 A view of the human biliary system (illustration after Schatz 2020).

41–60 years, respectively, while for women, the lengths are 24.8 ± 7.6, 28.0 ± 3.6 and 29.1 ± 6.2 mm in the same age groups (Nahar et al. 2011). Interestingly, the mean CD diameters of men are 1.47 ± 0.50, 2.07 ± 0.56 and 2.73 ± 0.59 mm, but the mean diameters of women are 1.70 ± 0.34, 2.20 ± 0.64 and 2.95 ± 0.60 mm in the same age groups, respectively (Nahar et al. 2011).

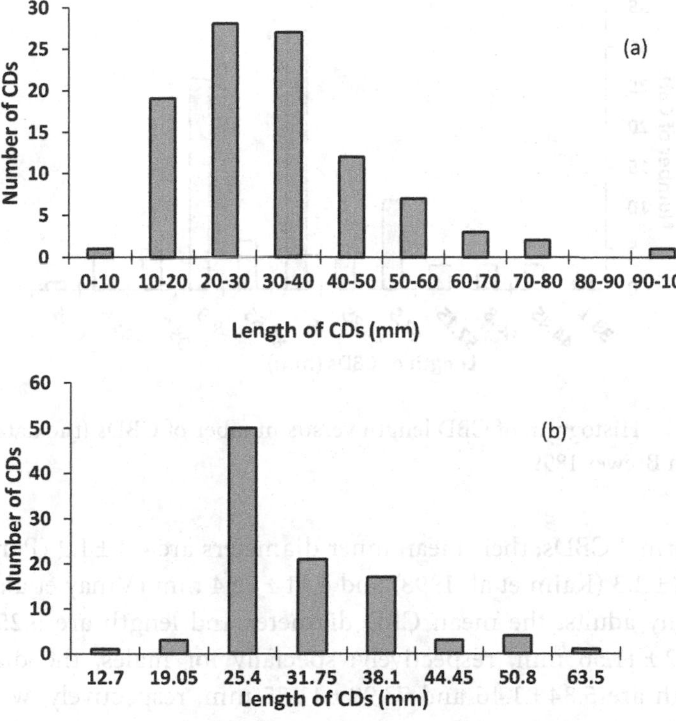

FIGURE 1.2 Histograms of CD length versus number of CDs: (a) the data from Lichtenstein and Ivy (1937) and (b) the data from Brewer (1899).

Two histograms of CD length are illustrated in Figure 1.2 based on observation data made by Lichtenstein and Ivy (1937) as well as Brewer (1899). Mostly, CD length is in a range of 20–40 mm.

The common duct is about 10–15 cm long and 5–15 mm wide, in which the CHD is ~4 cm long (Dodds et al. 1989). The CBD penetrates the wall of the duodenum to meet pancreatic duct at the duodenal ampulla. Early measurements illustrated normal CBD outer diameter was in a range of 4–17 mm with a mean of 8.85 mm (Ferris and Vibert 1959), and a slightly late examination indicated the CBD outer diameter was in a range of 4–12 mm with a mean of 7.39 mm and the CBD wall thickness varied from 0.8 to 1.5 mm with an average of 1.1 mm (Mahour et al. 1967). A histogram of CDB length versus the number of CBDs observed is illustrated in Figure 1.3.

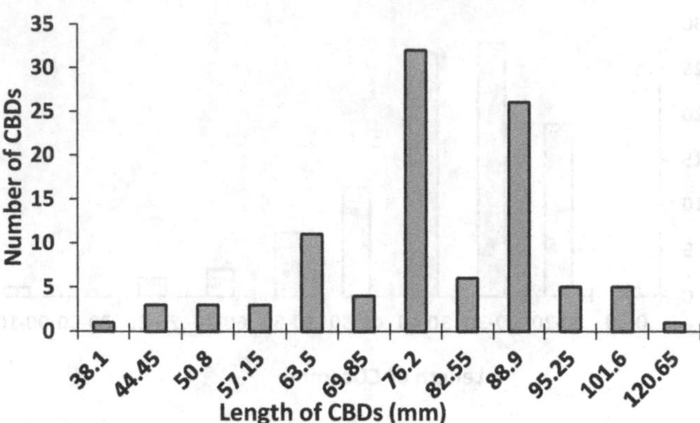

FIGURE 1.3 Histogram of CBD length versus number of CBDs (the data in figure is from Brewer 1899).

For normal CBDs, their mean inner diameters are 4.1 ± 1.2 (Parulekar 1979), 6.2 ± 2.3 (Kaim et al. 1998) and 4.11 ± 1.54 mm (Vinay et al. 2013). For healthy adults, the mean CBD diameter and length are 5.25 ± 1.28 and 72.02 ± 11.56 mm, respectively; specially for males, the diameter and length are 5.34 ± 1.46 and 73.90 ± 11.55 mm, respectively, while for females they are 5.10 ± 0.90 and 68.96 ± 11.00 mm (Blidaru et al. 2010). Mostly, CBD mean diameter can be correlated to age, when CBDs are in either healthy or diseased conditions (Adibi and Givechian 2006; Benjaminov et al. 2013; Chen et al. 2012; Lal et al. 2014; Perret et al. 2000; Nalaini et al. 2017; Peng et al. 2015; Senturk et al. 2012; Wu et al. 1984; Bachar et al. 2003). For example, $d_{\text{CBD}} = 0.0185 \times \text{age} + 2.42$, $\text{age} \in [60, 95]$ (Perret et al. 2000); $d_{\text{CBD}} = 0.055 \times \text{age} + 2.12$, $\text{age} \in [30, 80]$ (Nalaini et al. 2017); $d_{\text{CBD}} = 0.033 \times \text{age} + 2.624$, $\text{age} \in [5, 87]$ (Peng et al. 2015); $d_{\text{CBD}} = 0.06 \times \text{age} + 2.72$, $\text{age} \in [2, 80]$ (Wu et al. 1984); $d_{\text{CBD}} = 0.0262 \times \text{age} + 2.19$, $\text{age} \in [19, 90]$ (Ahmed 2017). However, two series of CBD measurements confirm mean CBD diameter have no correlation with age (Horrow et al. 2001; Karamanos et al. 2017).

Very recent work showed mean healthy CBD diameter is $d_{\text{CHD}} = 3.64 \pm 0.93$ mm for the Ethiopian group, and the diameter exhibits a nearly linear trend with age but is not statistically significant between male and female (Worku et al. 2020).

In childhood, the CBD diameter is also dependent on children aged between 1 and 14 years with mean values of 2–4.9 mm (Witcombe and Cremin 1978).

For CHDs, their mean diameter is also related linearly with age such as $d_{\text{CHD}} = 0.044 \times \text{age} + 3.93, \text{age} \in [2, 80]$ (Takahashi et al. 1985).

The Function of Bile

Bile is synthesised in the liver and excreted into the lumen of the duodenum when a fatty meal or any other sort of meal is consumed, or even water is drunk. Bile consists mostly of water with minor amount of ions, bilirubin (a pigment derived from haemoglobin), cholesterol and an assortment of lipids – the bile salts. The water and ions assist in the dilution and buffering of acids in chyme as it enters the small intestine. The large and water-soluble drops, containing a variety of lipids, are created by mechanical processing in the stomach. The enzymes from the pancreas are not lipid-soluble; consequently, they only interact with lipids at the surface of the lipid drop. Bile salts break the large lipid drops apart. The tiny droplets, with coating of bile salts, increase the surface area available for enzymatic attack. Furthermore, the layer coated by bile salts allows the interaction between the lipids and enzymes to be easier. After lipid digestion has been completed, bile salts promote the absorption of lipids by the intestinal epithelium. More than 90% of the bile salts are themselves reabsorbed primarily in the ileum as lipid digestion is finished. The reabsorbed bile salts enter the hepatic portal circulation and are collected and recycled by the liver (Martini 2001).

Physiology of the GB

The GB has two major functions: (1) bile storage and (2) bile modification. Although liver cells produce around 1 litre of bile each day, the sphincter of Oddi remains closed until chyme enters the duodenum. In this case, bile cannot flow along the CBD; it has to enter the CD for storage within the expandable GB. This kind of bile movement is named as refilling. When bile stays in the GB, much of the water in bile is absorbed, and bile salts and other components become increasingly concentrated. Consequently, the composition of bile gradually changes. Although bile is secreted continuously, bile release into the duodenum occurs only under the stimulation of cholecystokinin (CCK). In the absence of CCK, the sphincter of Oddi is closed and bile produced by the liver reaches the GB through both the CHD and the CD. When chyme enters the duodenum, CCK is released, but the amount secreted highly depends on the amount of lipids contained in chyme. CCK relaxes the sphincter of Oddi and stimulates contractions in the walls of the GB. These contractions push bile into the duodenum (small intestine) (Martini 2001). This process is called emptying.

Recently, a different view on GB function emerges (Turumin et al. 2013). The view is based on the fact that the bile is emptied from the GB in 5–20 min after the food arrives at the gut, while the gastric chyme moves into the duodenum from the gut in 1–3 h later, suggesting the bile may be insignificant for food digestion. The bile may stimulate intestinal peristalsis and promote the intestine to clean the gastric chyme. The role of the GB is to protect the liver, gut mucosa and colon from hepatotoxic and hydrophobic bile acids, and regulate the serum lipid level.

CLINICAL PHYSIOLOGY OF THE SYSTEM

Gallstones and the Other Diseases

As mentioned earlier, bile is composed of three major components: cholesterol, bile salts and bilirubin. When the GB is not functioning properly, the components of the bile become out of balance leading to the formation of solid crystals. The majority of stones (80%) are composed of cholesterol, and the remainder are pigmented stones consisting of bilirubin. Stones can be large or small, single or multiple (Figure 1.4). These factors do not necessarily

FIGURE 1.4 Four hundred and sixty-eight gallstones found in a Chinese woman GB (the picture is after Lo 2018).

predict the frequency of symptoms or the severity of the disease. In many cases, GB symptoms are caused by the dysfunctional GB that is forming stones rather than the stones themselves. The exception to this is when stones block off the CD or CBD. There are three kinds of biliary disease:

(a) Cholecystitis

A more serious form of GB disease, cholecystitis is an infection or inflammation of the GB often caused by obstruction of the CD. The symptoms are similar to biliary colic but more prolonged. Patients can also have fever, chills and an elevated white blood cell count.

(b) Choledocholithiasis (CBD stones)

Stones can drop out of the GB into the CBD. These stones often pass into the intestines without incident. Sometimes, they can cause obstructions in the bile duct leading to jaundice and life-threatening infections of the bile ducts.

(c) Biliary pancreatitis

When stones pass by the pancreatic duct, the pancreas can be irritated leading to this potentially serious condition. Symptoms usually consist of mid-abdominal pain radiating to the back with nausea and vomiting.

Archaeologists were astonished to find human gallstones originating from the 17th century B.C. in Mycenae, Greece, and this confirms that mankind has been suffering from this disease for at least 4000 years (Sandor et al. 1996).

Gallstone disease became much more common in Britain in the last quarter of the 20th century than ever before (Bateson 2000); consequently, British prevalence of gallstones is so high that it has ranked the seventh position in the list of international prevalence of gallstones (Bateson 1999). Cholecystectomy GB removal is the most commonly performed abdominal operation in the West, with some 60,000 operations estimated in England and Wales every year (Lam et al. 1996), costing the National Health Service (NHS) approximately £60 million/year (Calvert et al. 2000). In the United States, approximately 700,000 cholecystectomies are performed and medical expenses for the treatment of gallstone disease were over $6 billion (Wang et al. 2009).

Gallstone Formation

In these GB diseases, the gallstone formation is the key issue. The bile is a solution of solid substances of bile salts, cholesterol and lecithin in water. The bile salt is most soluble in water, but the cholesterol is least soluble in water. The state of three substances in bile can be represented by triangular coordinates in terms of per cent each substance, which is expressed by the ratio of one substance millimoles mass to the total millimoles mass of three substances (Hofmann and Small 1967; Admirand and Small 1968), as shown in Figure 1.5. The continuous line in the figure is the maximal cholesterol solubilisation curve determined by *in vitro* experiments.

If the cholesterol amount secreted by the liver is exactly on the curve, the cholesterol is completely dissolved in the water to form a saturated bile. If the cholesterol amount ratio is above the curve, the extra cholesterol dissociated from the bile solution to form a supersaturated bile with cholesterol crystals. In the figure, the normal bile is under the curve, but

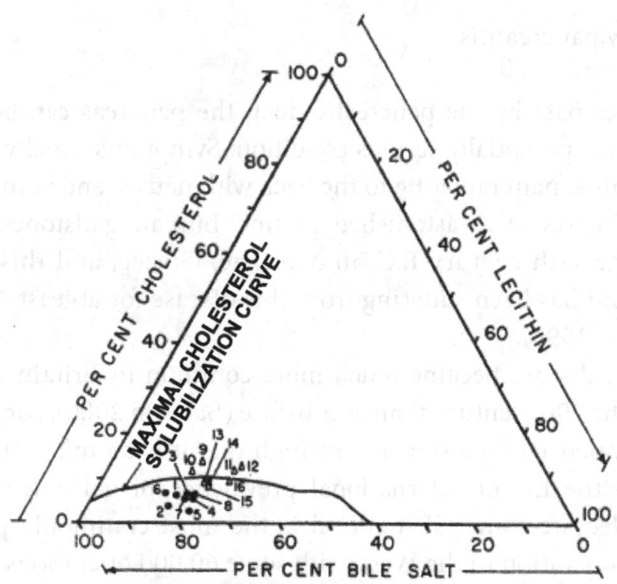

FIGURE 1.5 The maximal cholesterol solubilisation curve determined by *in vitro* experiments and the bile compositions of GB bile from gallstone patients and normal subjects are plotted in a triangular coordinates (the picture is after Admirand and Small 1968; 1–8 for normal bile and 9–16 for abnormal bile).

the bile from the gallstone patients is on or above the curve. The maximal cholesterol solubilisation curve provided in a later study (Holzbach et al. 1973) is slightly lower than that shown in Figure 1.5. Additionally, the absolute concentration of total bile acid also plays a very important role in the dissolution ability (Matsushiro et al. 1981).

The gallstone formation depends on two necessary conditions: (1) cholesterol crystal nucleation and (2) cholesterol crystal growth. The necessary physical-chemical factor for crystal nucleation or growth is cholesterol supersaturation in bile (Admirand and Small 1968; Brenneman et al. 1972; Antsaklis et al. 1975; Holzbach et al. 1973; Carey and Small 1978; Holzbach 1984a,b; Catnach et al. 1992).

The bile with cholesterol supersaturation is called lithogenic bile. The bile stasis in the GB may cause cholesterol supersaturation and in turn allow cholesterol to form stones (Thureborn 1965). This assumption was made in 1856 by Meckel von Helmbach and has gained support in a great deal via medical and chemical facts. An additional study shows that hypersecretion of GB mucus is also an important factor in gallstone formation (Doty et al. 1983a). Human GB mucin can accelerate nucleation of cholesterol (Levy et al. 1984; LaMont et al. 1984). It has been found that increased glycoproteins in bile precede cholesterol saturation and crystallisation (Zak et al. 1984).

Interestingly, small gallstones or cholesterol crystals are frequently found in groups along the wall of the GB where bile flow is particularly larger due to absorption, suggesting cholesterol concentration is higher near the wall than the position off the wall (Cussler et al. 1970). Based on this observation, a mathematical model was established to quantitatively estimate the variation of cholesterol concentration as a function of distance from the GB wall by assuming GB is a sphere. The cholesterol precipitation effects were assessed, and predictions of the concentration were compatible with qualitative clinical observations (Cussler et al. 1970).

Another piece of work established that the early stage of gallstone formation may be associated with reduced GB sensitivity to CCK based on the guinea pig model owing to lowered concentration of CCK receptors on GB smooth muscle (Poston et al. 1992).

It has been found that the CD flow resistance is increased before gallstone formation (Pitt et al. 1981; Jazrawi 2000). The increased flow resistance may be owing to a higher bile viscosity and narrowed CD inner diameter. This suggests that the gallstone formation may have a closed

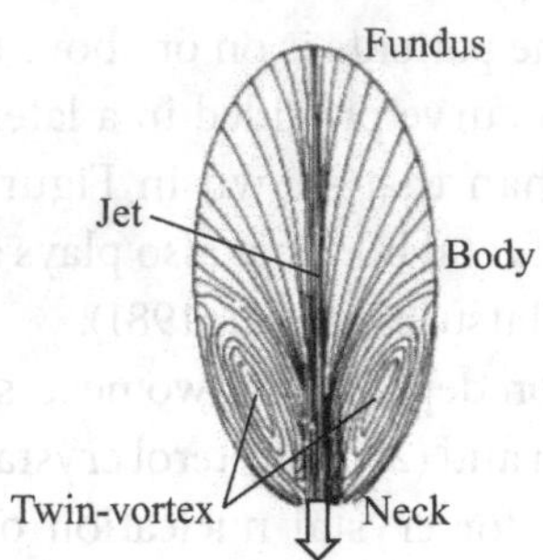

FIGURE 1.6 Twin-vortex in a GB predicted with numerical fluid mechanics (the streamline picture is adopted from Long et al. 1997).

link with the fluid dynamics because the flow resistance in CD represents the hydraulic feature of bile flow in the human biliary system.

Furthermore, the gallstone formation might have a link with bile flow pattern inside GB. A snapshot of the twin-vortex pattern of bile laminar flow during emptying is illustrated in Figure 1.6, which is predicted by numerical fluid mechanics methods such as finite-element method (Xie et al. 1990; Zeng et al. 1992) or finite-volume method (Long et al. 1997) based on a revolutionary ellipsoidal GB model. The GB model was generated by using time-sequence computed tomography (CT) images taken after the patient consumed fired eggs. There is a jet flow between the two vortices. The tiny cholesterol crystals can be trapped in the centres of the vortices and are allowed to grow into gallstones.

Acalculous Biliary Pain

The acalculous biliary pain (ABP) or functional biliary pain or stoneless GB disease is defined as steady pain located in the epigastrium and right upper quadrant in the absence of gallstones or when other structural abnormalities exist in biliary tract (Delgado-Aros et al. 2003; Shaffer 2003). The frequency of acalculous biliary pain may be as high as 7.6% in men and 20.7% in women, or as low as 2.4% overall (Corazziazri et al. 1999). The issue of acalculous biliary pain has drawn attention since 1920s (Meyer 1921; Muller 1927; Stanton 1932; Graham and Mackey 1934; Brown 1938). Currently, acalculous biliary pain is an increasingly interesting topic in recent years because there exists difficulty to diagnose the acalculous biliary pain. As a result, several comprehensive reviews on this topic are available now (Canfield et al. 1998; Corazziazri et al. 1999; Shaffer 2003; Delgado-Aros et al. 2003; Rastogi et al. 2005). The major points are illustrated in the following sections.

As mentioned in *Gallstones and the Other Diseases* section, the dysfunctional GB or impaired motor function of the GB links to gallstone formation, but it also has long been suspected a major factor to contribute to acalculous biliary pain. The presumed mechanism for biliary pain is obstruction leading to distension and inflammation. The obstruction might result from in-coordination between the GB and either the CD or the sphincter of Oddi due to increased flow resistance or tone (Corazzizri et al. 1999). Following this mechanism of pain, pain provocation test has been used as a diagnostic tool to select patients with impaired GB motor function who respond to therapy. In the test, CCK is injected to stimulate GB to contract and the GB emptying performance and the biliary pain can be observed. The bile ejection fraction (EF) in 30 min is applied to identify the emptying performance of the GB. When a GB EF <35% (Pickleman et al. 1985; Fink-Bennett et al. 1991; Kmiot et al. 1994; Michail et al. 2001; Misra et al. 1991; Ozden and DiBaise 2003; Bingener et al. 2004; Riyad et al. 2007) or 40% (Yap et al. 1991; Ponce et al. 2004), the GB emptying is abnormal and the GB motor function is impaired; otherwise, it is normal. The biliary pain of some patients has been alleviated after their GB with impaired motor function is removed (Pickleman et al. 1985; Fink-Bennett et al. 1991; Misra et al. 1991; Kmiot et al. 1994; Michail et al. 2001; Ozden and DiBaise 2003; Bingener et al. 2004; Brosseuk and Demetrick 2003; Yap et al. 1991; Ponce et al. 2004). However, the pain of the others still remains (Sunderland and Carter 1988; Smythe et al. 1998, 2004; Rastogi et al. 2005). These conflicting reports suggest that impaired GB motor function is incomplete to justify the acalculous biliary pain.

The acalculous biliary pain in the GB and biliary tract is one kind of visceral pains and results from obstruction of the CD or CBD. The obstruction elevates pressure within the biliary system. Because of being associated with pressure wave within the biliary system, the acalculous biliary pain is directly related to intraluminal pressure of the biliary tract (Ness and Gebhart 1990). Gaensler (1951) examined the pain threshold of CBD for 40 patients before and after GB removal with injection of bile and saline. It was found that the pain threshold varied from 14.7 to 59 mmHg (mean: 40 mmHg). Csendes et al. (1979) illustrated that the pain threshold is in the range of 15–60 mmHg (mean: 30 mmHg). Middelfart et al. (1998)

showed that the pain threshold of the GBs of 12 patients varied from 4 to 58 mmHg (mean: 23 mmHg). Obviously, stresses are experienced in the walls of GB and biliary tracts under the increased intraluminal pressure due to GB contraction. Therefore, these experimental observations seem to suggest that the muscle stress components may be involved in the acalculous biliary pain. Definitely, this new pain mechanism needs to be confirmed.

CRITICAL FACTORS RELEVANT TO DISEASES

GB Motor Function

The GB is able to react to the stimulus of eating by contracting and discharging bile into the duodenum. In contracting, the GB acts as a motor driving the enterohepatic circulation (Low-Beer et al. 1971). GB motor functions describe the behaviours in emptying during a meal and in refilling during fasting for a GB. They mainly include three aspects: (1) refilling with hepatic bile in fasting, (2) emptying/ejection of concentrated bile into the duodenum in a meal and (3) mixing of GB contents (Lanzini and Northfield 1989). There are two factors, namely EF and time to half-maximal emptying ($t_{½}$), as shown in Figure 1.7, to assess GB emptying characteristics (Brugge 1991).

Currently, GB is regarded as a 'slow pump', where emptying or refilling is operated separately; its volume change is related to inner pressure with a compliance (Ryan and Cohen 1976; Schoetz et al. 1981; Middelfart et al. 1998). Even though this relation is nonlinear, the linear one is applied in

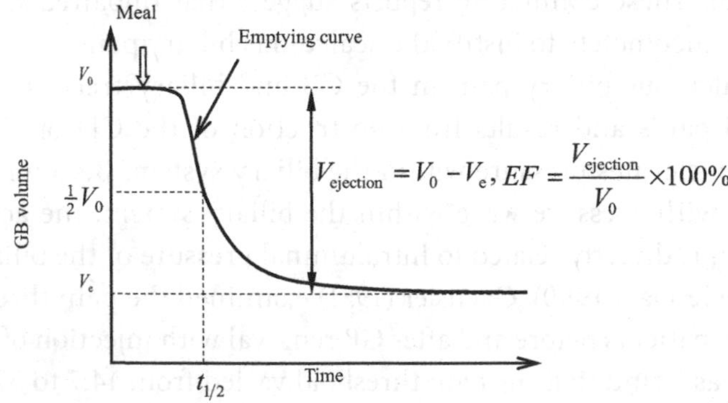

FIGURE 1.7 Emptying/ejection curve of a GB.

practice. Thus, a newly nonlinear relationship between compliance and GB volume is desirable.

There are two kinds of methods to measure GB motor functions (Lanzini and Northfield 1989; Brugge 1991). The first kind is ultrasonography, and the second is cholescintigraphy. In ultrasonography, GB longitudinal and transverse ultrasound images are taken first, and the GB bile volume is calculated by using the sum-of-cylinders (Everson et al. 1980; Hopman et al. 1985) or ellipsoid (Dodds et al. 1985) techniques. The method can reflect GB volume change during emptying and refilling phases. If images are taken at time interval longer than 10 min, the GB emptying will vary smoothly with time; however, if the time interval is at less than 10 min, the GB emptying curve can exhibit significantly fluctuations in the volume as shown in Figure 1.8 (Horrow et al. 1991). The fresh hepatic bile that went into the GB can generate a washout effect to make GB bile concentration dilute and dismiss a potential bile stasis in the GB.

This suggests that bile refilling can take place during GB bile emptying phase, presenting a refilling and emptying alternative pattern. The reason for this effect is that GB smooth muscle contraction is intermediate and fluctuating rather than in a smooth pattern (Takahashi et al. 1982).

In cholescintigraphy, a radioactive chemical or tracer, such as ^{99m}technetium-labelled hepato imino diacetic acid (^{99m}Tc-HIDA), is injected intravenously and the tracer is taken up by the liver and excreted by it into the GB with bile flow. After 120 min, when GB ^{99m}Tc-HIDA

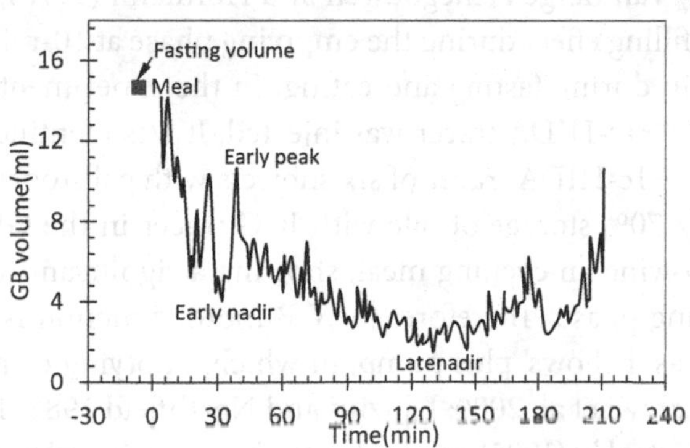

FIGURE 1.8 An instant GB volume versus time curve during emptying phase observed on a normal subject by Horrow et al. (1991) with ultrasound at 1-min time interval.

uptake phase is completed, a meal or CCK dose is applied to imitate emptying phase, and then GB images are monitored and the number of isotopes is counted in the GB with time by using an external gamma camera with a pinhole collimator. The isotope counts are in a linear relationship with GB volume (Krishnamurthy et al. 1981). Thus, GB EF can be figured out based on the isotope counts at different time instants (Bobba et al. 1984; Krishnamurthy et al. 1982; Pellegrini et al. 1986). Cholescintigraphy can result in nearly the same EF value as ultrasonography (Krishnamurthy et al. 1981; Kronert et al. 1989). In the early trails, iopanoic acid labelled with radioactive iodine (I^{131}) has been used ever in cholescintigraphy (Englert and Chiu 1966).

In comparison with ultrasonography, cholescintigraphy can measure isotope counts in bile flows in CHD and CBD, suggesting cholescintigraphy being able to provide quantitative biliary dynamics. It was identified that in patients with cholellthiasis, the fraction of hepatic bile flowing into the GB was normal, but GB EF was significantly reduced (Krishnamurthy et al. 1983).

Interestingly, in the study by Lanzini and Northfield (1989), a CCK dose was given before completion of GB tracer uptake phase and the tracer counts exhibited a fluctuation pattern similar to that in Figure 1.8. This suggested that cholescintigraphy still could demonstrate refilling effect during the emptying phase.

In Lanzini et al.'s (1987) study, except GB ^{99m}Tc-HIDA tracer, another tracer, such as indocyanine green (ICG), which had been used to label the hepatic bile by van Berge Henegouwen and Hofmann (1978), was utilised to quantity refilling effect during the emptying phase at 10-min time intervals for 60 min during fasting and eating. In the experiment, an intravenous bolus of ^{99m}Tc-HIDA tracer was injected. It was identified that mean GB EF of GB ^{99m}Tc-HIDA tracer of six subjects with gallstones was 43% by accompanying 70% storage of bile with ICG tracer in the GB during the first hour following an evening meal, showing a significant refilling effect in the emptying phase. Therefore, the GB motor function is that the GB is considered as 'bellows' plus pump, in which emptying or refilling is an alternative (Jazrawi et al. 2000; Lanzini and Northfield 1988, 1989).

In Jazrawi et al.'s (1995) study, both cholescintigraphy and ultrasonography were applied simultaneously in 14 patients with gallstones and 11 healthy controls when they were fasting and at 10-min time intervals after a standard meal for 90 min. In the experiment, an intravenous bolus of ^{99m}Tc-HIDA tracer was injected as well. The instant bile postprandial refilling volume was calculated by the two emptying curves measured by cholescintigraphy and ultrasonography; subsequently, the cumulative GB bile refilling volume at 90 min, i.e., bile turnover, was worked out. The turnover index which is defined as the ratio of the turnover to the GB fasting volume was assessed. The mean turnover index of the patients was 1.8 in comparison with 3.5 of the control subjects.

CCK stimulations not only cause GB to contract but also allow CD (Courtney et al. 1983) and CBD (Severi et al. 1988) to contract. However, the sensitivity to CCK decreases from GB to CBD. This sensitivity gradient may act to facilitate GB emptying in response to hormonal and neural stimulation (Severi et al. 1988). Experiments show the changes in canine GB volume have no relation with CD resistance and the CD does not demonstrate sphincter mechanism at physiological GB volumes (Sharp et al. 1990). No motor activity was found in CBD and pancreatic duct (Csendes et al. 1979).

The poor GB motor function can lead to an abnormal GB emptying and is related to gallstones. Studies have found that GB emptying is abnormal in patients with gallstones (Shaffer et al. 1980; Fisher et al. 1982; Thompson et al. 1982; Forgacs et al. 1984; Pomeranz and Shaffer 1985; Sylvestrowics and Shaffer 1988; Masclee et al. 1989; Spengler et al. 1989; Festi et al. 1990; Van Erpecum et al. 1992; Portincasa et al. 1997; Zhu et al. 2005,). These studies cannot distinguish whether abnormal GB emptying is caused by the presence of gallstones or vice versa. However, several reports proposed that abnormal GB emptying precedes the developments of gallstones (La Morte et al. 1979; Fridhandler et al. 1983; Pellegrini et al. 1986; Doty et al. 1983b). The GB motor functions are instrumental in the pathogenesis of GB diseases (Everson et al. 1980).

GB emptying and refilling experimental data can be fitted with mathematical models. There are two kinds of mathematical models: the first is analytical model and the second is first-order ordinary differential equation. In the first kind model, the mathematical model is expressed by using a power-exponential function (Elashoff et al. 1982; Jonderko 1989) as follows:

$$V = V_0 e^{-(t/t_{1/2})^a} \tag{1.1}$$

where a is a positive model constant determined with experimental data.

In the second kind model, the model is in terms of two first-order ordinary differential equations for the emptying and refilling processes. If the emptying and refilling occur in a time sequence, the two processes are described mathematically by the following equations (Zulpo et al. 2018):

$$\begin{cases} \dfrac{\mathrm{d}V}{\mathrm{d}t} = k_{\mathrm{g}} V \left(\dfrac{V_{\mathrm{total}} - V}{V_{\mathrm{total}}} \right) \\ \dfrac{\mathrm{d}V}{\mathrm{d}t} = -V_{\mathrm{g}} \sin^2(\omega t) \end{cases} \tag{1.2}$$

where the first equation is for refilling, but the second one is for emptying, k_g is the GB refilling rate constant (min^{-1}), V_{total} is the total volume secreted by the liver (ml), V_g is the bile volume discharged per GB contraction (ml/min) and ω is the GB contraction frequency (min^{-1}). The first equation in Eq. (1.2) is the standard Bernoulli differential equation and its general solution exists, the second equation Eq. (1.2) is a first-order linear differential equation and its general solution is available as well. The two general solutions of Eq. (1.2) can be written as follows:

$$\begin{cases} V = \dfrac{V_{\mathrm{total}} \mathrm{e}^{k_{\mathrm{g}} t}}{c_1 V_{\mathrm{total}} + \mathrm{e}^{k_{\mathrm{g}} t}} \\ V = -V_{\mathrm{g}} \left[\dfrac{t}{2} - \dfrac{1}{4\omega} \sin(2\omega t) \right] + c_2 \end{cases} \tag{1.3}$$

where the model constants k_g, V_g, ω, c_1 and c_2 are decided by using GB refilling and emptying experimental data.

If the emptying and refilling emerge alternatively, the two processes are described mathematically by the following equation (Zulpo et al. 2018):

$$\frac{\mathrm{d}V}{\mathrm{d}t} = -V_{\mathrm{g}} \sin^2(\omega t) + k_{\mathrm{g}} (V_{\mathrm{total}} - V) \tag{1.4}$$

Eq. (1.4) is a first-order linear differential equation and the general solution reads as follows:

$$V = V_{\mathrm{total}} - \frac{V_{\mathrm{g}}}{2} \left[\frac{1}{k_{\mathrm{g}}} - \frac{2\omega \sin(2\omega t) + k_{\mathrm{g}} \cos(2\omega t)}{4\omega^2 + k_{\mathrm{g}}^2} \right] + c_3 \mathrm{e}^{-k_{\mathrm{g}} t} \tag{1.5}$$

where the model constants k_g, V_g, ω and c_3 are determined by means of GB refilling and emptying experimental data.

Effects of Bile Rheology

Clinical measurements have demonstrated that the density of GB bile is very close to 1000 kg/m^3 of water density at room temperature, but the dynamic viscosity differs greatly from that of water (Ooi 2004). The dynamic viscosity of GB bile is 1.77–8.0 mPa·s (Ooi 2004). Tera (1963) and Thureborn (1966) found that the normal GB bile was layered. For example, the dynamic viscosity of top thinnest layer is 2.0 mPa·s, but the thickest layer is 2.2 mPa·s after 2 hours of sedimentation (Tera 1963). Bouchier et al. (1965) reported that the dynamic viscosity of bile in pathological GB bile is greater than normal GB bile, and both are more viscous than hepatic tube bile. The concentration of normal GB bile is a major factor determining the viscosity. In pathological and hepatic duct bile, the content of mucous substances was the major factor determining the viscosity. Cowie and Sutor (1975) illustrated that the mean viscosity of bile from GBs with stones was greater than that from healthy ones. The presence of mucus in GBs with stones is likely to account for the differences in viscosity (Doty et al. 1983a). Jungst et al. (2001) showed that the biliary viscosity was markedly higher in GB bile of patients with cholesterol (5.0 mPa·s) and mixed stones (3.5 mPa·s) compared with hepatic duct bile (0.2 mPa·s). A positive correlation between mucin and viscosity was found in GB bile but not in hepatic duct bile.

Gottschalk and Lochner (1990) measured the postoperative bile viscosity sampled by T-drainages of 29 patients with a modified horizontal capillary viscometer from the day of operation to the ninth day and confirmed that the viscosity is time-dependent and the bile behaves like a Maxwell fluid (one simple kind of viscoelastic fluids). The dynamic viscosity of 33 bile specimens showed shear-thinning non-Newtonian behaviour by using a Contraves low shear viscometer. The bile dynamic viscosity decreases from 5 mPa·s at the shear rate of 0.1 s^{-1} to 1.5 mPa·s at 2.0 s^{-1}.

Saida (1992) measured the viscosity of the bile of 15 healthy GBs (10 male and 5 female adults) and the bile of five dogs. Table 1.1 lists the mean experimental data of bile viscosity of the human and dog bile. Clearly, the bile is shear thinning since the viscosity is decreased with an increasing shear rate.

TABLE 1.1 Viscosity of the human and dog bile (mean ± standard deviation).

Human	Shear rate, γ (s^{-1})	37.5	75	150	375	750
	Viscosity, μ (mPa·s)	2.55±0.51	2.49±0.47	2.44±0.46	2.41±0.44	2.39±0.43
Dog	Shear rate γ (s^{-1})	18.75	37.5	75	150	375
	Viscosity, μ (mPa·s)	5.33±1.35	5.22±1.24	5.04±1.08	4.83±0.93	4.53±0.79

The bile is non-Newtonian fluid, and its shear stress can be best fitted with a power of shear rate, i.e.,

$$\tau = \mu_c \gamma^a, 0<a<1 \tag{1.6}$$

where τ is the shear stress, μ_c is the non-Newtonian coefficient of viscosity, γ is the shear rate and a is the viscosity index. For the human bile, $\mu_c = 2.63 \pm 0.68$ mPa·s and $a = 0.98 \pm 0.01$; for the dog bile, $\mu_c = 6.36 \pm 2.36$ mPa·s and $a = 0.95 \pm 0.04$.

For the human bile, the $\tau - \gamma$ relation can be represented by the well-known Casson model as well, namely

$$\sqrt{\tau} = \sqrt{\tau_0} + b\sqrt{\gamma} \tag{1.7}$$

where τ_0 is the yield shear stress, $\tau_0 = 0.2 \pm 0.1$ mPa for the human bile and b is the slope of the $\sqrt{\tau} - \sqrt{\gamma}$ curve, $b = 1.50$.

Coene et al. (1994) reported that the bile viscosity of gallstone patients as a function of shear rate based on the results by using Contraves low-shear 30 rotational viscometer. The bile dynamic viscosity decreases from 2.5 mPa·s at a shear rate of 0.1 s^{-1} to 1.5 mPa·s at 100 s^{-1}. These observations indicate that the bile in GBs with gallstones will be characterised by shear-thinning non-Newtonian behaviour at a low shear rate.

Ooi (2004) measured the bile dynamic viscosity of 59 patients (cholecystectomy) and illustrated that the bile rheology (the relation between shear stress and shear rate) of 20 patients was Newtonian, the bile of 22 patients was shear thickening at a high shear rate and shear thinning at a low shear rate. The bile of eight patients was shear thickening and that of four patients (with mucus) was shear-thinning non-Newtonian. Even though the results at low shear rate are not reliable due to viscometer limitation, these experiments suggest that GB bile really can demonstrate a complicated non-Newtonian behaviour. Obviously, this behaviour and the viscosity increased due to it may modify the flow features of bile in the human biliary system. An increased viscosity of GB bile has been considered an important factor in the pathogenesis of gallstone disease (Jungst et al. 2001), However, whether the bile rheology is changed during gallstone formation needs to be investigated.

Kuchunov et al. (2014) measured the shear stress-shear rate curves of the hepatic bile and the GB bile and identified the shear-thickening non-Newtonian behaviour. Both Casson and Carreau's rheological model parameters were extracted from the experimental data and applied in CFD simulations of non-Newtonian bile flow in a specific human biliary system.

Bile viscosity can be affected by nonsteroidal anti-inflammatory drugs. For example, indomethacin decreases viscosity of GB bile in patients with gallstones to 2.9 ± 0.6 from 5.6 ± 1.2 mPa·s (van Ritter et al. 1993). A possible reason may be attributed to an alternation in mucin macromolecular composition.

Effects of CD

The CD connects the neck of the GB to the CHD and CBD. The CD allows bile to go into and out of the GB in refilling and emptying in response to hormonal and neural stimuli (Dasgupta and Stringer 2005). Recent observations on the CD anatomy show that the duct has two kinds of structure: (1) the duct lumen wall has folds which look like the Heister's valve (Lichtenstein and Ivy 1937) and (2) the duct has a smooth lumen which develops in space in various patterns, such as spiral, winding, kinks, spiral incorporating M-shape loop and so on (Ooi 2004; Bird et al. 2006). The first kind of structure can raise the pressure drop or flow resistance significantly compared with the second one.

The relationships between CD geometry and cholelithiasis have been investigated quantitatively based on 250 patients with cholelithiasis and 250 healthy controls (Deenitchin et al. 1998). The patients with gallstones have significantly longer and narrower CDs (a mean of 48 and 4 mm in length and diameter, respectively) than those without stones (a mean of 28 and 7 mm in length and diameter, respectively). The results suggest that CD geometry is associated with cholelithiasis.

It is obvious that the flow resistance is affected by the CD geometry. As mentioned earlier, the CD has a complicated structure compared with the other bile ducts. Consequently, it inevitably makes a great contribution to the resistance of the whole biliary system. The pressure drop across the biliary system of a dog has been examined (Rodkiewicz et al. 1979). It was found that flow rate of bile in the CD and the whole system (including the sphincter of Oddi) tree was related to the associated pressure drop by a power law, which differed from that for Poiseuille's flow in a rigid tube. The power was in the range of 1.47–2.05 for bile. The bile flow in a long, circular, smooth and rigid tube also has been investigated by the same authors, and it was found that the flow rate-pressure drop relation was very close to Poiseuille's flow (Rodkiewicz and Otto 1979). These facts illustrate the complicated geometry can modify flow characteristics. Ooi (2004) and Ooi et al. (2004) applied a circular, straight and rigid duct with uniform baffles to model the real human CD and performed lots of CFD simulations in two-dimensional (2D) and three-dimensional (3D) cases to investigate the effect of baffle on flow resistance. The flow in a 2D model based on two patients' biliary system images was simulated too. It was shown that both the dimensionless baffle height and number of baffles, especially dimensionless baffle height, significantly affect flow resistance. The 3D laminar flow in three real 3D human CD casts has been done and increased pressure drop was found compared with that of a circular straight tube (Ooi 2004). The flow patterns in a circular, straight and rigid tube with uniform baffles were observed (Al-Atabi et al. 2005). The pressure drop was predicted and compared with the experimental data (Al-Atabi et al. 2006). The Reynolds number is more than 50 in these experiments (in human biliary system, the Reynolds number is less than 40). The flow patterns in 2D models on which were scaled based on the two patient's biliary system images illustrated by Ooi et al. (2004) were demonstrated (Al-Atabi et al. 2006, 2012).

The fluid-structure interaction (FSI) for 2D channel with uniform baffles has also been tackled (Ooi et al. 2003).

Effects of Sphincter of Oddi

The sphincter of Oddi is located at the termination of CBD and pancreatic duct. The human sphincter of Oddi segment is 6–15 mm in length and demonstrates a muscular structure. The physiological functions include the following: (1) to regulate bile flow into the duodenum, (2) to divert hepatic bile into the GB and (3) to prevent reflux of duodenal contents from the duodenum into the biliary tree (Dodds et al. 1989; Funch-Jensen 1995). The CCK released due to stimulation of ingestion of a fatty meal can cause the GB to contract and the sphincter of Oddi to relax for bile emptying. Coordination of GB and sphincter of Oddi function may also be influenced by nerve bundles which connect the GB and the sphincter of Oddi via the CD (Luman et al. 1997; Mawe 1998). The effect of enteric hormones on sphincter of Oddi motor function was determined by recording intraluminal pressure from the human sphincter of Oddi (Geenen et al. 1980). It was found that the human sphincter of Oddi demonstrated unique phasic contractions, which would have an important role in regulating biliary and pancreatic duct emptying (Geenen et al. 1980). In recent years, several studies on the opossum sphincter of Oddi, which is ~3 cm in length, have been conducted (Toouli et al. 1983; Liu et al. 1992; Calabuig et al. 1990; Grivell et al. 2004). The findings show that the sphincter of Oddi regulates bile as a peristaltic pump at low CBD pressure and as a resistor at higher CBD pressure. Funch-Jensen et al. (1981) observed that the CCK injection inhibited the activity of the sphincter of Oddi of dogs; as a result, bile flow was increased, but GB pressure remained unchanged. Furthermore, Funch-Jensen (1995) argued that the peristaltic pump activity of human sphincter of Oddi was probably less important since CCK inhibited phasic wave of sphincter of Oddi.

The basal pressure of human sphincter of Oddi is affected by gallstone disease. Cicala et al. (2001) investigated motor activity of sphincter of Oddi of 155 patients and found that gallstones are frequently associated with increased basal pressure in the sphincter of Oddi which might obstruct bile flow thus acting to facilitate GB stasis. The early work on pressure drop and flow resistance across a dog biliary duct illustrated that the resistance of sphincter of Oddi is at least about three times higher than CD (Otto et al. 1979). These results suggest that the fluid dynamics in human sphincter of Oddi should be investigated significantly in the future.

FLUID MECHANICS IN THE BILIARY SYSTEM

Fluid mechanics is the study that deals with the action of forces on fluids. It involves various properties of the fluid, such as velocity, pressure, density and temperature, as functions of space and time. The flow system of the human biliary system is composed of the GB, biliary tract and bile. The fluid velocity, pressure and the shear stress in fluid and at wet solid walls can be calculated by solving the flow governing equations in the system.

There has been interest in the flow details in the human biliary system, but also in the general hydraulic characteristics, for instance, thus far the relation of flow resistance across this system with the flow rate or the Darcy friction factor with Reynolds number have been tackled for rigid and compliant biliary tract models. CFD technique is increasingly applied in fluid mechanics in the biliary system. However, bile-gallstone two-phase flows have remained untreated presently.

SOLID MECHANICS IN THE BILIARY SYSTEM

The GB wall has four layers: serosa, perimuscularis, muscularis and mucosa (Cai and Gabella 1983; MacPherson et al. 1984; Dodds et al. 1989). The serosa envelopes the entire fundus but covers only the posterior surface of the body and neck. The permuscularis consists of connective tissue richly supplied with nutrient vessels. The GB muscularis forms a netlike arrangement of longitudinal, transverse and oblique smooth muscle bundles separated by connective tissue. The smooth muscle cells are small and loose with few intercellular contacts (Dodds et al. 1989). The mucosa consists of numerous elevated folds, covered with simple columnar epithelium.

The GB can contract under the control of nerve and hormone to empty bile (Mawe 1998). Contraction of the muscle fibre mesh generates a vector of force directed towards the centre of the GB lumen. The biomechanical properties of the GB wall include the following: (1) dynamic volume-pressure relation of the GB, (2) steady length-tension relation of strips of the GB and (3) dynamic length-tension relation or force-velocity of the strips.

Pressure-Volume Relations

GB pressure-volume curves have been measured *in vitro* by inflating a GB with saline in passive and active states (Miura and Saito 1967; Ryan and Cohen 1976; Schoetz et al. 1981; Brotschi et al. 1984; Matsuki 1985a,b; Borly et al. 1996). It was demonstrated that GB pressure-volume curves presented viscoelastic property (Miura and Saito 1967; Schoetz et al. 1981; Brotschi et al. 1984; Matsuki 1985a,b).

The relationship between the volume change ΔV and the differential pressure Δp in a GB can be expressed as follows:

$$\Delta V = C\Delta p \tag{1.8}$$

where C is the compliance of the GB, which can be determined from *in vitro* or *in vivo* experiments. If the C is known, the differential pressure Δp can be estimated from the volume change ΔV.

The pressure-volume response of the opossum was studied by Ryan and Cohen (1976) under basal condition and after the continuous intravenous infusion of gastrin I, secretin and CCK. The GB was forced to expand with a continuous-infusion pump in the experiments. It was found that the GB un-stimulated was capable of accommodation increases in volume with only slight changes in pressure and the CCK significantly increased the pressure in the GB. The compliances were 0.77, 0.38, 0.32 and 0.34 ml/mmHg in the basal condition and in CCK stimulations of 0.025, 0.25 and 2.5 μg/kg·h, respectively.

Schoetz et al. (1981) measured the dynamic pressure-volume relation of adult female baboons. When the cyclical infusion of bile into and withdrawal of bile from the GB were conducted, the continuous pressure was monitored in conditions without and with hormones, such as pilocarpine, histamine, atropine and CCK. The results showed that a hysteretic loop existed in the pressure-volume relations, especially under the stimulations of pilocarpine, histamine and CCK rather than atropine. This illustrates that the GB muscle possesses the feature of viscoelasticity. Table 1.2 illustrates the mean compliances during the experiments for guinea pig GBs. The values of this set of compliances are very close to that of Ryan and Cohen (1976).

Middelfart et al. (1998) illustrated the pressure-volume of the GBs of 11 patients with gallstones when the GBs were fed by saline via McGaham catheter under no CCK stimulation condition. The compliances have been obtained from the *in vitro* pressure-volume relations. It is found that the compliance of human GBs with gallstones depends upon the individual patient heavily and varies from 0.17 to 4.0 ml/mmHg. The mean value of the compliance is about 2.731 ml/mmHg.

Stress-Stretch Relations

The stress consists of passive and active stresses in the GB smooth muscle. The active stress is generated automatically under hormone stimulants, whereas the passive stress/tension is induced by stretching. In fact, the

TABLE 1.2 **Compliances in different conditions.**

C (ml/mmHg)	**Pilocarpine**	**Histamine**	**CCK**	**Atropine**
Basal	0.76	0.66	0.79	0.71
Stimulated	0.30	0.30	0.34	0.81

GB smooth muscle responding to hormone or contraction varies with time. Therefore, the stress-stretch relation of the muscle is unsteady. If we are just interested in the maximal stress generated at various stretches, the stress-stretch will be steady. Currently, most of the investigation into the GB smooth muscle is just associated with the steady stress-stretch relation.

Fifty strips of human GB from 25 specimens removed at operations were studied (Mack and Todd 1968). It was found that the human GB smooth muscle was capable of showing maintained tone *in vitro* and of contracting power in response to certain substances, especially acetylcholine (ACh) and CCK, and that the peak stress could be achieved in 3–5 min from when the hormone stimulants were applied. However, quantitative relations between length and tension are unavailable.

Washabau et al. (1991) obtained muscle strips from GBs of mature female guinea pigs (400–500 g) and measured the isometric stress under 10^{-8}–10^{-4} M ACh and 10–80 mM KCl stimulations. The passive tension increases with L/L_0, and the active stress achieves the maximal value at L_0, afterwards the active stress declines (Figure 1.9). However, the total tension (active + passive) rises with L/L_0.

The peak active stress links to the CCK dose strongly. The disagreement exists between various experimental data of the peak active stress

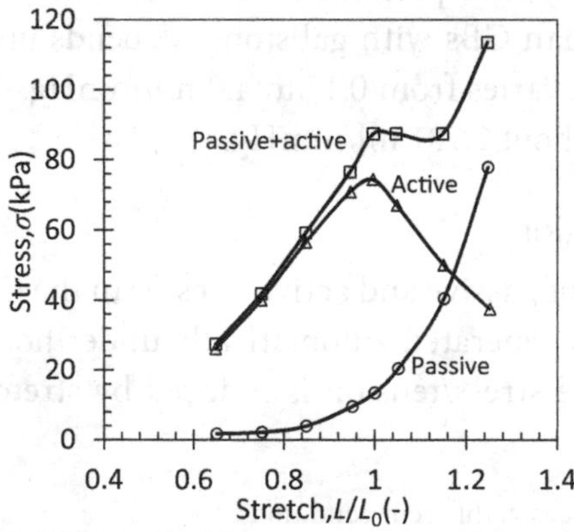

FIGURE 1.9 Passive, active and total isometric stress at various lengths of GB smooth muscle strips in ACh-treated (10^{-4}M), L_0 is a length at which a peak active stress is developed (Washabau et al. 1991).

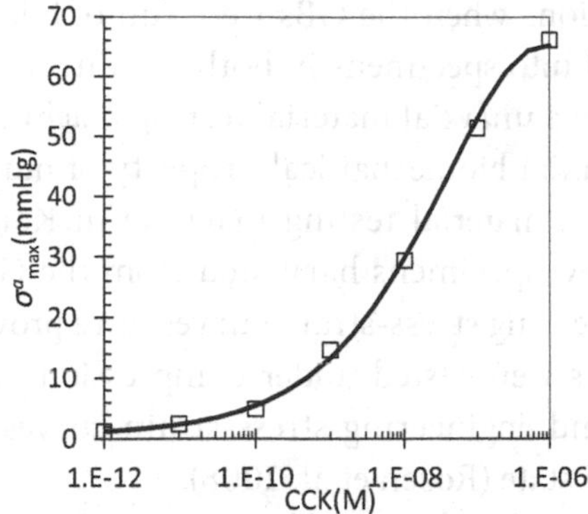

FIGURE 1.10 Peak active stress versus CCK dose. The stress rises with an increase of CCK dose. The symbols present the experimental data (the data is from van De Heijning et al. 1999).

(Portincasa et al. 1994; van De Heijning et al. 1999). Figure 1.10 shows a typical example of peak active stress for a strip of a GB.

Bird et al. (1996) obtained muscle strips from human GB removed at cholecystectomy in longitudinal, circular and oblique directions, respectively, and the other strips also were taken from the body and neck regions of the GBs. No differences were seen between samples taken from the longitudinal, circular and oblique axes. The samples from the body contracted more forcefully than those of neck tissue. Strips from the body are more sensitive to muscarinic stimulation than those from the neck. The authors failed to obtain the length-tension relation.

Ahmed et al. (2000) considered that the diagnosis and identification of patients with acalculous biliary pain remained a significant clinical problem and CCK provocation test helped diagnosis. Hence, the GB muscle strips from patients with acalculous biliary pain were taken to characterise the response of the strips to CCK-8 and carbachol, respectively, and to compare these with strips from GB of normal controls.

There is no difference in the CCK responses for the two groups, so it casts doubt over the effectiveness of the CCK test to diagnose acalculous biliary pain. However, if carbachol sensitivity is different in the two groups, then the muscarinic stimulation with something like carbachol would help in the diagnosis. Once again, length-tension relations are unavailable.

Xiong et al. (2013) harvested porcine GBs from a slaughterhouse and carried out indentation experiments along the circumferential and longitudinal directions when the GBs were full of bile. Additionally, the GBs were dissected into specimens in both the directions, and the specimens were tested on a uniaxial material testing machine.

The passive uniaxial biomechanical property of human GB walls was tested on a uniaxial material testing machine in Karimi et al.'s (2017) study based on a few specimens harvested from the GBs of corpses in a hospital. The engineering stress-strain curves were provided.

Porcine GB walls were tested under compression loads on a material testing machine, and engineering stress-strain curves were provided in passive compressed state (Rosen et al. 2008).

An organ inflating experimental rig was designed and built, and a lamb GB shape was measured *in vitro* when the GB was pressurised with phosphate-buffered solution in the study by Genovese et al. (2014). The passive biomechanical property constants were decided numerically with finite element analysis (FEA) based on the membrane mechanics model.

Force-Velocity Relations

In general, when the human GB smooth muscle contracts, the GB volume will begin to get small and the bile is emptied from the GB. Obviously, the stress in the muscle must vary with time during emptying, and then the stress-stretch relation is unsteady or dynamic. Besides, during the muscle contraction, the varied volume will generate a velocity, called contraction or shorting velocity. For skeletal muscle, the tension has a relation with the shortening velocity, called Hill's equation (Fung 1993), written as follows:

$$(v+b)(T+a)=b(T_0+a) \tag{1.9}$$

where T represents the tension in a muscle, v is the velocity of contraction, and a, b and T_0 are constants. While a, b, T_0 and v are available, T can be predicted by using Eq. (1.9).

Gordon and Siegman (1971) obtained the tension-length and force-velocity of rabbit taenia coli. The inverse relationship between the load or developed tension and the velocity could be described by the Hill's equation. The dynamic constants from Hill's equation were independent of the initial muscle length. The similarity of these mechanical properties and those of skeletal muscle suggests that a similar mechanism of contraction exists in taenia coli. Table 1.3 shows a summary of the constants in Hill's

TABLE 1.3 Comparison of force-velocity constants for various muscles.

Constants	a (g/cm³)	a/T_0	b (s^{-1})	T_0 (g/cm³)	v_{max} (s^{-1})
Sartorius, frog	399	0.257	244	2.0	1.29
Papillary muscle, cat	175	0.22	0.27	0.80	1.24
Uterus, rabbit	56	0.44	0.09	0.13	0.18
Trachealis, canine	244	0.21	0.04	1.17	0.17
Penis, tortoise	-	0.11	0.03	-	0.36
Duodenum, cat	220	0.524	0.1	0.42	0.20
Taenia coli, pig	255	0.17	0.05	1.5	0.3
Taenia coli, rabbit	295	0.331	0.010	0.89	0.031

equation that are available (Gordon and Siegman 1971). From this table, the constants for human GBs cannot be found.

The volume-pressure, stress-stretch and force-velocity relations can describe the mechanical property of human GB as a whole. However, there is limited information about the pressure-volume relation, only a few stress-stretch relations are known and presently there are not investigations into the force-velocity relation for the human GB at all.

Because there is a lack of mechanical property, it is impossible to build an exact model to deal with the contraction and emptying activity of the human GB. An experiment study of volume-pressure, stress-stretch and force-velocity relations for healthy GB, GB with gallstone or the GB diseases under basal and CCK stimulation conditions should be conducted to explore the differences in these relations. Experimental data and related clinical observations should be compared to ultimately establish the GB contractile activity.

Biliary Duct Biomechanical Property

In vivo experiments were performed on the canine CBD of six mongrel dogs to characterise the pressure response to an infusion of saline in passive state, and nonlinear pressure-saline volume curves were obtained (Slater et al. 1983).

Sixteen bile duct systems of healthy adult canines have been tested with well-devised apparatus (Jian and Wang 1991). In normal physiological conditions, the bile duct experiences an almost uniform circumferential and longitudinal stress. However, in disease cases, the stresses are much larger at inside wall than at outside. The elastic modulus becomes small from CBD to CHD and to hepatic duct. The modulus of CD is close to the hepatic duct (Jian and Wang 1991).

Eleven porcine bile ducts were examined in two locations, in the hepatic duct and the CBD (Duch et al. 1998, 2004). Consequently, the luminal cross-sectional area and circumferential tension-strain properties have been established during distension of the isolated CBD *in vitro*. The cross-sectional areas in the CBD are significantly higher than those in the hepatic duct in the pressure range of 0–8 kPa. The stress-strain relations at both locations are fitted with exponential equation.

The structural and mechanical changes in the CBD at different time intervals after acute obstruction were quantified (Duch et al. 2002, 2003). The duct was ligated in pigs, near the duodenum, and studied after 3 hours, 12 hours, 2 days, 8 days and 32 days (five groups). It was found that the diameter and wall thickness were increased with time, and the circumferential stress-strain relation differed from each group. A mathematical model was established to model remodelling of pig CBD post-obstruction based on a series of experimental remodelling data (Dang et al. 2004). Furthermore, four model parameters of Fung-type strain energy (Fung 1993) were determined inversely based on pig CBD post-obstruction experimental remodelling data (Dang et al. 2005). It was shown that the chronic remodelling process could lead to an initial circumferential softening followed by stiffening of the CBD.

Healthy CBD samples were harvested from fresh cadavers, their wall microstructure was observed and additional specimens were made and tested on a material testing machine by Girard et al. (2019). There are four layers in the CBD wall, namely epithelium (mucous layer), internal submucosal layer (longitudinal muscular and elastic fibres), external submucosal layer (circumferential muscular and elastic fibres) and adventitia (collagen fibres) from the inside to the outside. The biomechanical property is anisotropic and viscoelastic, especially, the longitudinal property is stiffer than the circumferential.

PROSPECTS AND CONTENT OF THE BOOK

Based on the clinical practice and investigation in the human biliary system mentioned earlier, nevertheless, a few prospects for the research in the system can be expected. Fluid mechanics of non-Newtonian bile flow in the GB and biliary tract needs to be investigated to provide insights into clinical physiology of gallstone formation. The study on gallstone-bile two-phase flow condition in the system should be launched as well. Biomechanics of intact GBs and GB wall tissue should be in line with soft bio-tissue context, especially constitutive laws of the GB wall in the passive state and viscoelastic property. The GB in active state should be modelled and is coupled with GB emptying-refilling physiological cycle. The relationships between the GB wall tissue biomechanical property and disease condition are worth being explored to improve the clinical diagnosis.

As an initial response to the prospects above, in the book, the following topics are covered in both biomechanical and applied mathematics context: (1) one-dimensional (1D) Newtonian and non-Newtonian bile flows in the biliary tract with rigid and compliant walls, (2) 3D FSI modelling of the CD, (3) biomechanical model for GB pain, (4) cross-bridges of human GB smooth muscle contraction, (5) nonlinear analysis of the anisotropic, heterogeneous biomechanical human GB wall, (6) constitutive laws of GB walls with damage effects, (7) viscoelasticity modelling of intact GBs and (8) GB 3D geometrical generation from ultrasound images and FEA analysis.

REFERENCES

Adibi, A., and B. Givechian. 2006. Diameter of common bile duct: What are the predicting factor? *Journal of Research in Medical Sciences* 12:121–124.

Admirand, W. H., and D. M. Small. 1968. The physicochemical basis of cholesterol gallstone formation in man. *Journal of Clinical Investigation* 47:1043–1052.

Ahmed, H. K. 2017. Measurement of normal common bile duct diameter using ultrasonography. Master's degree thesis, Sudan University of Science and Technology, Khartoum, Sudan.

Ahmed, R., N. Bird, R. Chess-Williams, et al. 2000. In vitro responses of gallbladder muscle from patients with acalculous biliary pain. *Digestion* 61:140–144.

Al-Atabi, M., S. B. Chin, and X. Y. Luo. 2005. Flow structure in circular tubes with segmental baffles. *Journal of Flow Visualization and Image Processing* 3:301–311.

Al-Atabi, M., S. B. Chin, and X. Y. Luo. 2006. Visualization experiment of flow structures inside two-dimensional human biliary system models. *Journal of Mechanics in Medicine and Biology* 6:249–260.

Al-Atabi, M., R. C. Ooi, X. Y. Luo, et al. 2012. Computational analysis of the flow of bile in human cystic duct. *Medical Engineering & Physics* 34:1177–1183.

Antonldio, G., M. R. Lewin., J. D. Sutor, et al. 1975. Gallbladder function, cholesterol stones, and bile composition. *Gut* 16:937–942.

Bachar, G. N., M. Cohen, A. Belenky, et al. 2003. Effect of aging on the adult extrahepatic bile duct. *Journal of Ultrasound Medicine* 22:879–882.

Bateson, M. C. 1999. Gallbladder disease. *British Medical Journal* 318:1745–1748.

Bateson, M. C. 2000. Gallstones and cholecystectomy in modern Britain. *Postgraduate Medical Journal* 76:700–703.

Benjaminov, F., G. Leichtman, T. Naftali, et al. 2013. Effects of age cholecystectomy on common bile duct diameter as measured by endoscopic ultrasonography. *Surgical Endoscopy* 27:303–307.

Bingener, J., M. L. Richards, W. H. Schwesinger, et al. 2004. Laparoscopic cholecystectomy for biliary dyskinesia. *Surgical Endoscopy* 18:802–806.

Bird, N. C., R. C. Ooi, X. Y. Luo, et al. 2006. Investigation of the functional three-dimensional anatomy of the human cystic duct: A single helix? *Clinical Anatomy* 19:528–534.

Bird, N. C., H. Wegstapel, R. Chess-Williams, et al. 1996. In vitro contractility of stimulated and non-stimulated human gallbladder muscle. *Neurogastroenterology & Motility* 8:63–68.

Blidaru, D., M. Blidaru, C. Pop, et al. 2010. The common bile duct: Size, course, relations. *Romanian Journal of Morphology and Embryology* 51:141–144.

Bobba, V. R., G. T. Krishnamurthy, E. Kingston, et al. 1984. Gallbladder dynamics induced by a fatty meal in normal subjects and patients with gallstones: Concise communication. *Journal of Nuclear Medicine* 25:21–24.

Borly, L., L. Hojgaard, S. Gronvall, et al. 1996. Human gallbladder pressure and volume: Validation of a new direct method for measurements of gallbladder pressure in patients with acute cholecystitis. *Clinical Physiology and Functional Imaging* 16:145–156.

Bouchier, I. A., S. R. Cooperband, and B. M. El-Kodsi. 1965. Mucous substances and viscosity of normal and pathological human bile. *Gastroenterology* 49:343–353.

Brenneman, D. E., W. E. Connor, E. L. Forker, et al. 1972. The formation of abnormal bile and cholesterol gallstones from dietary cholesterol in the prairie dog. *Journal of Clinical Investigation* 51:1495–1503.

Brewer, G. E. 1899. Preliminary report on the surgical anatomy of the gall-bladder and ducts from an analysis of one hundred dissections. *Annals of Surgery* 29:721–730.

Brosseuk, D., and J. Demetrick. 2003. Laparoscopic cholecystectomy for symptoms of biliary colic in the absence of gallstones. *American Journal of Surgery* 186:1–3.

Brotschi, E. A., W. W. Lamorte, and L. F. Williams. 1984. Effect of dietary cholesterol and indomethacin on cholelithiasis and gallbladder motility in guinea pig. *Digestive Diseases and Sciences* 29:1050–1056.

Brown, M. J. 1938. Non-calculous chronic gallbladder disease. *American Journal of Surgery* 41:238–254.

Brugge, W. R. 1991. Motor function of the gallbladder: Measurement and clinical significance. *Seminars in Roentgenology* 26(3):226–231.

Cai, W. Q., and G. Gabella. 1983. The musculature of the gall bladder and biliary pathways in the guinea-pig. *Journal of Anatomy* 136:237–250.

Calabuig, R., M. G. Ulrich-Baker, F. G. Moody, et al. 1990. The propulsive behaviour of the opossum sphincter of Oddi. *American Journal of Physiology* 258:G138–G142.

Calvert, N. W., G. P. Troy, and A. G. Johnson. 2000. Laparoscopic cholecystectomy: A good buy? A cost comparison with small-incision (mini) cholecystectomy. *European Journal of Surgery* 166:782–786.

Canfield, A., S. P. Hetz, J. P. Schriver, et al. 1998. Biliary dyskinesia: A study of more than 200 patients and review of the literature. *Journal of Gastrointestinal Surgery* 2:443–448.

Carey, M. C., and D. M. Small. 1978. The physical chemistry of cholesterol solubility in bile. *Journal of Clinical Investigation* 61:998–1026.

Caroli-Bosc, F. X., P. Pugliese, E. P. Peten, et al. 1999. Gallbladder volume in adults and its relationship to age, sex, body mass index, body surface area and gallstones. *Digestion* 60:344–348.

Castelain, M., C. Grimaldi, A. G. Harris, et al. 1993. Relationship between cystic duct diameter and the presence of cholelithiasis. *Digestive Diseases and Sciences* 38:2220–2224.

Catnach, S. M., P. D. Fairclough, R. C. Trembath, et al. 1992. Effect of oral erythromycin on gallbladder motility in normal subjects and subjects with gallstones. *Gastroenterology* 102:2071–2076.

Chen, T., C. R. Hung, A. C. Huang, et al. 2012. The diameter of the common bile duct in an asymptomatic Taiwanese population: Measurement by magnetic resonance cholangiopancreatography. *Journal of Chinese Medical Association* 75:384–388.

Cicala, M., F. I. Habib, F. Fiocca, et al. 2001. Increased sphincter of Oddi basal pressure in patients affected by gall stone disease: A role for biliary stasis and colicky pain? *Gut* 48:414–417.

Coene, P. P., A. K. Groen, P. H. Davids, et al. 1994. Bile viscosity in patients with biliary drainage. *Scandinavian Journal of Gastroenterology* 29:757–763.

Corazziazri, E., E. A. Shaffer, and W. J. Hogan. 1999. Functional disorders of the biliary tract and pancreas. *Gut* 45(Suppl II):II48–II54.

Courtney, D. F., A. S. Clanachan, and G. W. Scott. 1983. Cholecystokinin constricts the canine cystic duct. *Gastroenterology* 85:1154–1159.

Cowie, A. G. A., and D. J. Sutor. 1975. Viscosity and osmolality of abnormal bile. *Digestion* 13:312–315.

Csendes, A., A. Kruse, P. Funch-Jensen, et al. 1979. Pressure measurements in the biliary and pancreatic duct systems in controls and in patients with gallstones, previous cholecystectomy, or common bile duct stones. *Gastroenterology* 77:1203–1210.

Cussler, E. L., D. F. Evans, and R. G. DePalma. 1970. A model for gallbladder function and cholesterol gallstone formation. *Proceedings of National Academy of Sciences* 67:400–407.

Dang, Q., H. Gregersen, B. Duch, et al. 2004. Indicial response functions of growth and remodeling of common bile duct postobstruction. *American Journal of Physiology-Gastrointestinal and Liver Physiology* 286:G420–G427.

Dang, Q., H. Gregersen, B. Duch, et al. 2005. Remodeling of strain energy function of common bile duct post obstruction. *Mechanics & Chemistry of Biosystems* 2:53–61.

Dasgupta, D., and M. D. Stringer. 2005. Cystic duct and Heister's "valves". *Clinical Anatomy* 18:81–87.

Deenitchin, G. P., J. Yoshida, K. Chijiiwa, et al. 1998. Complex cystic duct is associated with cholelithiasis. *Hepato-Biliary-Pancreatic Surgery* 11:33–37.

Delgado-Aros, S., F. Cremonini, A. J. Bredenoord, et al. 2003. Systematic review and meta-analysis: Does gall-bladder ejection fraction on cholecystokinin cholescintigraphy predict outcome after cholecystectomy in suspected functional biliary pain? *Aliment Pharmacology* 18:167–174.

Dodds, W. J., W. J. Groh, R. M. Darweesh, et al. 1985. Sonographic measurement of gallbladder volume. *American Journal of Radiology* 145:1009–1101.

Dodds, W. J., W. J. Hogan, and J. E. Green. 1989. Motility of the biliary system. In *Handbook of Physiology*, ed. S. G. Schultz, 1055–1101. Bethesda: American Physiological Society.

Doty, J. E., H. A. Pitt, S. L. Kuchenbecker, et al. 1983a. Impaired gallbladder emptying before gallstone formation in the prairie dog. *Gastroenterology* 85:168–174.

Doty, J. E., H. A. Pitt, S. L. Kuchenbecker, et al. 1983b. Role of gallbladder mucus in the pathogenesis of cholesterol gallstone. *American Journal of Surgery* 145:54–61.

Duch, B. U., H. L. Andersen, and H. Gregersen. 2003. Morphometric and biomechanical remodelling following reopening of the obstructed bile duct. *Physiological Measurement* 24:N23–N34.

Duch, B. U., H. L. Andersen, and H. Gregersen. 2004. Mechanical properties of the porcine bile duct wall. *BioMedical Engineering OnLine* 3:1–8.

Duch, B. U., H. L. Andersen, J. Smith, et al. 2002. Structural and mechanical remodelling of the common bile duct after obstruction. *Neurogastroenterology & Motility* 14:111–122.

Duch, B. U., J. A. Petersen, and H. Gregersen. 1998. Luminal cross-sectional area and tension-strain relation of the porcine bile duct. *Neurogastroenterology & Motility* 10:203–209.

Elashoff, J. D., T. J. Reedy, and J. H. Meyer. 1982. Analysis of gastric emptying data. *Gastroenterology* 83:1306–1312.

Englert, E., and V. S. Chiu. 1966. Quantitative analysis of human biliary evacuation with radioisotopic technique. *Gastroenterology* 50:506–518.

Everson, G. T., D. Z. Braverman, M. L. Johnson, et al. 1980. A critical evaluation of real-time ultrasonography for the study of gallbladder volume and contraction. *Gastroenterology* 79:40–48.

Ferris, D. O., and J. C. Vibert. 1959. The common bile duct: Significance of its diameter. *Annals of Surgery* 149:249–251.

Festi, D., R. Frabboni, F. Bazzoli, et al. 1990. Gallbladder motility in cholesterol gallstone disease. *Gastroenterology* 99:1779–1785.

Fink-Bennett, D., P. DeRidder, W. Z. Kolozsi, et al. 1991. Cholecystokinin cholescintigraphy: Detection of abnormal gallbladder motor function in patients with chronic acalculous gallbladder disease. *Journal of Nuclear Medicine* 32:1695–1699.

Fisher, R. S., F. Stelzer, E. Rock, et al. 1982. Abnormal gallbladder emptying in patients with gallstones. *Digestive Diseases and Sciences* 27:1019–1024.

Forgacs, J. C., M. N. Maisey, G. M. Murphy, et al. 1984. Influence of gallstone and ursodeoxycholic acid therapy on gallbladder emptying. *Gastroenterology* 87:229–307.

Fridhandler, T. M., J. S. Davison, and E. A. Shaffer. 1983. Defective gallbladder contractility in the ground squirrel and prairie dog during the early stages of cholesterol gallstone formation. *Gastroenterology* 85:830–836.

Funch-Jensen, P. 1995. Sphincter of Oddi physiology. *Journal of Hepato-Biliary-Pancreatic Surgery* 2:249–254.

Funch-Jensen, P., H. Stodkilde-Jorgensen, K. Kraglund, et al. 1981. Biliary manometry in dogs. *Digestion* 22:89–93.

Fung, Y. C. 1993. *Biomechanics-Mechanical Properties of Living Tissues* (2nd edition). New York: Springer.

Gaensler, E. A. 1951. Quantitative determination of the visceral pain threshold in man. *Journal of Clinical Investigation* 30:406–420.

Geenen, J. E., W. J. Hogn, W. J. Dodds, et al. 1980. Intraluminal pressure recoding from human sphincter of Oddi. *Gastroenterology* 78:317–324.

Genovese, K., L. Casaletto, J. D. Humphrey, et al. 2014. Digital image correlation-based point-wise inverse characterization of heterogeneous material properties of gallbladder in vitro. *Proceedings of the Royal Society Series A* 470:20140152.

Girard, E., G. Chagnon, and E. Gremen, et al. 2019. Biomechanical behaviour of human bile duct wall and impact of cadaveric preservation processes. *Journal of Mechanical Behavior of Biomedical Materials* 98:291–300.

Gordon, A. R., and M. J. Siegamn. 1971. Mechanical properties of smooth muscle I: Length-tension and force-velocity relations. *American Journal of Physiology* 221:1243–1249.

Gottschalk, M., and A. Lochner. 1990. Behavior of postoperative viscosity of bile fluid from T-drainage. *Gastroentroloisches Journal* 50:65–67.

Graham, E. A., and W. A. Mackey. 1934. A consideration of the stoneless gallbladder. *JAMA* 103:1497–1499.

Grivell, M. B., C. M. Woods, A. R. Grivell, et al. 2004. The possum sphincter of Oddi pumps or resists flow depending on common bile duct pressure: A multilumen manometry study. *Journal of Physiology* 558:611–611.

Hofmann, A. F., and D. M. Small. 1967. Detergent properties of bile salts: Correlation with physiological function. *Annual Review of Medicine* 18:333–376.

Holzbach, R. T. 1984a. Effects of gallbladder function on human bile: Compositional and structural changes. *Hepatology* 4:57S–60S.

Holzbach, R. T. 1984b. Factors influencing cholesterol nucleation in bile. *Hepatology* 4:173S–176S.

Holzbach, R. T., M. Marsh, M. Olszewski, et al. 1973. Cholesterol solubility in bile. *Journal of Clinical Investigation* 52:1467–1479.

Hopman, W. P., W. F. Brouwer, G. Rosenbusch, et al. 1985. A computerized method for rapid quantification of gallbladder volume from real-time sonograms. *Radiology* 154:236–237.

Horrow, M. M., J. C. Horrow, A. Niakosari, et al. 2001. Is age associated with size of adult extrahepatic bile duct: Sonographic study. *Radiology* 221:411–414.

Horrow, P. J., G. M. Murphy, and R. H. Dowling. 1991. Gall bladder emptying patterns in response to a normal meal in healthy subjects and patients with gall stones: Ultrasound study. *Gut* 32:1406–1411.

Jazrawi, R. P., P. Pazzi, M. L. Petroni, et al. 1995. Postprandial gallbladder motor function: Refilling and turnover of bile in health and in cholelithiasis. *Gastroenterology* 102:582–591.

Jazrawi, R. P., P. Pazzi, M. L. Petroni, et al. 2000. Role of the gallbladder in the pathogenesis of gallstone disease. In *Bile Acid in Hepatology Disease*, ed. T. C. Northfield, H. A. Ahmed, and R. P. Jazrawi, 182–191. Dordrecht, the Netherlands: Kluwer Academic Publishers.

Jian, C. Y., and G. R. Wang. 1991. Biomechanical study of the bile duct system outside the liver. *Bio-Medical Materials and Engineering* 1:105–113.

Jonderko, K. 1989. Comparative analysis of quantitative gastric emptying indices and power-exponential modelling of gastric emptying curves. *Clinical Physics and Physiological Measurement* 10:161–170.

Jungst, D., A. Niemeyer, I. Muller, et al. 2001. Mucin and phospholipids determine viscosity of gallbladder bile in patients with gallstones. *World Journal of Gastroenterology* 7:203–207.

Kaim, A., K. Steinke, M. Frank, et al. 1998. Diameter of the common bile duct in the elderly patient: Measurement by ultrasound. *European Radiology* 8:1413–1415.

Karamanos, E., K. Inaba, R. J. Berg, et al. 2017. The relationship between age, common bile duct diameter and diagnostic probability in suspected choledocholithiasis. *Digestive Surgery* 34:421–428.

Karimi, A., A. Shojaei, and P. Tehrani. 2017. Measurement of the mechanical properties of the human gallbladder. *Journal of Medical Engineering & Technology* 41:541–545.

Kmiot, W. A., E. P. Perry, I. A. Donovan, et al. 1994. Cholesterolosis in patients with chronic acalculous biliary pain. *British Journal of Surgery* 81:112–115.

Krishnamurthy, G. T., V. R. Bobba, and E. Kingston. 1981. Radionuclide ejection fraction: A technique for quantitative analysis of motor function of the human gallbladder. *Gastroenterology* 80:482–490.

Krishnamurthy, G. T., V. R. Bobba, E. Kingston, et al. 1982. Measurement of gallbladder emptying sequentially using a single dose of ^{99m}Tc-labeled hepatobiliary agent. *Gastroenterology* 83:773–776.

Krishnamurthy, G. T., V. R. Bobba, D. McConnell, et al. 1983. Quantitative biliary dynamics: Introduction of a new noninvasive scintigraphic technique. *Journal of Nuclear Medicine* 24:217–223.

Kronert, K., Gotz, V., Reuland, P., et al. 1989. Gallbladder emptying in diabetic patients and controls subjects assessed by real-time ultrasonography and cholescintigraphy: A methodological comparison. *Ultrasound in Medicine and Biology* 15:535–539.

Kuchumov, A. G., V. Gilev, V. Popov, et al. 2014. Non-Newtonian flow of pathological bile in the biliary system: Experimental investigation and CFD simulations. *Korea-Australia Rheology Journal* 26:81–90.

Lal, N., S. Mehra, and V. Lal. 2014. Ultrasonographic measurement of normal common bile duct diameter and its correlation with age, sex and anthropometry. *Journal of Clinical and Diagnostic Research* 8:AC01–AC04.

Lam, C. M., F. E. Murray, and A. Cuschieri. 1996. Increased cholecystectomy rate after the introduction of laparoscopic cholecystectomy. *Gut* 38:282–284.

LaMont, J. T., B. F. Smith, and J. R. Moore. 1984. Role of gallbladder in pathophysiology of gallstones. *Hepatology* 4:51S–56S.

LaMorte, W. W., D. J. Shoetz, D. H. Birkett, et al. 1979. The role of the gallbladder in the pathogenesis of cholesterol gallstone. *Gastroenterology* 77:580–592.

Lanzini, A., R. P. Jazrawi, and T. C. Northfield. 1987. Simultaneous quantitative measurements of absolute gallbladder storage and emptying during fasting and eating in humans. *Gastroenterology* 92:852–861.

Lanzini, A., and T. C. Northfield. 1988. Gallbladder motor function in man. In *Bile Acid in Hepatology Disease*, ed. T. C. Northfield, P. Zentler-Munro, and R. P. Jazrawi, 83–96. Dordrecht, the Netherlands: Kluwer Academic Publishers.

Lanzini, A., and T. C. Northfield. 1989. Assessment of the motor functions of the gallbladder. *Journal of Hepatology* 9:383–391.

Levy, P. F., B. F. Smith, and J. T. LaMont. 1984. Human gallbladder mucin accelerates nucleation of cholesterol in artificial bile. *Gastroenterology* 187:270–275.

Lichtenstein, M. E., and A. C. Ivy. 1937. The function of the "Valve" of Heister. *Surgery* 1:38–52.

Liu, Y. F., T. P. Saccone, A. Thune, et al. 1992. Sphincter of Oddi regulates flow by acting as a variable resistor to flow. *American Journal of Physiology* 263:G683–G689.

Lo, T. 2018. Woman who loves deep-fried food has 468 gallstones removed from her body in one operation and doctors claim her diet could be the cause. *Daily Mail News*. www.dailymail.co.uk/news/china/article-5762167/Woman-loves-deep-fried-food-468-gallstones-removed-body-ONE-operation.html (accessed July 30, 2020).

Long, T. Y., R. F. Yang, and Y. P. Wu. 1997. Mathematical model of bile flow in a contracting gallbladder. *Journal of Chongqing University (Natural Science Edition)* 20:95–99.

Low-Beer, T. S., S. T. Heaton, and A. E. Read. 1971. Gallbladder inertia and sluggish enterohepatic circulation of bile salts in celiac disease. *Lancet* 1:991–994.

Luman, W., A. J. K. Williams, A. Pryde, et al. 1997. Influence of cholecystectomy on sphincter of Oddi motility. *Gut* 41:371–374.

Mack, A. J., and J. K. Todd. 1968. A study of human gall bladder muscle in vitro. *Gut* 9:546–549.

MacPherson, B. R., G. W. Scott, J. P. N. Chansouria, et al. 1984. The muscle layer of the canine gallbladder and cystic duct. *Acta Anatomy* 120:117–122.

Mahour, G. H., K. G. Wakim, and D. O. Ferris. 1967. The common bile duct in man: Its diameter and circumference. *Annals of Surgery* 165:415–419.

Martini, F. 2001. *Fundamentals of Anatomy & Physiology* (5th edition). Upper Saddle River, NJ: Prentice Hall.

Masclee, A. A., J. B. Jansen, L. M. Geuskens, et al. 1989. Plasma cholecystokinin and gallbladder responses to intraduodenal fat in gallstone patients. *Digestive Disease and Sciences* 34:353–359.

Matsuki, Y. 1985a. Spontaneous contractions and the visco-elastic properties of the isolated guinea-pig gall-bladder. *Japanese Journal of Smooth Muscle Research* 21:71–78.

Matsuki, Y. 1985b. Dynamic stiffness of the isolated Guinea-pig gallbladder during contraction induced by cholecystokinin. *Japanese Journal of Smooth Muscle Research* 21:427–438.

Matsushiro, T., H. Cho, H. Nagashima, et al. 1981. Factors affecting the cholesterol dissolution ability of human bile. *Tohoku Journal of Experimental Medicine* 135:51–61.

Mawe, G. M. 1998. Nerves and hormones interact to control gallbladder function. *News Physiological Science* 13:84–90.

Meyer, W. 1921. Chronic cholecystitis without stones: Diagnosis and treatment. *Annals of Surgery* 74:439–448.

Michail, S., D. Preud'Homme, J. Christian, et al. 2001. Laparoscopic cholecystectomy: Effective treatment for chronic abdominal pain in children with acalculous biliary pain. *Journal of Pediatric Surgery* 36:1394–1396.

Middelfart, H. V., P. Jensen, L. Hojgaard, et al. 1998. Pain patterns after distension of the gallbladder in patients with acute cholecystitis. *Scandinavian Journal of Gastroenterology* 33:982–987.

Misra, D. C., G. B. Blossom, D. Fink-Bennett, et al. 1991. Results of surgical therapy for biliary dyskinesia. *Archives of Surgery* 126:957–960.

Miura, K., and S. Saito. 1967. Visco-elastic properties of the gallbladder in rabbit and guinea-pig. *Journal of Showa Medical Association* 27:135–138.

Muller, G. P. 1927. The noncalculous gallbladder. *JAMA* 89:786–789.

Nahar, N., S. Ara, M. Rahman, et al. 2011. Length and diameter of the cystic duct: A post-mortem study. *Bangladesh Journal of Anatomy* 9:89–92.

Nalaini, F., M. G. Salehi, and N. Farshchian. 2017. Relationship between age and common bile duct diameter in adults: Ultrasonographic study. *International Journal of Life Science and Pharma Research* 7:P20–P23.

Ness, T. J., and G. F. Gebhart. 1990. Visceral pain: A review of experimental studies. *Pain* 41:167–234.

Oluseyi, K. Y. H. 2018. Ultrasound determination of gall bladder size and wall thickness in normal adults in Abuja North Central Nigeria. *Archives of International Surgery* 6(4):214–218.

Ooi, R. C. 2004. The flow of bile in human cystic duct. PhD diss., University of Sheffield, Sheffield, UK.

Ooi, R. C., X. Y. Luo, S. B. Chin, et al. 2004. The flow of bile in the human cystic duct. *Journal of Biomechanics* 37:1913–1922.

Ooi, R. C., X. Y. Luo, S. B. Chin, et al. 2003. Fluid-structure interaction (FSI) simulation of the human cystic duct. *Summer Bioengineering Conference*, 25–29 June, Key Biscayne, FL. www.tulane.edu/~sbc2003/pdfdocs/0811.PDF.

Otto, W. J., G. W. Scott, and C. M. Rodkiewicz. 1979. A comparison of resistances to flow through the cystic duct and the sphincter of Oddi. *Journal of Surgical Research* 27:68–72.

Ozden, N., and J. K. DiBaise. 2003. Gallbladder ejection fraction and symptom outcome in patients with acalculous biliary-like pain. *Digestive Diseases and Sciences* 48:890–897.

Palasciano, G., G. Serio, P. Portincasa, et al. 1992. Gallbladder volume in adults, and relationship to age, sex, body mass index, and gallstones: A sonographic population study. *American Journal of Gastroenterology* 87:493–497.

Parulekar, S. G. 1979. Ultrasound evaluation of common bile duct size. *Radiology* 133:703–707.

Pellegrini, C. A., T. Ryan, W. Broderick, et al. 1986. Gallbladder filling and emptying during cholesterol gallstone formation in the prairie dog. *Gastroenterology* 90:143–149.

Peng, R., L. Zhang, X. M. Zhang, et al. 2015. Common bile diameter in an asymptomatic population: A magnetic resonance imaging study. *World Journal of Radiology* 7:501–508.

Perret, R. S., G. D. Sloop, and J. A. Borne. 2000. Common bile duct measurements in an elderly population. *Journal of Ultrasound Medicine* 19:727–730.

Pickleman, J., R. l. Peiss, R. Henkin, et al. 1985. The role of sincalide cholescintigraphy in the evaluation of patients with acalculous gallbladder disease. *Archives of Surgery* 120:693–697.

Pitt, H. A., J. J. Roslyn, S. L. Kuchenbecker, et al. 1981. The role of cystic duct resistance in the pathogenesis of cholesterol gallstones. *Journal of Surgical Research* 30:508–514.

Pomeranz, I. S., and E. A. Shaffer. 1985. Abnormal gallbladder emptying in a subgroup of patients with gallstones. *Gastroenterology* 85:787–791.

Ponce, J., V. Pons, R. Sopena, et al. 2004. Quantitative cholescintigraphy and bile abnormalities in patients with acalculous biliary pain. *European Journal of Nuclear Medicine and Molecular Imaging* 31:1160–1165.

Portincasa, P., A. Di Ciaula, G. Baldassarre, et al. 1994. Gallbladder motor function in gallstone patients: Sonographic and *in vitro* studies on the role of gallstones, and smooth muscle function and gallbladder wall inflammation. *Journal of Hepatology* 21:430–440.

Portincasa, P., A. Di Ciaula, V. Palmieri, et al. 1997. Impaired gallbladder and gastric motility and pathological gastro-oesophageal reflux in gallstone patients. *European Journal of Clinical Investigation* 27:653–661.

Poston, J. G., P. Singh, E. Draviam, et al. 1992. Early stages of gallstone formation in guinea pig are associated with decreased biliary sensitivity to cholecystokinin. *Digestive Diseases and Sciences* 37:1236–1244.

Rastogi, A., A. Slivka, A. J. Moser, et al. 2005. Controversies concerning pathophysiology and management of acalculous biliary-type abdominal pain. *Digestive Diseases and Sciences* 50:1391–1401.

Riyad, K., C. R. Chalmers, A. Aldouri, et al. 2007. The role of ^{99m}technetium-labelled hepato imino diacetic acid (HIDA) scan in the management of biliary pain. *HPB* 9:219–224.

Rodkiewicz, C. M., and W. J. Otto. 1979. On the Newtonian behaviour of bile. *Journal of Biomechanics* 12:609–612.

Rodkiewicz, C. M., W. J. Otto, and G. W. Scott. 1979. Empirical relationships for the flow of bile. *Journal of Biomechanics* 12:411–413.

Rosen, J., J. D. Brown, S. De, et al. 2008. Biomechanical properties of abdominal organs in vivo and postmortem under compression loads. *Journal of Biomechanical Engineering* 130:021020.

Ryan, J., and S. Cohen. 1976. Gallbladder pressure-volume response to gastrointestinal hormones. *American Journal of Physiology* 230:1461–1463.

Saida, Y. 1992. Clinical and experimental evaluations of bile viscosity in the gallbladder. *Japanese Journal of Gastroenterological Surgery* 25:2129–2138.

Sandor, J., A. Sandor, A. Zaborszky, et al. 1996. Why laparoscopic cholecystectomy today? *Surgery Toady* 26:556–560.

Schatz, P. 2020. Gallbladder and bile ducts – structure and function. www.lecturio.com/magazine/gall-bladder-bile-ducts/. https://d3uigcfkiiww0g.cloudfront.net/wordpress/blog/pics-en/uploads/Accessory-Organs.jpg (accessed August 1, 2020).

Schoetz, D. J., W. W. LaMorte, W. E. Wise, et al. 1981. Mechanical properties of primate gallbladder: Description by a dynamic method. *American Journal of Physiology* 241:G376–G381.

Senturk, S., T. C. Miroglu, A. Bilici, et al. 2012. Diameter of the common bile duct in adults and postcholecystecomy patients: A study with 64-slice CT. *European Journal of Radiology* 81:39–42.

Severi, C., J. R. Grider, and G. M. Makhlouf. 1988. Functional gradient in muscle cells isolated from gallbladder, cystic duct and common bile duct. *American Journal of Physiology* 255:G647–G652.

Shaffer, E. A. 2003. Acalculous biliary pain: New concepts for an old entity. *Digestive and Liver Disease* 35(Suppl):S20–S25.

Shaffer, E. A., P. McOrmond, and H. Duggan. 1980. Quantitative cholescintigraphy: Assessment of gallbladder filling and emptying and duodenogastric reflux. *Gastroenterology* 79:899–906.

Sharp, K. W., C. B. Ross, V. N. Tillman, et al. 1990. Changes in gallbladder do not affect cystic duct resistance. *Archives of Surgery* 125:460–462.

Slater, G., P. I. Tartter, D. Dreiling, et al. 1983. Resistance of the canine common bile duct. *Bulletin of New York Academy of Medicine* 59(8):711–720.

Smythe, A., R. Ahmed, M. Fitzhenry, et al. 2004. Bethanechol provocation testing does not predict symptom relief after cholecystectomy for acalculous biliary pain. *Digestive and Liver Disease* 36:682–686.

Smythe, A., A. W. Majeed, M. Fitzhenry, et al. 1998. A requiem for the cholecystokinin provocation test? *Gut* 43:571–574.

Spengler, U., M. Sackmann, T. Sauerbruch, et al. 1989. Gallbladder motility before and after extracorporeal shock-wave lithotripsy. *Gastroenterology* 96:860–863.

Stanton, E. M. 1932. The stoneless gallbladder: A study of operative cases. *American Journal of Surgery* 18:246–250.

Sundeland, G. T., and D. C. Carter. 1988. Clinical application of the cholecystokinin provocation test. *British Journal of Surgery* 75:444–449.

Sylvestrowics, T. A., and E. A. Shaffer. 1988. Gallbladder function during gallstone dissolution. *Gastroenterology* 95:740–748.

Takahashi, I., M. Nakaya, T. Suzuki, et al. 1982. Postprandial changes in contractile activity and bile concentration in gallbladder of the dog. *American Journal of Physiology-Gastrointestinal and Liver Physiology* 243:G365–G371.

Takahashi, Y., T. Takahashi, W. Takahashi, et al. 1985. Morphometrical evaluation of extrahepatic bile ducts in reference to their structural changes with aging. *Tohoku Journal of Experimental Medicine* 147:301–309.

Tera, H. 1963. Sedimentation of bile constituents. *Annals of Surgery* 157:467–472.

Thompson, J. C., G. M. Fried, W. D. Ogden, et al. 1982. Correlation between release of cholecystokinin and contraction of the gallbladder in patients with gallstones. *Annals of Surgery* 195:670–676.

Thureborn, E. 1965. Formation of gallstone in man. *Archive of Surgery* 91:952–957.

Thureborn, E. 1966. On the stratification of human bile and its importance for the solubility of cholesterol. *Gastroenterology* 50:775–780.

Toouli, J., W. J. Doods, R. Honda, et al. 1983. Motor function of the opossum sphincter of Oddi. *Journal of Clinical Investigation* 71:208–220.

Turumin, T. L., V. E. Shanturov, and H. E. Turumina. 2013. The role of the gallbladder in humans. *Revista Gastroenterologia de Mexico* 78:177–187.

van Berge Henegouwen, G. P., and A. F. Hofmann. 1978. Nocturnal gallbladder storage and emptying in gallstone patients and healthy subjects. *Gastroenterology* 75:879–885.

van de Heijning, B. J., P. C. van de Meeberg, P. Portincasa, et al. 1999. Effects of ursodeoxycholic acid therapy on in vitro gallbladder contractility in patients with cholesterol gallstones. *Digestive Diseases and Sciences* 44:190–196.

van Erpecum, K. J., G. P. van Berge Henegouwen, M. F. Stolk, et al. 1992. Fasting gallbladder volume, postprandial emptying and cholecystokinin release in gallstone patients and normal subjects. *Journal of Hepatology* 14:194–202.

van Ritter, C., A. Niemeyer, V. Lange, et al. 1993. Indomethacin decreases viscosity of gallbladder bile in patients with cholesterol gallstone disease. *Clinical Investigator* 71:928–932.

Vinay, S., A. Manali, A. Pathak, et al. 2013. Ultrasonic measurement of common bile duct diameter in an Indian population. *Journal of Advance Researches in Biological Sciences* 5:221–223.

Wang, D. Q. H., D. E. Cohen, and M. C. Carey. 2009. Biliary lipids and cholesterol gallstone disease. *Journal of Lipid Research* 50(Suppl):S406–S411.

Washabau, R. J., M. B. Wang, C. L. Dorst, et al. 1991. Effect of muscle length on isometric stress and myosin light chain phosphorylation in gallbladder smooth muscle. *American Journal of Physiology* 260:G920–G924.

Witcombe, J. B., and B. J. Cremin. 1978. The width of the common bile duct in childhood. *Pediatric Radiology* 7:147–149.

Worku, M. G., E. F. Enyew, Z. T. Desita, et al. 2020. Sonographic measurement of normal common bile duct diameter and associated factors at the University of Gondar comprehensive specialized hospital and selected private imaging center in Gondar town, North West Ethiopia. *PLoS One* 15(1):e0227135. https://doi.org/10.1371/journal.pone.0227135.

Wu, C. C., Ho, Y. H., and C. Y. Chen. 1984. Effect of aging on common bile duct diameter: A real-time ultrasonographic study. *Journal of Clinical Ultrasound* 12:473–478.

Xie, G., G. H. Wu, P. Chen, et al. 1990. Kinetic pattern and bile flow analysis in vivo gallbladder. *Journal of Third Military Medical University* 12:70–75.

Xiong, L., C. K. Chui, and C. L. Teo. 2013. Reality based modelling and simulation of gallbladder shape deformation using variational methods. *International Journal of Computer Assisted Radiology and Surgery* 8:857–865.

Yap, L., A. G. Wycherley, A. D. Morphett, et al. 1991. Acalculous biliary pain: Cholecystectomy alleviates symptoms in patients with abnormal cholescintigraphy. *Gastroenterology* 101:786–793.

Zak, R. A., P. G. Frenkiel, J. W. Marks, et al. 1984. Cyclic nucleotides and glycoproteins during formation of cholesterol gallstones in prairie dogs. *Gastroenterology* 87:263–269.

Zeng, X. J., R. F. Yang, G. Xie, et al. 1992. The application of finite-element methods for the analysis of gallbladder bile flow in vivo. *Journal of Chongqing University* 15:91–98.

Zhu, J., T. Q. Han, S. Chen, et al. 2005. Gallbladder motor function, plasma cholecystokinin and cholecystokinin receptor of gallbladder in cholesterol stone patients. *World Journal of Gastroenterology* 11:1685–1689.

Zulpo, M., J. Balbus, P. Kuropka, et al. 2018. A model of gallbladder motility. *Computers in Biology and Medicine* 93:139–148.

CHAPTER 2

1D Models of Newtonian Bile Flow in the Biliary Tract

FLOW IN THE BILIARY TRACT

The biliary tract/tree consists of CD, CHD and CBD (Figure 2.1). Biliary diseases such as cholelithiasis and cholecystitis necessitate surgical removal of the GB, which is the most performed abdominal operation in the world. In this chapter, attention will be paid to flow resistance in the biliary tract since the resistance is related to bile stasis in the GB, which can initiate gallstone formation.

Currently, only a little is known about fluid mechanics in the biliary tract. Tera (1963) measured the dynamic viscosity of GB bile by using eight 8-cm-long capillary tubes with a diameter of 0.2 mm. He found that the normal GB bile was layered and the relative viscosity of the top, thinnest layer was 2.1 and the bottom thickest layer was 5.1. The relative viscosity is defined as the dynamic viscosity of the investigated fluid compared with that of distilled water, both at the same temperature.

Bouchier et al. (1965) also reported that the relative viscosity, determined by a capillary flow viscometer, was greater in pathological GB bile than normal GB bile and both were more viscous than hepatic duct bile. Although the concentration of normal GB bile affected the bile viscosity,

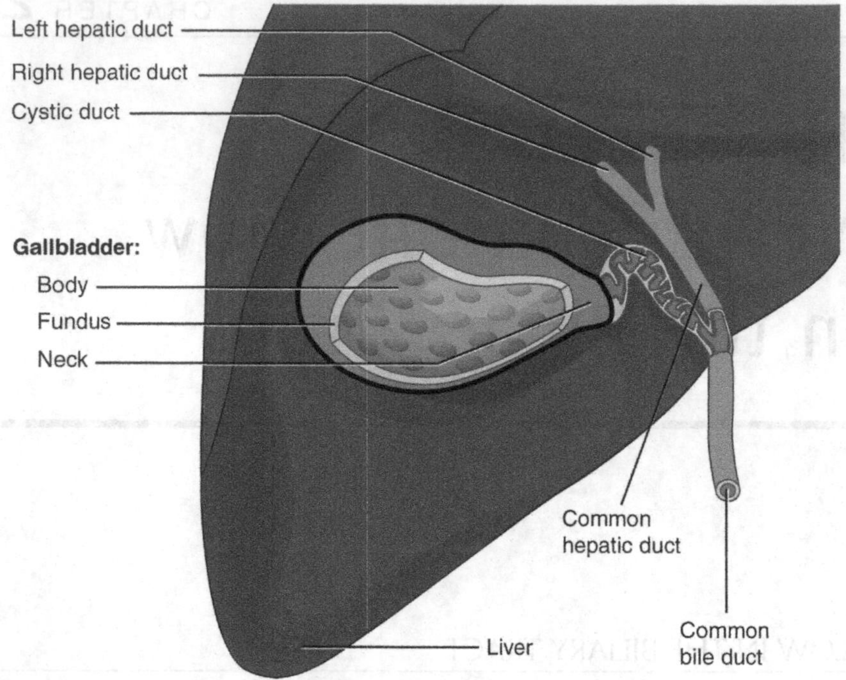

FIGURE 2.1 The biliary duct/tract (adapted from Schatz 2020).

in pathological and hepatic bile, the content of mucous was the major factor determining the viscosity.

Cowie and Sutor (1975) showed that the mean viscosity of bile from GBs containing stones was greater than that from healthy ones. The presence of mucous in GBs with stones was likely to account for the differences in viscosity based on the viscosity results using a Cannon-Fiske capillary viscometer at room temperature. The experimental work by Rodkiewicz and Otto (1979) showed that bile behaved like a Newtonian fluid. The bile samples of the growing pigs exhibited 1.41–1.45 mPa·s dynamic viscosity and 1009–1010 kg/m^3 density (Sambrook 1981).

Ooi (2004) measured the bile dynamic viscosity at various shear rates with the Brookfield rotational viscometer and found that the bile rheology of one-third of 59 patients was Newtonian. Reinhart et al. (2010) examined bile samples from the CBD of 138 patients by using a rotational viscometer at 37°C and in a range of 0.08–65.9 s^{-1} shear rates. The bile of the majority of the patients (64.5%) behaved like Newtonian fluid.

The complicated geometry of the biliary tree makes it difficult to estimate the flow resistance during bile emptying using the Poiseuille formula. Rodkiewicz et al. (1979) found that the flow of bile in the extrahepatic biliary tree of the dog was related to the flow resistance by a power law that differed from that for laminar flow in a rigid tube. Pitt et al. (1981) observed that the CD resistance was increased prior to gallstone formation; in consequence, the GB emptying function was impaired, suggesting that an increased flow resistance could play an aetiological role in the pathogenesis of gallstones.

Deenitchin et al. (1998) checked the relationships between CDs and cholelithiasis in 250 patients with cholelithiasis and 250 healthy controls. It was found that the patients with gallstones had significantly longer and narrower CDs than those without gallstones. The results suggested that complex geometry of the CDs might play an important role in cholelithiasis. An increase in the CD resistance has been shown to result in sludge formation and eventually stones in the GB (Pitt et al. 1982; Brugge et al. 1986; Soloway et al. 1977; Wolpers and Hofmann 1993).

Prolonged stasis of bile in the GB is a significant contributing factor to gallstone formation, suggesting that fluid mechanics, especially the pressure drop required to overcome the resistance experienced by bile flow during emptying, may play an important role in gallstone formation. Excessive flow resistance in bile emptying may result in incomplete bile emptying, leading to stasis and increasing the likelihood of forming cholesterol crystals as well as subsequent gallstone formation, but also high GB pressures, which could be a cause for acute pain observed *in vivo*.

The thin smooth muscle layer in CDs and CBD can respond to hormonal and neural stimuli (Burden 1925; Courtney et al. 1983; Daniels et al. 1961; Hauge and Mark 1965; Severi et al. 1988; Scott et al. 1979; Walsh 1979). However, the effect of the active contraction on the flow resistance across CDs and CBDs has been neglected in existing studies. In this context, the CD is considered either a rigid or passive compliant tube.

Ooi et al. (2004) performed a detailed numerical study on flow in 2D and 3D rigid CD models. The models were generated from patients' operative cholangiograms and acrylic casts. The flow resistances in these models were compared with those of an idealised straight duct with regular baffles or spiral structures. The influences of different baffle heights, numbers and Reynolds numbers on the flow resistance were investigated.

They found that the idealised CD duct model can provide qualitative results agreed with the realistic cast models from two different patients. Experimental work has been carried out to validate the CFD predictions in those simplified rigid CD models (Al-Atabi et al. 2005; Al-Atabi, Chin, and Luo 2006; Al-Atabi, Al-Zuhair et al. 2006).

The cholescintigraphy images of biliary tract were taken after a radioactive tracer, i.e., ^{99m}Tc-diisopropyl iminodiacetic acid was injected venously and then analysed by using optical flow method (dynamic 2D image analysis) in the study by Lo et al. (2015). The bile velocity and direction were captured when passing through the pylorus to the stomach.

CFD simulations have been performed in realistic CDs (Al-Atabi et al. 2012). 3D fluid domains were generated by scanning resin casts of real patients' CD lumen. The 3D flow structures and flow resistance across the CDs were obtained and assessed. However, the CFD modelling was limited to rigid CD models. Moreover, CFD simulations of flow in the whole biliary system in emptying phase (Kuchumov et al. 2014) and refilling phase (Kuchumov 2019) with a rigid GB and biliary tract were conducted in ANSYS CFX. Nevertheless, CFD simulation of a whole biliary system is time-consuming.

CHARACTERISTICS OF GEOMETRY

Anatomical descriptions of the human biliary tract/system date back to the 18th century when Heister (Dodds et al. 1989) reported spiral features in the lumen of the CD and called them 'valves'. Although later researchers doubted the valvular function, the term 'Heister's valve' is still in use today. The valves of Heister are projections of mucous membrane into lumen of the CD and can be folds or spiral leaflets (Mentzer 1927), as shown Figure 2.2.

The patency experiments illustrated that either folds or leaflets did not prevent the egress of bile from the GB and the bile could pass either direction through the CD with equal facility (Mentzer 1927). The patency of the human CD was experimented *in vitro* on cadavers and a flow resistance of 1–8 cm of water was found from the CD to the GB (Johnston and Brown 1932). The function of the valves of Heister was examined once more by observing CD anatomical structure and conducting patency experiments in the study by Lichtenstein and Ivy (1937). It was shown that the CD did not check bile flow in either direction. The function of the folds or leaflets was to prevent distension or collapse of the CD when it

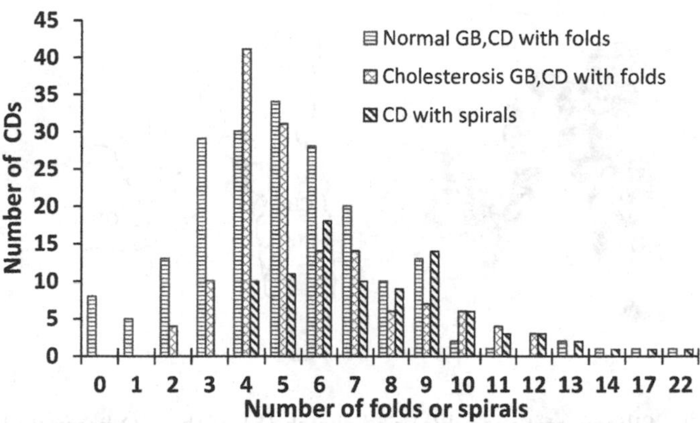

FIGURE 2.2 Number of folds or spirals in CDs (the data is from Mentzer 1927).

was subject to sudden pressure in the GB or CBD, and the twisting of CD to form folds was possibly due to the erect posture has limited the space in the upper abdomen for accommodation of the liver (Lichtenstein and Ivy 1937).

Recently, Dasgupta and Stringer (2005) debuted that the internal spiral folds in the CD might maintain its patency rather than regulate bile flow as unidirectional valves (Otto et al. 1979), even though the duct could facilitate GB emptying.

The complex shape of folds in the human CD can be observed by using cast models of the CDs harvested from autopsies or patients; Figure 2.3 is such an example. In the figure, there are at least five folds in the CD and the spiral structure is not clear.

Bird et al. (2006) have observed the effects of different geometries and their anatomical functions of the CDs based on acrylic resin casts of the neck and first part of the CD in GBs removed from gallstone diseases and patients undergoing partial hepatectomy for metastatic diseases. It was identified that in over half of the casts, spiral folds were not a dominant feature of the CD that could regulate bile flow in both directions rather than like unidirectional valves.

More recently, Pina et al. (2015) observed the cross sections of CDs harvested from 46 adult human cadavers. There existed 32 CDs with folds, and these folds were distributed uniformly along the CD in a discontinuous manner, like a number of isolated baffles.

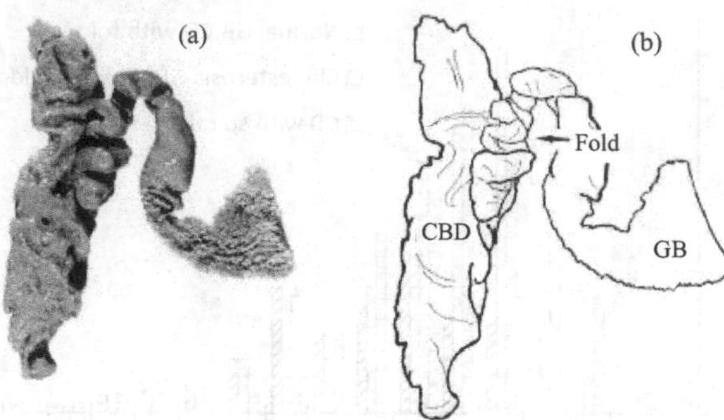

FIGURE 2.3 Silicon cast model (a) and sketch (b) of the CD harvested from an autopsy viewed from dorsal side (the picture is after Tominaga et al. 1995).

In this chapter, two 1D models of the human biliary tract are going to be proposed, one with a rigid wall and the other with an elastic wall. Both refilling and emptying phases are modelled, and the bile flow in the CHD and CBD is also taken into account. These models are based on the 3D straight duct with regular baffles used by Ooi et al. (2004). The rigid model is validated against the 3D simulations, and the differences between the elastic and rigid models are discussed. Using these models, the effects of physical parameters such as the CD diameter, dimensionless baffle height, number of baffles, the Young's modulus and the bile viscosity, on the pressure drop are studied in detail. The model details can be found in Li et al.'s (2007) study.

THE SIMPLIFIED BILIARY TRACT

The human biliary tract begins from the GB neck that funnels into the CD. Mucous membrane folds are generally prominent in the proximal part of the CD, which then smooths out to form a circular lumen at the distal end and joints with CBD and CHD. Furthermore, the actual geometry of the CD, CHD and CBD is curved, complicated and subject-dependent. To obtain a global flow resistance feature, the biliary tract is schematically reflected in Figure 2.4 and all the ducts are within the circular lumen.

The emptying phase begins immediately after a meal for about half an hour, but the refilling phase usually takes place after the emptying and lasts for several hours until a next meal.

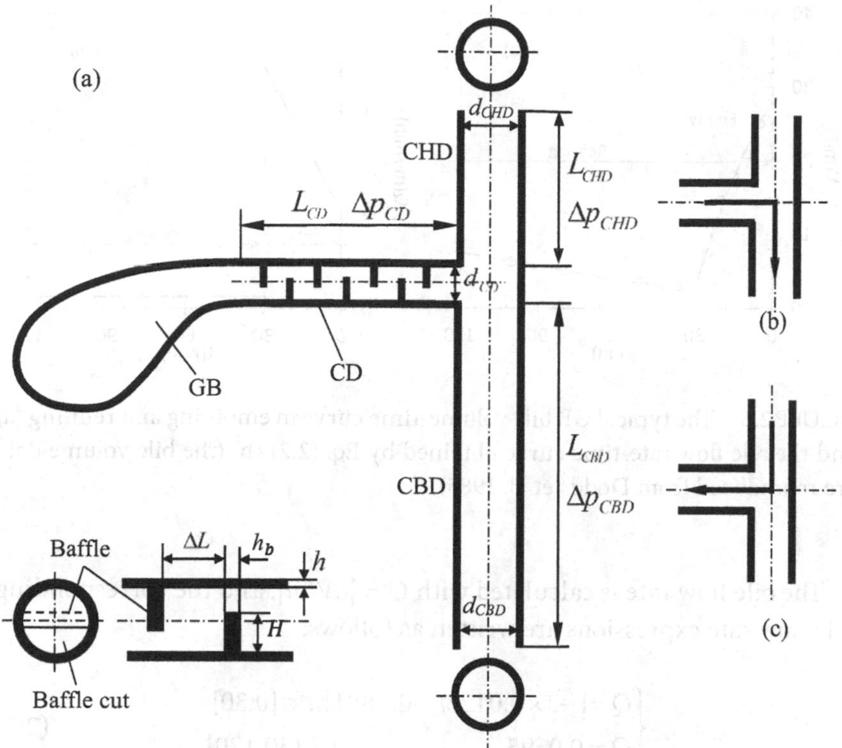

FIGURE 2.4 Schematic geometry model of human biliary system (a), the CD with a few baffles and the bile flow directions in emptying (b) and refilling (c) phases are shown.

The typical GB bile volume V and flow rate Q in both emptying and refilling phases are shown in Figure 2.5 in terms of time t (Dodds et al. 1985). The GB bile volume data in the emptying and refilling phases can be best fitted by using parabolic and linear equations, respectively. The fitted equations read as follows:

$$\begin{cases} V = -0.0123t^2 - 0.4881t + 27.9263, t \in [0,30] \\ V = 0.0595t + 0.3533, \quad t \in (30,120] \end{cases} \tag{2.1}$$

where the first formula is for the emptying and the second one is for the refilling.

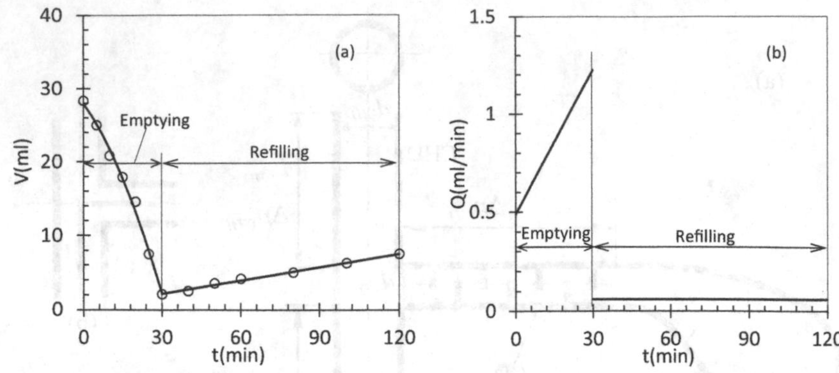

FIGURE 2.5 The typical GB bile volume-time curve in emptying and refilling (a) and the bile flow rate-time curve obtained by Eq. (2.2) (b) (the bile volume data are reproduced from Dodds et al. 1985).

The bile flow rate is calculated with $Q = |dV/dt|$, and the corresponding bile flow rate expressions are written as follows:

$$\begin{cases} Q = |-2\times 0.0123t - 0.4881|, t \in [0.30] \\ Q = 0.0595 \qquad\qquad t \in (30,120] \end{cases} \tag{2.2}$$

Based on Figure 2.5, the Reynolds number Re_{CD} and the scale of temporal acceleration of bile $\rho\partial u/\partial t$ can be estimated, where u is the mean bile velocity in a CD. For a healthy GB, the range of diameter of the CD is roughly d_{CD} = 1–4 mm (Bird et al. 2006). The maximum Reynolds number $Re_{\mathrm{CD}} = 4Q/\pi\nu d_{\mathrm{CD}} \approx 40$ at d_{CD} = 1 mm and the bile kinematic viscosity $\nu = 1.0\ \mathrm{mm}^2/\mathrm{s}$ in the emptying, and even smaller in the refilling, as such that the bile flow is laminar in the biliary tract.

The bile average density ρ is about 1000 kg/m^3. The temporal acceleration of bile $\rho\partial u/\partial t \approx 10^{-3}\ \mathrm{m/s}^2$ in the emptying but $\rho\partial u/\partial t \approx 10^{-4}\ \mathrm{m/s}^2$ in the refilling. The temporal acceleration can therefore be ignored in the model. For healthy GBs, the bile is considered as a Newtonian fluid.

1D MODELS

Since the pressure drop or flow resistance across the biliary tract in emptying may have a link with gallstone formation in the GB (Deenitchin et al. 1998), the pressure drop will be predicted with 1D mathematical models.

The CD makes the greatest contribution to the pressure drop in comparison with the CHD and CBD. Hence, the principal modelling work will be focused on the CD, while the Poiseuille flow is assumed in the other two biliary ducts. Furthermore, the effects of the baffles in the CD are taken into account in determining the equivalent diameter and length. The effects of the elastic wall are then involved by a straight smooth tube model of the CD by using the equivalent diameter and length.

The Rigid Wall Model

For a given bile flow rate, the flow resistance across a duct is defined as the pressure drop required to drive the flow along the duct. The pressure drop includes skin friction loss and any local flow separation or vortex loss.

Equivalent Diameter and Length

The CBD and CHD are assumed to be straight circular tubes and join at the T-junction (Figure 2.4). As done by Ooi et al. (2004), the baffles are arranged in a simple manner as shown in Figure 2.6. Unlike in the straight

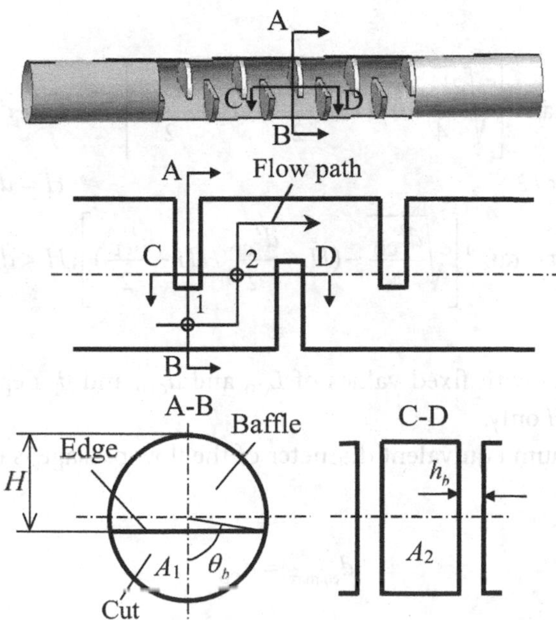

FIGURE 2.6 The baffle and cross sections of CD: A_1 is the cross-sectional area of flow at point 1 and A_2 is the cross-sectional area of the flow at point 2.

tube, the bile flow in the CD needs to negotiate its way around the baffles as shown by the arrow in the figure. Thus, the key issue in the modelling work is estimating the equivalent length, L_{eq}, and the equivalent diameter, d_{eq}, then treating the CD as an equivalent straight tube. Eventually, the pressure drop in the CD is calculated straightforward by assuming Poiseuille flow.

The equivalent diameter for the CD, d_{eq}, is dependent on the number of baffles, n, as well as the baffle height, H. In Figure 2.6, the bile flow travels twice the distance from points 1 to 2 between any two baffles in the duct, A_1 and A_2 are the corresponding cross-sectional areas at points 1 and 2. The lumen area A_1 can be easily calculated from the following equation:

$$A_1 = \frac{d^2_{CD}\theta_1}{4} - \left(H - \frac{d^2_{CD}}{2}\right)\sqrt{\frac{d^2_{CD}}{4} - \left(H - \frac{d_{CD}}{2}\right)^2} \quad (2.3)$$

where θ_b is half of the centre angle of the baffle cut, and is written as follows:

$$\theta_b = \begin{cases} \tan^{-1}\left[\sqrt{\frac{d^2_{CD}}{4} - (H - \frac{d^2_{CD}}{2}} / (H - \frac{d_{CD}}{2})\right], & H > d_{CD}/2 \\ \pi/2 & H = d_{CD}/2 \\ \pi + \tan^{-1}\left[\sqrt{\frac{d^2_{CD}}{4} - (H - \frac{d^2_{CD}}{2}} / (H - \frac{d_{CD}}{2})\right], & H < d_{CD}/2 \end{cases} \quad (2.4)$$

for a given CD with fixed values of L_{CD} and d_{CD}, and θ_b depends on the baffle height H only.

The maximum equivalent diameter of the flow passage is equal to d_{CD}, i.e.,

$$d_{eq,max} = d_{CD} \quad (2.5)$$

and the minimum equivalent diameter of the flow passage is associated with A_1. Here, θ_b represents the area of the square intersection formed by the plane through the central line of CD and paralleling the baffle edge with two successive baffles as well as the CD wall, i.e.,

$$d_{\text{eq,min}} = \sqrt{A_1 / \pi} \tag{2.6}$$

Now it is assumed that the equivalent diameter of CD varies linearly with the number of baffles between $d_{\text{eq,min}}$ and $d_{\text{eq,max}}$.

$$d_{\text{eq}} = d_{\text{eq,min}} + \left(d_{\text{eq,max}} - d_{\text{eq,min}}\right)\left(1 - \frac{n}{n_{\text{c}}}\right) \tag{2.7}$$

where n_c is the critical number of baffles in which $A_1 = A_2$. A_2 can be written as follows:

$$A_2 = d_{\text{CD}} \Delta L \tag{2.8}$$

The space ΔL between the two successive baffles is given as follows:

$$\Delta L = \frac{L_{\text{CD}} - n h_{\text{b}}}{n - 1} \tag{2.9}$$

For given values of L_{CD} and h_b, ΔL or A_2 varies with number of baffles n only. The values of A_2 are calculated and plotted as a function of number of baffles n in Figure 2.7. As given in the study by Ooi et al. (2004),

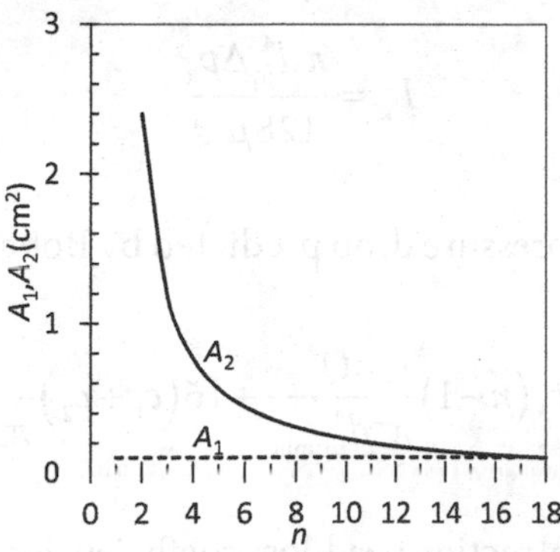

FIGURE 2.7 Variations of A_1 and A_2 with the number of baffles n.

the typical geometric parameters representing the average human CD are $L_{CD} = 50$ mm, $d_{CD} = 5$ mm, $n = 0$–14, $h = 1$ mm and $h_b = 1$ mm for the plot.

The value of A_1, which is independent of n, see Eq. (2.3), is also shown. As n increases, A_2 decreases towards A_1. When $A_1 = A_2$, the number of baffles will be as follows:

$$n_c = \frac{L_{CD} + \frac{A_1}{d_{CD}}}{h_b + \frac{A_1}{d_{CD}}} \tag{2.10}$$

Using the parameters given above, Eq. (2.10) predicts $n_c = 18$. For the range of parameters interested, A_1 is always smaller than A_2. Therefore, as it is very rare for human CDs to have more than the equivalent of 18 baffles, in the model, we assume that, $A_2 > A_1$, so that the minimum equivalent diameter is always estimated from A_1.

The equivalent length of the CD is determined from the actual length of the flow passage along the duct plus an extra length due to the complicated flow pattern, namely,

$$L_{eq} = H(n-1) + L_{CD} + L_k \tag{2.11}$$

where L_k denotes the extra length due to the minor pressure drop of local vortex from the cross-sectional area expansion, contraction and the flow path bending in the baffle zone. It can be estimated from White (2011) that

$$L_k = \frac{\pi d_{eq}^4 \Delta p_k}{128 \mu Q} \tag{2.12}$$

where Δp_k is the local pressure drop predicted by Bober and Kenyon (1980)

$$\Delta p_k = 16c_3(n-1)\frac{\rho Q^2}{\pi^2 d_{eq,min}^4} + 16(c_1 + c_2)\frac{\rho Q^2}{\pi^2 d_{eq,min}^4}. \tag{2.13}$$

Here, the sudden contraction head-loss coefficient is $c_1 = 0.42(1 - A_1/A_{CD})$, and the sudden expansion head-loss coefficient is $c_2 = (1 - A_1/A_{CD})^2$ for the CD (White 2011). The head-loss coefficient c_3 depends on the flow bending angle around baffles which is related to the number of baffles, n, and the dimensionless baffle height ξ ($=H/d_{CD}$).

Usually, the averaged flow bending angle for different n is less than 90°. For a 90° bend, c_3 has been measured to be 0.75 (Bober and Kenyon 1980). In the model, the angle through which the flow bends around a baffle should largely depend on the dimensionless baffle height, ξ, and to a lesser extent, on the number of baffles too. For simplicity, however, it is assumed that the angle is a linear function of ξ, i.e., c_3 = constant × ξ, where the constant is chosen to be 0.85. Thus, for $\xi = 0$ (straight tube flow), $c_3 = 0$, and for $\xi = 0.9$, where 3D simulations typically show that flow turns through 90° around baffles, $c_3 = 0.75$.

The Emptying Phase

The pressure drop in the CD in the emptying phase for a given number of baffles can be estimated for the Poiseuille flow (White 2011) as follows:

$$\Delta p_{\text{CD}} = \frac{128\mu Q}{\pi d_{\text{eq}}^4} L_{\text{eq}} \tag{2.14}$$

For the CBD, in the emptying phase, the pressure drop can be written as follows:

$$\Delta p_{\text{CBD}} = \frac{128\mu Q}{\pi d_{\text{CBD}}^4} L_{\text{CBD}} + \Delta p_{\text{te}} \tag{2.15}$$

where Δp_{te} accounts for the pressure drop owing to the T-junction which consists of one 90° bend and one expansion, and is given by

$$\Delta p_{\text{te}} = 16c_4 \frac{\rho Q^2}{\pi^2 d_{\text{CD}}^4} + 16c_2 \frac{\rho Q^2}{\pi^2 d_{\text{CD}}^4} \tag{2.16}$$

The coefficients $c_4 = 0.75$ for 90° bend and c_2 may be treated in the same manner as those for Eq. (2.13). Thus, the total pressure drop in the biliary system during the emptying phase is given as follows:

$$\Delta p_{\text{EM}} = \frac{128\mu Q}{\pi d_{\text{eq}}^4} L_{\text{eq}} + \frac{128\mu Q}{\pi d_{\text{CBD}}^4} L_{\text{CBD}} + \Delta p_{\text{te}} \tag{2.17}$$

The Refilling Phase

Likewise, during the refilling phase, the pressure drop in the CBD is expressed by Eq. (2.15), and the pressure drop in the CHD is given as follows:

$$\Delta p_{\mathrm{CHD}} = \frac{128\mu Q}{\pi d_{\mathrm{CHD}}^4} L_{\mathrm{CHD}} + \Delta p_{\mathrm{th}} \tag{2.18}$$

where

$$\Delta p_{\mathrm{th}} = 16 c_4 \frac{\rho Q^2}{\pi^2 d_{\mathrm{CHD}}^4} + 16 c_1 \frac{\rho Q^2}{\pi^2 d_{\mathrm{CHD}}^4} \tag{2.19}$$

and the total pressure drop during the refilling is

$$\Delta p_{\mathrm{RF}} = \frac{128\mu Q}{\pi d_{\mathrm{eq}}^4} L_{\mathrm{eq}} + \frac{128\mu Q}{\pi d_{\mathrm{CHD}}^4} L_{\mathrm{CHD}} + \Delta p_{\mathrm{th}} \tag{2.20}$$

The Elastic Wall Model

In order to obtain a more realistic description for the pressure drop in the human biliary tract, an elastic wall model will be considered. The wall of CDs is a nonlinear soft tissue, i.e., the Young's modulus varies with the internal pressure (Jian and Wang 1991; Duch et al. 1998; Duch et al. 2002; Duch et al. 2004; Slater et al. 1983). In the model, however, the CD wall is assumed to be a linear, homogeneous, isotropic, elastic material with a uniform wall thickness and without active contraction.

Additionally, the CHD and CBD are assumed to be rigid and without active contraction as well. The reasons for those assumptions are that the Young's modulus of these ducts is larger than that of the CD (Jian and Wang 1991) and the deformation of the ducts is also much smaller than the CD.

The elastic CD is modelled as an equivalent tube with an equivalent length, L_{eq}, and a diameter, d_{eq}. The effects of baffles on the flow come implicitly through L_{eq} and d_{eq}. The initial cross section of the CD is circular. When the GB starts to contract and the sphincter of Oddi opens, the bile pressure rises in the GB but reduces in the CBD. As a result, the bile flow out of the GB is initiated and the emptying phase begins. A negative pressure gradient is established along the CD, i.e., the bile pressure decreases downstream in the CD. Thus, the transmural pressure (internal pressure minus external pressure) will become negative downstream the CD during emptying and the CD becomes partially collapsed downstream. This belongs an FSI problem, which can be modelled with well-known work on collapsible tube flows (Flaherty et al. 1972; Cancelli and Pedley 1985; Pedley et al. 1996).

The Emptying Phase

A partially collapsed CD is shown schematically in Figure 2.8, where p_e is the external pressure, which is equal to the pressure in the chest, $p_e = 1.5$ kPa (Dodds et al. 1989) (above one atmospheric pressure). A 1D coordinate system x originating from point O is introduced. As the bile flows down the CD, the internal pressure decreases due to viscous losses, causing a decrease in transmural pressure, $p - p_e$, from the inlet (A_{in}) to the outlet (A_{out}). The governing equations for the flow in the elastic CD read as follows (Pedley et al. 1996):

$$Q = Au \tag{2.21a}$$

$$\rho u \frac{du}{dx} = -\frac{dp}{dx} - \frac{8\pi\mu Q}{A^2} \tag{2.21b}$$

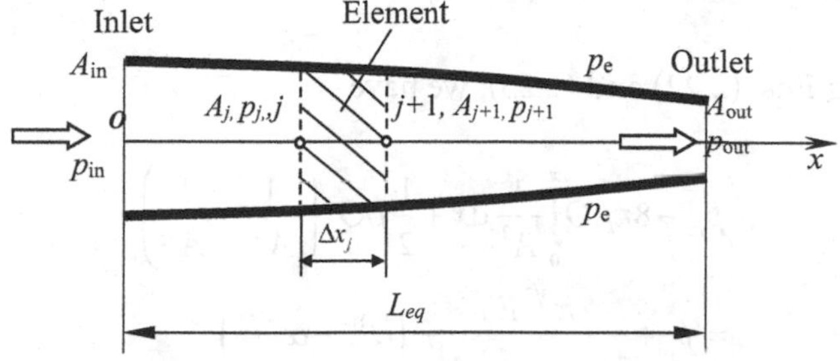

FIGURE 2.8 A CD tube with elastic wall in the emptying phase, the tube is circular at the inlet initially, but it collapses partially downstream as bile flows due to the bile pressure reduction along the tube.

The bile pressure at the inlet is chosen as the reference pressure. At a given flow rate, the corresponding pressure in the duct is derived by integrating Eq. (2.21b)

$$p = p_{\text{in}} - 8\pi\mu Q \int_0^x \frac{1}{A^2}\,\mathrm{d}x + \frac{1}{2}\rho Q^2 \left(\frac{1}{A_{\text{in}}^2} - \frac{1}{A^2}\right). \tag{2.22}$$

The constitutive equation for the duct with an elastic wall obeys the tube law for linear, homogeneous, isotropic, elastic materials (Flaherty et al. 1972)

$$p - p_{\text{e}} = \frac{Eh^3}{12\left(1-\upsilon^2\right)r^3} F(\alpha) \tag{2.23}$$

where $\alpha = A/A_{\text{eq}}$, $A_{\text{eq}} = \pi d_{\text{eq}}^2/4$, E is Young's modulus, υ is the Poisson's ratio of elastic CD wall, h is the thickness of wall and r is the inner diameter of the duct. $F(\alpha)$ is usually determined by experiments. For veins, the tube law can be expressed as follows (Cancelli and Pedley 1985; Pedley et al. 1996):

$$F(\alpha) = \alpha^{10} - \alpha^{-3/2} \tag{2.24}$$

Since there are not experimental data for the CD, here it is assumed that the tube law for the CD obeys Eq. (2.24). The fluid pressure estimated by using Eq. (2.23) is given as follows:

$$p = p_{\text{e}} + \frac{\pi^{3/2} Eh^3}{12\left(1-\upsilon^2\right)A^{3/2}}\left(\alpha^{10} - \alpha^{-3/2}\right) \tag{2.25}$$

Combining Eqs. (2.22) and (2.25), we have

$$\begin{aligned} p_{\text{in}} - 8\pi\mu Q \int_0^x \frac{1}{A^2}\,\mathrm{d}x + \frac{1}{2}\rho Q^2 \left(\frac{1}{A_{\text{in}}^2} - \frac{1}{A^2}\right) \\ = p_{\text{e}} + \frac{\pi^{3/2} Eh^3}{12\left(1-\upsilon^2\right)A^{3/2}}\left(\alpha^{10} - \alpha^{-3/2}\right) \end{aligned} \tag{2.26}$$

Equation (2.26) represents a 1D boundary value problem, which is solved for $A(x)$ using a finite difference method. The duct is divided into J elements

(J is chosen to be >300); a typical element extending from node j to $j+1$ is illustrated in Figure 2.8. At the (j+1)th node,

$$\begin{aligned} & p_j + \frac{1}{2}\rho Q^2\left(\frac{1}{A_j^2} - \frac{1}{A_{j+1}^2}\right) - 8\pi\mu Q\left(\frac{1}{A^2}\right)_{j+1/2} \\ & \Delta x_j = p_e + \frac{\pi^{3/2}Eh^3}{12\left(1-\upsilon^2\right)A_{j+1}^{3/2}}\left[\left(\frac{A_{j+1}}{A_{eq}}\right)^{10} - \left(\frac{A_{j+1}}{A_{eq}}\right)^{-3/2}\right] \end{aligned} \tag{2.27}$$

where

$$p_j = p_{in} - 8\pi\mu Q\int_0^{x_j}\frac{1}{A^2}\,\mathrm{d}x + \frac{1}{2}\rho Q^2\left(\frac{1}{A_{in}^2} - \frac{1}{A_j^2}\right)$$

$$\left(\frac{1}{A^2}\right)_{j+1/2} = \frac{1}{2}\left(\frac{1}{A_j^2} + \frac{1}{A_{j+1}^2}\right)$$

and p_j is known. Expressing $(1/A^2)_{j+1/2}$ in terms of A_j and A_{j+1}, Eq. (2.27) can also be written as follows:

$$\begin{aligned} & p_j + \frac{1}{2}\rho Q^2\left(\frac{1}{A_j^2} - \frac{1}{A_{j+1}^2}\right) - 4\pi\mu Q\left(\frac{1}{A_j^2} + \frac{1}{A_{j+1}^2}\right)\Delta x_j \\ & = p_e + \frac{\pi^{3/2}Eh^3}{12\left(1-\upsilon^2\right)A_{j+1}^{3/2}}\left[\left(\frac{A_{j+1}}{A_{eq}}\right)^{10} - \left(\frac{A_{j+1}}{A_{eq}}\right)^{-3/2}\right] \end{aligned} \tag{2.27a}$$

The bisection method is employed to solve Eq. (2.27a) to find unknown A_{j+1} in the region $A_{j+1} \in \left[0.1A_{eq}\, 2A_{eq}\right]$ iteratively. After A_{j+1} is obtained, the pressure at node $j+1$ will be given by

$$p_{j+1} = p_j + \frac{1}{2}\rho Q^2\left(\frac{1}{A_j^2} - \frac{1}{A_{j+1}^2}\right) - 8\pi\mu Q\left(\frac{1}{A^2}\right)_{j+1/2}\Delta x_j \tag{2.28}$$

The boundary conditions are applied at the inlet (node 1)

$$\begin{cases} \alpha_{in} = A_{in}/A_{eq} \\ p_{in} = p_e + \dfrac{\pi^{3/2}Eh^3}{12\left(1-\upsilon^2\right)A_{in}^{3/2}}\left(\alpha_{in}^{10} - \alpha_{in}^{-3/2}\right) \end{cases} \tag{2.29}$$

If α_{in}, then $p_{in} = p_e$; else if $\alpha_{in} > 1$, then $p_{in} > p_e$. The maximum pressure drop in the CD is thus $\Delta p_{CD} = p_{in} - p_{out}$, and the total pressure drop occurring during emptying is

$$\Delta p_{EM} = \Delta p_{CD} + \frac{128\mu Q}{\pi d_{CBD}^4} L_{CBD} + \Delta p_{te}. \tag{2.30}$$

The Refilling Phase

Because the bile flow rate is very small during refilling phase and the refilling time is at least three times longer than the emptying time, the CD wall can be regarded as rigid during this phase. Equations (2.18)–(2.20) in the rigid model are applied to calculate the pressure drop.

PARAMETER SET AND MODEL VALIDATION

Parameter Set

The parameters used in the models are listed in Table 2.1. Most of these are taken from the statistics of human CDs given by Deenitchin et al. (1998). The range of values for ξ, n and d_{CD} is chosen to be the same as in the 3D models of Ooi et al. (2004). The GB bile flow rate is shown in Figure 2.5(b).

The range of the Young's modulus used is based on the measurements of Jian and Wang (1991), where the biliary ducts of 16 healthy adult dogs were tested with a pressure ranging from 4.7 to 8 kPa. The physiological internal pressure is normally around 1.5 kPa in the human biliary tract outside the pressure range used in Jian and Wang's (1991) study. To obtain meaningful results, the Young's modulus for the pressure around 1.5 kPa was estimated by extrapolating the data in the study by Jian and Wang (1991). The Young's modulus chosen is therefore in the range of 100–1000 Pa for the internal pressure varying from 1.03 to 1.9 kPa.

TABLE 2.1 A summary of parameters for the human biliary tract.

Duct and fluid	Parameter
CD	d_{CD} = 1–6 mm, L_{CD} = 40 mm, $h = h_b$ = 0.5 mm, ξ = 0.3–0.7, n = 0–18, υ = 0.5, E = 100–1000 Pa
CBD	d_{CBD} = 6 mm, L_{CBD} = 100 mm
CHD	d_{CHD} = 6 mm, L_{CHD} = 40 mm
Bile	ρ = 1000 kg/m^3, v = 1–3 mm^2/s, $\mu = \rho v$

The Model Validation

Since several assumptions have been adopted in deriving the equivalent diameter and length for the 1D model, the model has to be compared with the 3D rigid CD models solved with the numerical methods by using a general purpose CFD software FLUENT 6.2.

The mesh corresponding to a mesh-independent pressure drop is determined at first. The mesh sizes 0.5, 0.4 and 0.3 mm as well as the first- and second-order upwind schemes respectively are used to calculate a steady, laminar flow of Newtonian bile in an ideal 3D CD in Fluent 6.2. The mesh size is measured by the edge of a tetrahedral cell, which is applied to discretise the flow domain. The CD geometry parameters are: L_{CD} = 50 mm, d_{CD} = 5 mm, n = 6, ξ = 0.5, $h = h_b$ = 1 mm, and the Newtonian bile: ρ = 1000 kg/m³, ν = 1 mm²/s (Ooi et al. 2004). The effect of the mesh size on the pressure drop is illustrated in Figure 2.9.

The pressure drop at the mesh size 0.5 mm (about 63,000 cells) differs from that at the 0.3 mm (about 220,000 cells) significantly in the first-order upwind scheme. In the second-order upwind scheme, however, the difference in the pressure drop at the mesh sizes 0.5 and 0.3 mm is very small. Thus, the mesh size 0.5 mm and the second-order upwind scheme are used in the CFD computations.

The pressure drop curves with Reynolds number are illustrated in Figure 2.10 with the rigid model for the CD with or without baffles. The geometry and bile parameters are L_{CD} = 50 mm, d_{CD} = 5 mm, n = 0, 2,

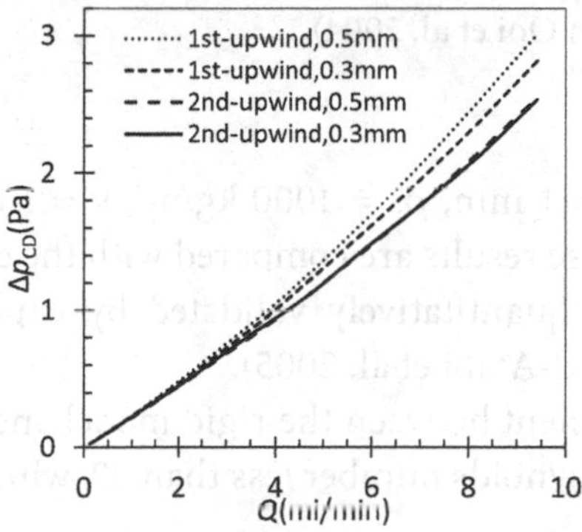

FIGURE 2.9 The pressure drops of the Newtonian bile in the rigid CD, the results were obtained by using FLUENT 6.2 with mesh sizes 0.3, 0.4 and 0.5 mm and the first- and second-order upwind schemes.

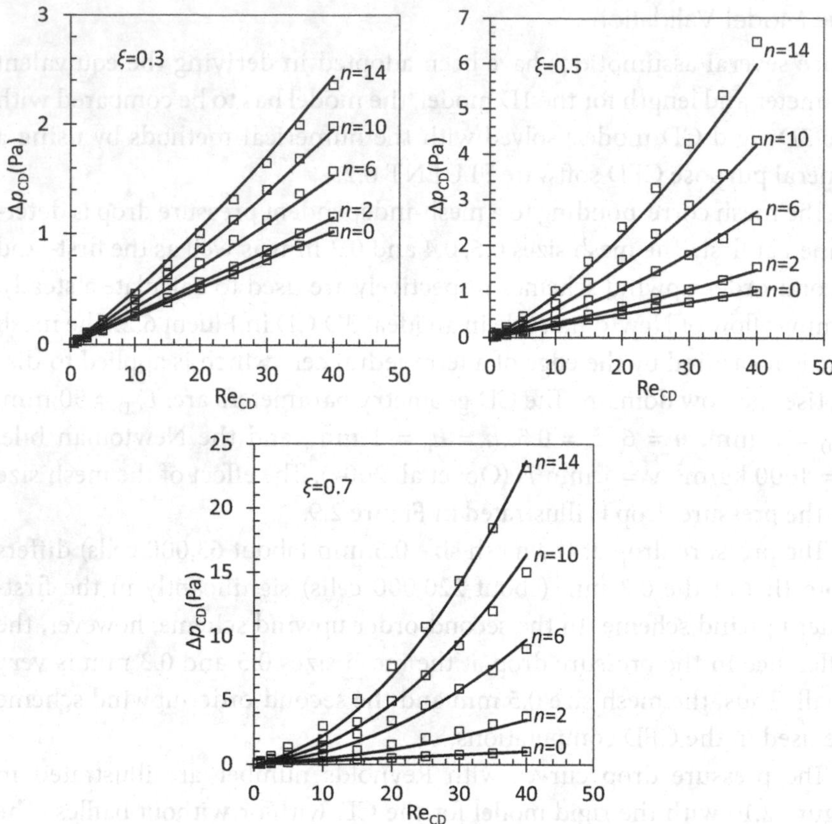

FIGURE 2.10 Comparison of the pressure drop estimated using the 1D rigid model (solid line) and the 3D numerical simulations (symbols) (the 3D geometry of the CD is taken from Ooi et al. 2004).

6, 10 and 14, $h = h_b$ = 1 mm, ρs = 1000 kg/m^3, ν = 1 mm^2/s, respectively (Ooi et al. 2004). These results are compared with the corresponding CFD results, which were quantitatively validated by experiments at higher Reynolds numbers (Al-Atabi et al. 2005).

Clearly, the agreement between the rigid model and 3D CFD results is good, especially at Reynolds number less than 20, which is in the range of interest.

The elastic model is derived for a straight circular tube with equivalent diameter and length to the duct with baffles, and the tube law is based on the experimental curve for a straight rubber tube (Pedley et al. 1996).

Therefore, if the rigid model with the correct equivalent diameter and length is acceptable, then the elastic model is likely to be satisfactory.

Pressure Drop for the Reference Parameter Set

There are a few parameters presented in the models above, and each can vary within its own physiological range. To clarify the effect of each individual parameter on pressure drop, a reference parameter set is specified, which is based on the averaged values of the normal human CDs. The reference parameter set is: $n = 7$, $\xi = 0.5$, $\nu = 1.275$ mm²/s, $d_{CD} = 1$ mm, $L_{CD} = 40$ mm, $E = 300$ Pa, $\alpha_{in} = 1$ and $Q = 1$ ml/min. The effect of a particular parameter on the pressure drop is determined by varying this parameter while keeping all the other parameters fixed.

The predicted pressure drops in the human biliary tract for the reference parameter set using the rigid and elastic models in the emptying and refilling phases are illustrated in Figure 2.11. Two cases $\alpha_{in} = 1$ and 1.2 are considered. $\alpha_{in} = 1$ is the case where the inlet of the CD is not expanded, while $\alpha_{in} = 1.2$ indicates a duct expansion because this has been observed clinically. For $\alpha_{in} = 1$, the elastic model predicts a greater pressure drop in the emptying phase due to the collapse of the CD. It is also noted that the maximum pressure drop agrees with the typical physiological observation of 20–200 Pa (Dodds et al. 1989; Doty et al. 1983).

The ratio of the total pressure drop in the CBD or CHD to the pressure drop in the CD can indicate the importance of the pressure drop across

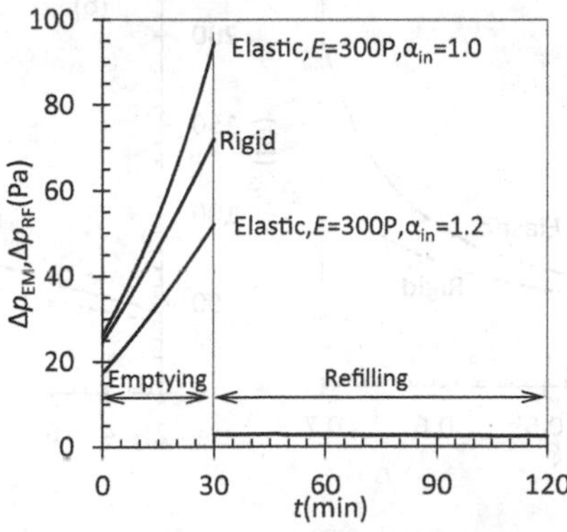

FIGURE 2.11 The pressure drop curves with time predicated using both the rigid and the elastic 1D models, the parameters in the reference parameter set are used.

the CD in the human biliary tract. The results show that the pressure drop in the CBD is less than 1.5%, and in the CHD less than 0.15% only, compared with that in the CD. This justifies estimating the pressure drop in the human biliary tract only from the CD model.

In the following sections, only the pressure drop in the CD is presented. All the parameters used below are in those in the reference parameter set unless otherwise stated.

EFFECTS OF PARAMETER ON PRESSURE DROP

The effects of the dimensionless baffle height ξ $(= H/d_{\text{CD}})$ and number of baffles n on the pressure drop are shown in Figure 2.12. The pressure drop predicted by the elastic model is also compared with the corresponding rigid model. The pressure drop rises as ξ increases since the greater the dimensionless baffle height, the narrower the equivalent diameter. As ξ varies from 0.3 to 0.7, the pressure drop increases from 50 to 100 and 200 Pa for the rigid and elastic ducts, respectively.

The pressure drop increases as both n and ξ rises. The change in ξ leads to a more significant variation in pressure drop for both rigid and elastic ducts (from approximately 50 to 200 Pa) than n does (from approximately 50 to 100 Pa). Also, the pressure drop predicted by the elastic wall model is always higher than that estimated by the rigid wall model for all values of ξ and n owing to the duct collapse downstream (Figure 2.14).

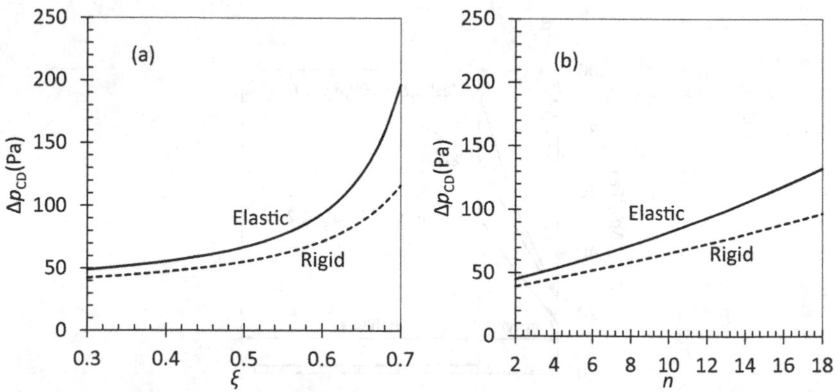

FIGURE 2.12 The pressure drop curves with dimensionless baffle height ξ (a) and number of baffles n (b), all the other parameters are in the reference parameter set.

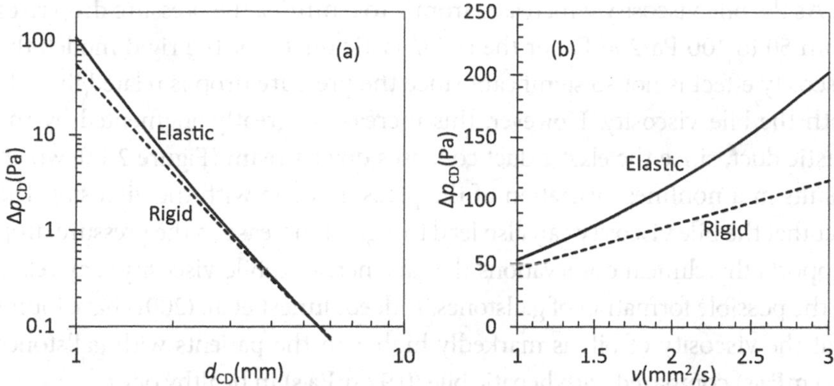

FIGURE 2.13 The pressure drop curves with CD diameter d_{CD} (E = 100 Pa) (a) and bile viscosity ν (E = 300 Pa) (b), all the other parameters are those in the reference parameter set.

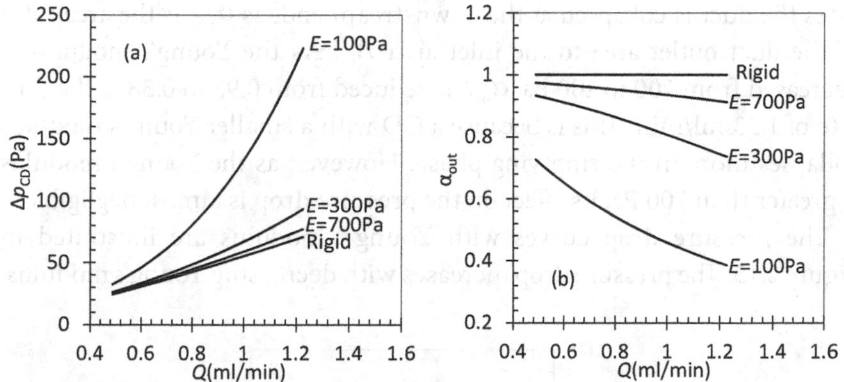

FIGURE 2.14 The variation of pressure drop Δp_{CD} (a) and the area ratio α_{out} (b) with flow rate at various Young's moduli, all the other parameters are those in the reference parameter set.

The pressure drop curves with the CD diameter d_{CD} and bile viscosity ν are demonstrated in Figure 2.13. The diameter has the strongest effect on the pressure drop, i.e., the pressure drop is almost proportional to d_{CD}^{-4} (this is strictly true from the Poiseuille flow in a duct without baffles), thus a narrow diameter causes a dramatic increase in pressure drop, as shown in Figure 2.13, and 1% decrease in d_{CD} gives rise to 4–4.4% increment in the pressure drop.

As the bile viscosity v increases from 1 to 3 mm²/s, the pressure drop rises from 50 to 100 Pa/200 Pa for the rigid/elastic duct. For the rigid model, the viscosity effect is not so significant since the pressure drop is related linearly with the bile viscosity. However, this increase is greatly augmented by the elastic duct, since the elastic duct collapses downstream (Figure 2.14), which results in a nonlinear variation of the pressure drop with the viscosity. The fact that the bile viscosity can also lead to a great increase in the pressure drop supports the clinical observations that an increased bile viscosity may relate to the possible formation of gallstones. Indeed, Jungst et al. (2001) have found that the viscosity of bile is markedly higher in the patients with gallstones (5.0 mPa·s) compared with hepatic bile (0.92 mPa·s) in healthy ones.

The effects of the Young's modulus on the CD and the bile flow rate are shown in Figure 2.14, where the pressure drop and the minimum cross-sectional area ratio are plotted against flow rate for various Young's moduli. A lower Young's modulus (i.e., a more compliant duct) produces a greater pressure drop and a more dominant reduction in α_{out}. A value of α_{out} indicates the duct is collapsed at the downstream end, as α_{out} is the area ratio of the duct outlet area to the inlet area A_{in}. As the Young's modulus is decreased from 700 to 100 Pa, α_{out} is reduced from 0.92 to 0.38 at the flow rate of 1.23 ml/min. This is because a CD with a smaller Young's modulus collapses more in the emptying phase. However, as the Young's modulus is greater than 700 Pa, its effect on the pressure drop is almost negligible.

The pressure drop curves with Young's modulus are illustrated in Figure 2.15. The pressure drop increases with decreasing Young's modulus.

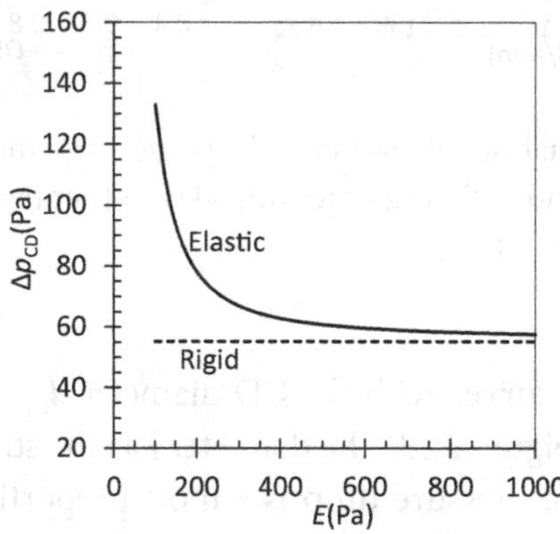

FIGURE 2.15 The variation of pressure drop Δp_{CD} with Young's modulus, all the other parameters are in the reference parameter set.

As the modulus varies from 1000 to 100 Pa, the pressure drop increases from 64 to 130 Pa.

THE DARCY FRICTION FACTOR

To gain more understanding to the results obtained, the Darcy friction factor (White 2011) is selected as a dimensionless parameter to show the effects of the parameters on the pressure drop.

For the steady fully developed laminar flow in a circular straight pipe with an inner diameter d and length L, the pressure drop across the pipe Δp is related to the flow shear stress at the wall τ_w in the following expression (White 2011):

$$\Delta p = 4\tau_w \left(\frac{L}{d}\right) \tag{2.31}$$

The result of the dimensional analysis of the pressure drop (flow resistance) of flow in this pipe is given as follows:

$$\frac{\Delta p}{\frac{1}{2}\rho\left(\frac{4Q}{\pi d^2}\right)^2} = f\left(\frac{L}{d}, \frac{4\rho Q}{\pi \mu d}\right) \tag{2.32}$$

where Q and μ are the flow rate and dynamic viscosity of the fluid, respectively, and f is a dimensionless parameter that is related to the Reynolds number $4\rho Q/\pi\mu d$. Since Eq. (2.31) indicates Δp is proportional to L/d, Eq. (2.32) can be simplified as follows:

$$\frac{\Delta p}{\frac{1}{2}\rho\left(\frac{4Q}{\pi d^2}\right)^2} = \left(\frac{L}{d}\right) f\left(\frac{4\rho Q}{\pi \mu d}\right) \tag{2.33}$$

Thus, the dimensionless parameter is

$$f = \frac{\Delta p}{\frac{1}{2}\rho\left(\frac{L}{d}\right)\left(\frac{4Q}{\pi d^2}\right)^2} \tag{2.34}$$

This parameter f is called the Darcy friction factor, after Henry Darcy (1803–1858), a French engineer. f represents the dimensionless pressure drop of flow due to the friction at the pipe wall.

The Darcy factor varies with the geometrical similarity and Reynolds number of flows. If the geometry of two flow systems is similar, the friction factor of the two systems is the same at any Reynolds number. Otherwise, it will differ from each other. When the baffle height and number of baffles vary, the equivalent diameter and length will be different, then the geometrical similarity of the CD will be destroyed; as a result, the corresponding friction factor will change its value.

Similarly, the Darcy friction factor for the models with baffles is defined as follows:

$$f = \frac{\Delta p_{\mathrm{CD}}}{\frac{1}{2}\rho\left(\frac{L_{\mathrm{eq}}}{d_{\mathrm{eq}}}\right)\left(\frac{4Q}{\pi d_{\mathrm{eq}}^2}\right)^2} \tag{2.35}$$

For a straight circular tube with diameter d_{eq} and length L_{eq}, it is assumed that the Poiseuille flow exists in the tube without baffles, then the Darcy friction factor, denoted by f_0, is given by the following equation:

$$f_0 = \frac{\dfrac{128\mu Q L_{\mathrm{eq}}}{\pi d_{\mathrm{eq}}^4}}{\frac{1}{2}\rho\left(\frac{L_{\mathrm{eq}}}{d_{\mathrm{eq}}}\right)\left(\frac{4Q}{\pi d_{\mathrm{eq}}^2}\right)^2} \tag{2.36}$$

The friction factor ratio is then,

$$f/f_0 = \Delta p_{\mathrm{CD}} \Bigg/ \left(\frac{128\mu Q L_{\mathrm{eq}}}{\pi d_{\mathrm{eq}}^4}\right) \tag{2.37}$$

The ratio f/f_0 indicates now the pressure drop in a CD with baffles differing from that in an ideal pipe. When Δp_{CD} yields Poiseuille's formula, $f/f_0 = 1$, otherwise $f/f_0 > 1$.

Figure 2.16 illustrates the friction factor ratio f/f_0 variation with the Reynolds number ($Re = 4\rho Q/\pi\mu d_{\mathrm{eq}}$) for the CD with both rigid and elastic walls for two different values of n and ξ.

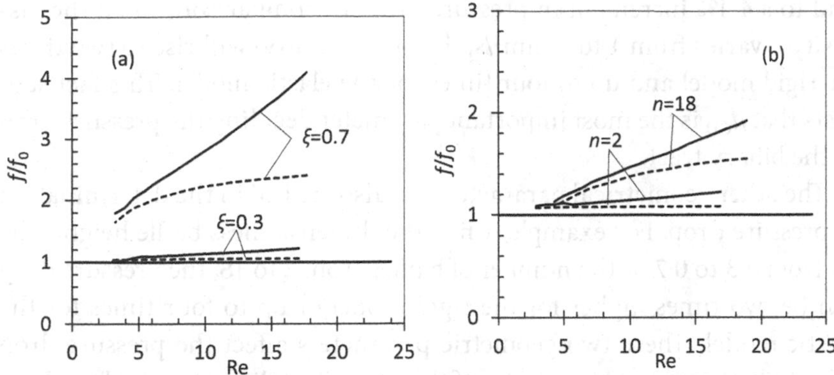

FIGURE 2.16 The variation of friction factor ratio f/f_0 with Reynolds number Re in the CDs with rigid and elastic walls at $\xi = 0.3$ and 0.7 (a) and $n = 2$ and 18 (b), all the other parameters are in the reference parameter set.

TABLE 2.2 Equivalent diameter and length.

	Case 1, $n=7$			Case 2, $\xi = 0.5$	
ξ	d_{eq}(mm)	L_{eq}(mm)	n	d_{eq}(mm)	L_{eq}(mm)
0.3	0.97–5.48	44.1–52.7	2	0.98–5.79	43.2–45.4
0.5	0.95–5.18	53.0–64.7	7	0.95–5.18	53.1–64.7
0.7	0.94–5.13	105.9–107.4	18	0.88–3.89	68.3–96.7

All the curves are above 1, usually more baffles or higher baffles result in a greater friction factor ratio, which is augmented further by having an elastic wall, especially at a large Reynolds number. The friction factor ratio owing to n variation is less than 1.2 for the rigid duct and less than 2.0 for the elastic wall duct. Whilst the change in the ratio caused from ξ variation is less than 2.0 and 4.0 for the rigid and elastic ducts, respectively. The reason for this fact is that the baffle height can lead to more significant changes in both the equivalent diameter and the equivalent length than the number of baffles. The detailed information on the equivalent diameter and length at various heights and numbers of baffles is listed in Table 2.2.

CLINICAL SIGNIFICANCE

The most significant parameter for the pressure drop across the biliary tract is the CD diameter d_{CD}. It is shown that a 1% decrement in d_{CD} can lead to a 4.4% increment in pressure drop. In comparison, when the viscosity v varies from 1 to 3 mm²/s, the pressure drop will rise to two times for rigid model and up to four times for the elastic model. This fact suggests that d_{CD} is the most important parameter deciding the pressure drop in the biliary tract.

The other geometrical parameters are also critical in the determination of pressure drop. For example, when the dimensionless baffle height varies from 0.3 to 0.7 or the number of baffles from 2 to 18, the pressure drop can be two times higher for the rigid model or up to four times for the elastic model. These two geometric parameters affect the pressure drop effectively through the changing of the equivalent diameter and length, as specified by Eqs. (2.7) and (2.11), respectively.

The elasticity of the CD also plays an important role. For instance, when the Young's modulus decreases from 1000 to 100 Pa, the pressure drop can be as high as three times. However, when the Young's modulus of the CD is larger than 700 Pa, the elastic-wall model and the rigid-wall model provides nearly the same pressure drop.

Gallstone formation is closely related to the CD resistance or the pressure drop (Pitt et al. 1982) because the resistance or the drop can cause bile stasis in GBs. Deenitchin et al. (1998) has illustrated that the patients with gallstones tend to have long and narrow CDs. The long and narrow CD should contribute a larger pressure drop across the CD and results in a bile stasis in the GB to prompt gallstone formation. The pressure drops presented in terms of CD diameter, dimensionless baffle height and number of baffles in section *Effects of Parameter on Pressure Drop* support that observation.

It was found that GB stasis is also related to hypersecretion of GB mucus from the liver, which contributes to an increase in bile viscosity in the GB (Jungst et al. 2001). The increased bile viscosity in the GB can raise the pressure drop across the CD. As a result, a bile stasis is established easily in the GB. The pressure drops against bile viscosity in *Effects of Parameter on Pressure Drop* section agree with that fact.

In this chapter, bile is assumed to be a Newtonian fluid, i.e., its viscosity is independent of the shear rate. However, recent experimental studies suggest that bile may display non-Newtonian behaviour such as shear thinning (Jungst et al. 2001; Kuchumov et al. 2014). In addition, tests that were carried out in the laboratory on fresh human bile after operations seem to suggest that the degree of the non-Newtonian behaviour of bile is not only subject-dependent, but also serves as an indication of whether crystals are present in the biliary system.

As the main purpose of this chapter is to identify possible indicators of gallstone formation for initially healthy GBs before any pathological changes have emerged, it is simple and reasonable to use a Newtonian fluid to represent bile. Nonetheless, it is important that the non-Newtonian properties of the bile or a two-phase flow with gallstone crystals are modelled to develop an even more suitable flow model for the diagnosis of individual patients in the future.

REFERENCES

Al-Atabi, M., S. Al-Zuhair, S. B. Chin, et al. 2006. Pressure drop in laminar and turbulent flows in circular pipe with baffles 1/m an experimental and analytic study. *International Journal of Fluid Mechanics Research* 33:303–319.

Al-Atabi, M., S. B. Chin, and X. Y. Luo. 2005. Flow structure in circular tubes with segmental baffles. *Journal of Flow Visualization and Image Processing* 12:301–311.

Al-Atabi, M., S. B. Chin, and X. Y. Luo. 2006. Visualization experiment of flow structures inside two-dimensional human biliary system models. *Journal of Mechanics in Medicine and Biology* 6:249–260.

Al-Atabi, M., R. C. Ooi, X. Y. Luo, et al. 2012. Computational analysis of the flow of bile in human cystic duct. *Medical Engineering & Physics* 34:1177–1183.

Bird, N. C., R. C. Ooi, X. Y. Luo, et al. 2006. Investigation of the functional three-dimensional anatomy of the human cystic duct: A single helix? *Clinical Anatomy* 19:528–534.

Bober, W., and R. A. Kenyon. 1980. *Fluid Mechanics*. New York: John Wiley & Sons.

Bouchier, I. A., S. R. Cooperband, and B. M. El-Kodsi. 1965. Mucous substances and viscosity of normal and pathological human bile. *Gastroenterology* 49:343–353.

Brugge, W. R., D. L. Brand, H. L. Atkins, et al. 1986. Gallbladder dyskinesia in chronic acalculous cholecystitis. *Digestive Diseases and Science* 31:461–468.

Burden, V. G. 1925. Observations on the histologic and pathologic anatomy of the hepatic, cystic, and common bile ducts. *Annals of Surgery* 82:584–597.

Cancelli, C., and T. J. Pedley. 1985. A separated-flow model for collapsible-tube oscillations. *Journal of Fluid Mechanics* 157:375–404.

Courtney, D. F., A. S. Clanachan, and G. W. Scott. 1983. Cholecystokinin constricts the canine cystic duct. *Gastroenterology* 85:1154–1159.

Cowie, A. G., and D. J. Sutor. 1975. Viscosity and osmolality of abnormal bile. *Digestion* 13:312–315.

Daniels, B. T., F. B. McGlone, H. Job, et al. 1961. Changing concepts of common bile duct anatomy and physiology. *Journal of American Medical Association* 178:120–123.

Dasgupta, D., and M. D. Stringer. 2005. Cystic duct and Heister's "valves". *Clinical Anatomy* 18:81–87.

Deenitchin, G. P., J. Yoshida, K. Chijiiwa, et al. 1998. Complex cystic duct is associated with cholelithiasis. *HPB Surgery* 11:33–37.

Dodds, W. J., W. J. Groh, R. M. Darweesh, et al. 1985. Sonographic measurement of gallbladder volume. *American Journal of Roentgenology* 145:1009–1011.

Dodds, W. J., W. J. Hogan, and J. E. Green. 1989. Motility of the biliary system. In *Handbook of Physiology*, ed. S. G. Schultz, 1055–1101. Bethesda: American Physiological Society.

Doty, J. E., H. A. Pitt, S. L. Kuchenbecker, et al. 1983. Role of gallbladder mucus in the pathogenesis of cholesterol gallstone. *American Journal of Surgery* 145:54–61.

Duch, B. U., H. L. Andersen, and H. Gregersen. 2003. Morphometric and biomechanical remodelling following reopening of the obstructed bile duct. *Physiological Measurement* 24:N23–34.

Duch, B. U., H. L. Andersen, and H. Gregersen. 2004. Mechanical properties of the porcine bile duct wall. *BioMedical Engineering OnLine* 3:1–8.

Duch, B. U., H. L. Andersen, J. Smith, et al. 2002. Structural and mechanical remodelling of the common bile duct after obstruction. *Neurogastroenterology and Motility* 14:111–122.

Duch, B. U., J. A. Petersen, and H. Gregersen. 1998. Luminal cross-sectional area and tension-strain relation of the porcine bile duct. *Neurogastroenterology and Motility* 10:203–209.

Flaherty, J. E., J. B. Keller, and S. I. Rubinow. 1972. Post buckling behavior of elastic tubes and rings with opposite sides in contact. *SIAM Journal of Applied Mathematics* 23:446–455.

Hauge, C. W., and J. B. Mark. 1965. Common bile duct motility and sphincter mechanism. *Annals of Surgery* 162:1028–1038.

Jian, C. Y., and G. R. Wang. 1991. Biomechanical study of the bile duct system outside the liver. *Bio-Medical Materials and Engineering* 1:105–113.

Johnston, C. G., and C. E. Brown. 1932. Studies of gall-bladder function: III—A study of the alleged impediment in the cystic duct to the passage of fluids. *Surgery, Gynecology and Obstetrics* 54:477–485.

Jungst, D., A. Niemeyer, I. Muller, et al. 2001. Mucin and phospholipids determine viscosity of gallbladder bile in patients with gallstones. *World Journal of Gastroenterology* 7:203–207.

Kuchumov, A. G. 2019. Biomechanical model of bile flow in the biliary system. *Russian Journal of Biomechanics* 23:224–248.

Kuchumov, A. G., V. Gilev, V. Popov, et al. 2014. Non-Newtonian flow of pathological bile in the biliary system: Experimental investigation and CFD simulation. *Korea-Australia Rheology Journal* 26:81–90.

Li, W. G., X. Y. Luo, A. G. Johnson, et al. 2007. One-dimensional models of the human biliary system. *ASME Journal of Biomechanical Engineering* 129:164–173.

Lichtenstein, M. E., and A. C. Ivy. 1937. The function of the "valve" of Heister. *Surgery* 1:38–52.

Lo, R. C., W. L. Huang, and Y. M. Fan. 2015. Evaluation of bile reflux in HIDA images based on fluid mechanics. *Computers in Biology and Medicine* 60:51–65.

Mentzer, S. H. 1927. The valves of Heister. *Archives of Surgery* 13:511–522.

Ooi, R. C. 2004. The flow of bile in human cystic duct. PhD diss., University of Sheffield, Sheffield, UK.

Ooi, R. C., X. Y. Luo, S. B. Chin, et al. 2004. The flow of bile in the human cystic duct. *Journal of Biomechanics* 37:1913–1922.

Otto, W. J., G. W. Scott, and C. M. Rodkiewicz. 1979. A comparison of resistances to flow through the cystic duct and the sphincter of Oddi. *Journal of Surgical Research* 27:68–72.

Pedley, T. J., B. S. Brook, and R. S. Seymour. 1996. Blood pressure and flow rate in the giraffe jugular vein. *Philosophical Transactions-Biological Sciences* 351:855–866.

Pina, L. N., F. Samoilovich, S. Urrutia, et al. 2015. Surgical considerations of the cystic duct and Heister valves. *Surgery Journal* 1:23–27.

Pitt, H. A., J. E. Doty, L. DenBesten, et al. 1982. Stasis before gallstone formation: Altered gallbladder compliance or cystic duct resistance? *American Journal of Surgery* 143:144–149.

Pitt, H. A., J. J. Roslyn, S. L. Kuchenbecker, et al. 1981. The role of cystic duct resistance in the pathogenesis of cholesterol gallstones. *Journal of Surgical Research* 30:508–514.

Reinhart, W. H., G. Naf, and B. Werth. 2010. Viscosity of human bile sampled from the common bile duct. *Clinical Hemorheology and Microcirculation* 44:177–182.

Rodkiewicz, C. M., and W. J. Otto. 1979. On the Newtonian behaviour of bile. *Journal of Biomechanics* 12:609–612.

Rodkiewicz, C. M., W. J. Otto, and G. W. Scott. 1979. Empirical relationships for the flow of bile. *Journal of Biomechanics* 12(6):411–413.

Sambrook, I. E. 1981. Studies on the flow and composition of bile in growing pigs. *Journal of Science of Food and Agriculture* 32:781–791.

Schatz, P. 2020. Gallbladder and bile ducts – structure and function. www.lecturio.com/magazine/gall-bladder-bile-ducts/. https://d3uigcfkiiww0g.cloudfront.net/wordpress/blog/pics-en/uploads/Gallbladder1.jpg (accessed August 1, 2020).

Scott, G. W., W. J. Otto, and C. M. Rodkiewicz. 1979. Resistance and sphincter-like properties of the cystic duct. *Surgery, Gynecology and Obstetrics* 149:177–182.

Severi, C. J. R. Grider, and G. M. Makhlouf. 1988. Functional gradient in muscle cells isolated from gallbladder, cystic duct and common bile duct. *American Journal of Physiology* 255:G647–G652.

Slater, G., P. I. Tartter, D. Dreiling, et al. 1983. Resistance of the canine common bile duct. *Bulletin of New York Academy of Medicine* 59:711–720.

Soloway, R. D., B. W. Trotman, and J. D. Ostrow. 1977. Pigment gallstones. *Gastroenterology* 72:167–182.

Tera, H. 1963. Sedimentation of bile constituents. *Annals of Surgery* 157:467–472.

Tominaga, K., K. Arai, H. Ishii, et al. 1995. Pathological study of inflammatory changes of the cystic duct. *Journal of Showa Medical Association* 55:253–261.

Walsh, T. H. 1979. The muscle content and contractile capability of the common bile duct. *Annals of Royal College of Surgeons of England* 61(3):206–209.

White, F. M. 2011. *Fluid Mechanics* (7th edition). New York: McGraw-Hill Companies Inc.

Wolpers, C., and A. F. Hofmann. 1993. Solitary versus multiple cholesterol gallbladder stones: Mechanism of formation and growth. *Clinical Investigation* 71:423–434.

CHAPTER 3

1D Models of Non-Newtonian Bile Flow in the Biliary Tract

BILE RHEOLOGY

Cholelithiasis/gallstone is one of the most common biliary diseases. Reduction in GB motility leading to prolonged stasis of bile in the GB is one of the essential factors in the pathogenesis of cholelithiasis (Holzbach et al. 1973; Catnach et al. 1992), but also increased flow resistance/pressure drop to the bile flow in the CD is aetiologically related to bile stasis (Pitt et al. 1982). As such that an understanding of fluid mechanics of bile flow in the biliary tract may contribute to the elucidation of the aetiology of cholelithiasis.

Bile rheology (the relation between shear stress and shear rate) is the essential of bile flow in CDs and has experienced significant experimental investigations since the 1930s. Tera (1963) and Thureborn (1966) found that the normal GB bile was layered. The dynamic viscosity of the top thinnest layer is 2.0 mPa·s, but the thickest layer is 2.2 mPa·s after 2 hours of sedimentation (Tera 1963). Bouchier et al. (1965) reported that the dynamic viscosity of pathological GB bile is greater than normal GB bile and both are more viscous than hepatic bile. The concentration of normal GB bile is a major factor determining the viscosity. In pathological bile and hepatic bile, the content of mucous substances is the major factor determining the viscosity. Cowie and Sutor (1975) illustrated that the mean viscosity of bile from GBs with gallstones is greater than that from healthy ones.

The presence of mucus in GBs with gallstones is likely to account for the differences in viscosity (Doty et al. 1983). Jungst et al. (2001) showed that the biliary viscosity was markedly higher in GB bile of patients with cholesterol (5.0 mPa·s) and mixed stones (3.5 mPa·s) compared with hepatic bile (0.2 mPa·s). A positive correlation between mucin and viscosity was found in GB bile but not in hepatic bile. An increased viscosity of GB bile has been considered an important factor in the pathogenesis of gallstone disease (Jungst et al. 2001).

More recent measurements with concentric cylinder viscometers have shown that GB bile may become non-Newtonian at low shear rate. Gottschalk and Lochner (1990) measured the postoperative bile viscosity sampled by T-drainages of 29 patients with a modified horizontal capillary viscometer from the day of operation to the ninth day and confirmed that the viscosity is time-dependent and the bile behaves like a Maxwell fluid (one simple kind of viscoelastic fluids). The dynamic viscosity of 33 bile specimens showed shear-thinning non-Newtonian behaviour by using a Contraves low shear viscometer. The bile dynamic viscosity decreases from 5 mPa·s at a shear rate of 0.1 s^{-1} to 1.5 mPa·s at 2.0 s^{-1}.

Saida (1992) tested the viscosity of the bile of 15 healthy GBs (10 male and 5 female adults at a shear rate of 37.5–750 s^{-1}) and the bile of five dogs (at a shear rate of 18.75–375 s^{-1}). The bile is shear-thinning non-Newtonian fluid, and its shear stress-shear rate relationship can be represented by using power function or the well-known Casson model.

Coene et al. (1994) indicated that the bile viscosity of gallstone patients is a function of shear rate based on the results by using the Contraves low-shear 30 rotational viscometer. The bile dynamic viscosity decreases to 1.5 mPa·s at 100 s^{-1} from 2.5 mPa·s at a shear rate of 0.1 s^{-1}. These observations established that the bile in a GBs with gallstones can be characterised by shear-thinning non-Newtonian behaviour at low shear rate.

Ooi (2004) measured the bile dynamic viscosity of 59 patients (cholecystectomy) and illustrated that the bile rheology of 20 patients' was Newtonian, 22 patients' bile is both shear-thickening and shear-thinning non-Newtonian, 8 patients' bile is shear-thickening non-Newtonian and 9 patients' bile (with mucus) is shear-thinning non-Newtonian. Even though the results at low shear rate are not reliable due to viscometer limitation, these experiments suggest that GB bile really can demonstrate a complicated non-Newtonian behaviour. Note that the density of GB bile is 965.9–1014.5 kg/m^3 and it is very close to water density of 1000 kg/m^3 at 20° (Ooi 2004).

Recently, the shear stress-shear rate curves of the hepatic bile and the GB bile were measured, and the shear-thickening non-Newtonian behaviour was identified. Both Casson and Carreau's rheological model parameters were extracted from the experimental data and applied in CFD simulations of non-Newtonian bile flow in a few specific human biliary systems (Kuchunov et al. 2014; Kuchunov 2019).

In this chapter, the idealised CD geometry in Chapter 2 is adapted to account for elastic wall and non-Newtonian bile flow, with the viscosity described by the Carreau's equation. The effects of CD geometry, wall elasticity and rheological properties on pressure drop are examined in terms of the Darcy friction factor. A 3D CFD simulation of non-Newtonian bile flow in the CD was also carried out to obtain the flow structure and to provide partial validation of the 1D model results. It was found that the most significant geometric parameter on non-Newtonian bile flow resistance is the baffle height, too. However, compared with the Newtonian bile, the elastic wall of the CD can produce large flow resistance to the non-Newtonian bile due to shear-thinning effect. The detailed models are referred to Li et al. (2008).

1D MODELS FOR THE BILIARY TRACT

The Geometry

The 1D models for the human biliary tract shown schematically in Figure 2.4 were described in Chapter 2 for Newtonian bile. Both the CBD and the CHD are represented by straight circular rigid tubes. These tubes are connected to the GB by an idealised CD at a T-junction. The valves of Heister in the CD are replaced by equally spaced and staggered baffles as illustrated in Figure 3.1. Two models of the idealised CD were used: one with rigid wall and the other with compliant wall. The directions of the bile flow during GB emptying and refilling are shown in Figure 2.4(b) and (c), respectively.

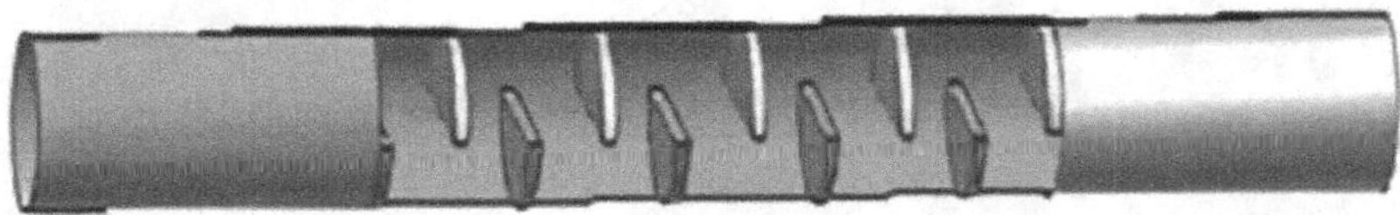

FIGURE 3.1 The valves of Heister in the CD are replaced by equally staggered baffles as in Chapter 2.

The Non-Newtonian Bile

Viscosity-shear rate relations of bile are few. Bile is a shear-thinning non-Newtonian fluid as the shear rate is at 0.1–10 (Gottschalk and Lochner 1990; Coene et al. 1994) and 37.5–750 s^{-1} (Saida 1992), respectively. The two sets of data (Gottschalk and Lochner 1990; Coene et al. 1994) were best fitted with the well-known Curreau model to obtain a more general viscosity-shear rate relation for bile. The model includes four parameters undetermined. With these data, the bile viscosity is calculated from the following Carreau model

$$\mu = \mu_\infty + \left(\mu_0 - \mu_\infty\right)\left(1 + m_1^2\gamma^2\right)^{\frac{m_2-1}{2}} \tag{3.1}$$

where μ_0 is the bile dynamic viscosity at zero shear rate, μ_∞ is the viscosity at infinite shear rate, m_1 is the time constant, γ is the shear strain rate and m_2 is the power. Figure 3.2 shows the experimental bile viscosity data and the best-fitted curve with $\mu_0 = 10^{-2}$ Pa·s, $\mu_\infty = 10^{-3}$ Pa·s, $m_1 = 160.5742$ s and $m_2 = 0.4843$. The GB bile density ρ is 1000 kg/m^3.

A reference Newtonian fluid is required to examine the effect of non-Newtonian behaviour of bile on the pressure drop in the human biliary tract. Usually, this reference Newtonian fluid should have a dynamic viscosity that is close to or equal to the viscosity at an infinite shear rate (Gijsen et al. 1999; Chen and Lu 2004; Chen et al. 2006; O'Callaghan et al.

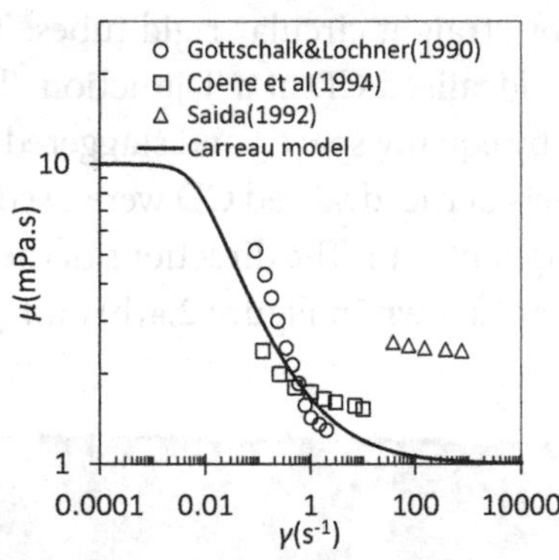

FIGURE 3.2 The Carreau model that fits the experimental bile viscosities measured by Coene et al. (1994) and Gottschalk and Lochner (1990) (the data of Said 1992 are shown for comparison).

2006). In this context, a Newtonian GB bile with $\mu = \mu_\infty$ and $\rho = 1000$ kg/m^3 is specified as the reference Newtonian fluid to compare with the results of the non-Newtonian bile in this chapter.

An Equivalent Duct for CD

The effects of the equally spaced and staggered baffles on the flow in the idealised CD are expressed in terms of equivalent diameter, d_{CD}, and equivalent length, L_{eq}. The equivalent diameter is calculated from the diameter of the CD, d_{CD}, the baffle cut area, A_1, and the number of baffles, n, as shown in Chapter 2

$$d_{eq} = d_{CD} + \left(d_{CD} - d_{min}\right)\left(1 - \frac{n}{n_c}\right) \tag{3.2}$$

where $d_{min} = 2\sqrt{A_1/\pi}$ (see Figure 2.6) and n_c is the critical number of baffles, which is calculated by

$$n_c = \frac{L_{CD} + \dfrac{A_1}{d_{CD}}}{h_b + \dfrac{A_1}{d_{CD}}} \tag{3.3}$$

The equivalent tube length is determined from the actual distance travelled by the bile as it negotiates the baffles and the extra length corresponding to minor losses in the baffle zone, i.e.,

$$L_{eq} = H\left(n - 1\right) + L_{CD} + L_k \tag{3.4}$$

where H is the baffle height (Figure 2.4), L_{CD} is the length of the CD and L_k is the length representing flow minor losses due to sudden expansions and contractions of the flow across the baffles. The length L_k is calculated from the following equations (Wilkinson 1960)

$$L_k = \frac{r_{eq}\Delta p_k}{2\tau_w} \tag{3.5a}$$

$$\frac{Q}{\pi r_{eq}^3} = \frac{1}{\tau_w^3}\int_0^{\tau_w} \gamma \tau^2 d\tau \tag{3.5b}$$

where $r_{eq} = 0.5d_{eq}$ and the τ_w is the shear stress at wall, but τ is the shear stress of flow, Δp_k is the pressure losses due to bends, sudden expansion and contraction and is given by the following expression (Bober and Kenyon 1980)

$$\Delta p_k = 16c_3(n-1)\frac{\rho Q^2}{\pi^2 d^4_{eq,min}} + 16n(c_1 + c_2)\frac{\rho Q^2}{\pi^2 d^4_{eq,min}} \tag{3.6}$$

where the sudden contraction head-loss coefficient $c_1 = 0.42(1 - A_1/A_{CD})$, the sudden expansion head-loss coefficient $c_2 = (1 - A_1 / A_{CD})^2$ (White 2011), the head-loss coefficient due to bends $c_3 = 0.85\xi$, where ξ is the dimensionless baffle height, Q is the bile flow rate which is known. The τ_w should be determined from Eq. (3.5b), and the detail will be mentioned in *The Shear Stress on the Wall* section.

FSI in the Emptying

Pressure drops in the biliary tract in the emptying phase are shown in Figure 2.4 that for the non-Newtonian bile flowing through the elastic CD and the rigid CBD. As the emptying phase starts, the CD cross section is circular under an external pressure p_e (chest pressure). During the emptying, the CD may be partially collapsed towards the downstream end owing to a decreased transmural pressure as the bile flows. The governing equations for the fully developed, steady laminar flow in the elastic CD are written as

$$Q = Au \tag{3.7a}$$

$$\rho u \frac{du}{dx} = -\frac{dp}{dx} - \frac{2\tau_w}{r} \tag{3.7b}$$

$$\frac{Q}{\pi r^3} = \frac{1}{\tau_w^3}\int_0^{\tau_w} \gamma \tau^2 d\tau \tag{3.7c}$$

where u is the bile velocity, p is the internal pressure acting on the CD, r is the internal radius of the CD and $0 < r \le r_{eq}$, $0 \le x \le L_{eq}$. The inlet pressure is selected as the reference pressure. At a given flow rate, the pressure in the CD is obtained by integrating Eq. (3.7b) along the duct from the inlet and is written as

$$p = p_{in} - 2\int_0^x \frac{\tau_w}{r} dx + \frac{1}{2}\rho Q^2 \left(\frac{1}{A_{in}^2} - \frac{1}{A^2} \right) \tag{3.8}$$

The constitutive equation for the elastic CD wall yields the tube law for homogeneous, isotropic, elastic materials (Cancelli and Pedley 1985; Pedley et al. 1996)

$$p - p_e = \frac{Eh^3}{12\left(1-\upsilon^2\right)r^2}\left(\alpha^{10} - \alpha^{-3/2}\right) \tag{3.9}$$

where $\alpha = A/A_{eq}$ and υ is the Poisson's ratio of elastic wall and p_e is the external pressure. Coupling Eqs. (3.8) and (3.9), the FSI equation is resulted as follows:

$$p_{in} - 2\int_0^x \frac{\tau_w}{r}dx + \frac{1}{2}\rho Q^2\left(\frac{1}{A_{in}^2} - \frac{1}{A^2}\right) = p_e + \frac{Eh^3}{12\left(1-\upsilon^2\right)r^2}\left[\left(\frac{A}{A_{eq}}\right)^{10} - \left(\frac{A}{A_{eq}}\right)^{-3/2}\right] \tag{3.10}$$

with the boundary condition at the inlet

$$p_{in} = p_e + \frac{\pi^{3/2}Eh^3}{12\left(1-\upsilon^2\right)A_{in}^{3/2}}\left(\alpha_{in}^{10} - \alpha_{in}^{-3/2}\right) \tag{3.11}$$

where $\alpha_{in} = A_{in}/A_{eq}$ is the cross-sectional area ratio at the tube inlet. The cross-sectional area of the rigid tube is $A_{eq} = \pi d_{eq}^2/4$.

Equation (3.10) represents a boundary value problem, which is solved for $A(x)$ by using the finite difference method with the aid of Eq. (3.7c) for given p_e, E, υ, A_{eq}, Q, and so on. Then the bile pressure is obtained from Eq. (3.8). The pressure drop in the CD is calculated by

$$\Delta p_{CD} = p_{in} - p_{out} \tag{3.12}$$

The pressure drop for the rigid CBD in the emptying is given by

$$\Delta p_{CBD} = \frac{2L_{CBD}\tau_w}{r_{CBD}} + \Delta p_{te} \tag{3.13a}$$

$$\frac{Q}{\pi r_{CBD}^3} = \frac{1}{\tau_w^3}\int_0^{\tau_w} \gamma\tau^2 d\tau \tag{3.13b}$$

where Δp_{te} accounts for the pressure losses at the T-junction. These losses are due to the flow rounding a 90° bend and expanding from the smaller CD to the larger CBD. And it is estimated by (Bober and Kenyon 1980)

$$\Delta p_{te} = 16c_4 \frac{\rho Q^2}{\pi^2 d_{CD}^4} + 16c_2 \frac{\rho Q^2}{\pi^2 d_{CD}^4} \tag{3.14}$$

and $c_4 = 0.75$ (Bober and Kenyon 1980). Then, the total pressure drop in the biliary tract in the emptying phase reads

$$\Delta p_{EM} = \Delta p_{CD} + \Delta p_{CBD} \tag{3.15}$$

Pressure Drop in the Refilling

In the refilling phase, the bile flows from the liver to the CHD and enters the GB via the CD, as shown in Figure 2.4(c). Since the flow rate of hepatic bile is very low and the refilling time is at least three times longer than that of the emptying, the CD and CHD walls should be assumed to be rigid. The pressure drop in the biliary tract in the refilling phase is written as

$$\Delta p_{RF} = \Delta p_{CD} + \Delta p_{CHD} \tag{3.16}$$

where Δp_{CD} is the pressure drop across the CD with rigid wall and is determined by

$$\Delta p_{CD} = \frac{2L_{eq}\tau_w}{r_{eq}} \tag{3.17a}$$

$$\frac{Q}{\pi r_{eq}^3} = \frac{1}{\tau_w^3} \int_0^{\tau_w} \gamma \tau^2 d\tau \tag{3.17b}$$

and Δp_{CHD} is the pressure drop in the CHD and is calculated with

$$\Delta p_{CHD} = \frac{2L_{CBD}\tau_w}{r_{CBD}} + 16c_4 \frac{\rho Q^2}{\pi^2 d_{CD}^4} + 16c_2 \frac{\rho Q^2}{\pi^2 d_{CD}^4} \tag{3.18a}$$

$$\frac{Q}{\pi r_{CHD}^3} = \frac{1}{\tau_w^3} \int_0^{\tau_w} \gamma \tau^2 d\tau \tag{3.18b}$$

The Shear Stress on the Wall

For non-Newtonian bile, the pressure drop depends on the shear stress on walls τ_w, which appears in Eqs. (3.5b), (3.7c), (3.13b), (3.17b) and (3.18b). Note that the τ_w is related to the constitutive or rheological equation of bile. In our case, this equation is the Carreau model represented by Eq. (3.1). The integral $\int_0^{\tau_w} \gamma\tau^2 d\tau / \tau_w^3$ is a function of τ_w and can be obtained numerically by using Eq. (3.1).

First, the shear rate γ is assumed to be in the range of 10^{-9} –500 s^{-1}, which covers the fully physiological shear rate in the human biliary system. The interval of γ is divided into 5000 points, then the τ_w is calculated at each point by using Eq. (3.1).

Second, the integral $\int_0^{\tau_w} \gamma\tau^2 d\tau / \tau_w^3$ is calculated by using the well-known trapezoid formula. As a result, the relation $\int_0^{\tau_w} \gamma\tau^2 d\tau / \tau_w^3$ with τ_w is available, i.e.,

$$F(\tau_w) = \frac{1}{\tau_w^3} \int_0^{\tau_w} \gamma\tau^2 d\tau \tag{3.19}$$

The function $F(\tau_w)$ has been shown in Figure 3.3.

Finally, the τ_w corresponding to a known $Q/\pi r^3$ can be interpolated linearly from this relation and the pressure drop can be determined.

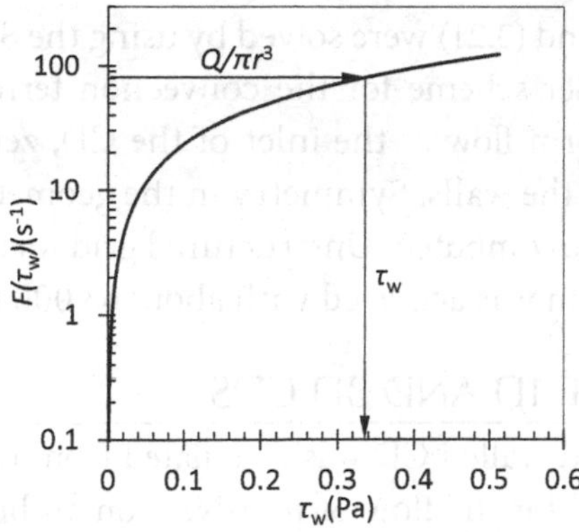

FIGURE 3.3 Integral $F(\tau_w) = \int_0^{\tau_w} \gamma\tau^2 d\tau / \tau_w^3$ as a function of τ_w with the Carreau model, for arbitrary known $Q/\pi r^3$ a definite τ_w can be determined with this curve.

3D CD Model

The detailed non-Newtonian flow structures in the CD can provide a benchmark to validate the 1D model for the rigid wall case. Thus, numerical simulations of 3D flow in a rigid CD model were performed by using the CFD software FLUENT 6.2. The governing equations of the steady, laminar flow of bile are

$$\frac{\partial u_i}{\partial x_j} = 0 \tag{3.20a}$$

$$\rho u_j \frac{\partial u_i}{\partial x_j} = -\frac{\partial p}{\partial x_j} + \frac{\partial \tau_{ij}}{\partial x_j} \tag{3.20b}$$

where τ_{ij} (i, $j = 1, 2, 3$) is the shear stress in the flow field and related to the shear strain rate, e_{ij}, and the dynamic viscosity of bile, μ. The constitutive equations are

$$\tau_{ij} = 2\mu e_{ij} \tag{3.21a}$$

$$e_{ij} = \frac{1}{2}\left(\frac{\partial u_i}{\partial x_j} + \frac{\partial u_j}{\partial x_i}\right) \tag{3.21b}$$

where for the Newtonian fluid, $\mu = \mu_\infty$, and for the non-Newtonian bile, the viscosity μ depends on the shear rate $\gamma \left(= \sqrt{2e_{ij}e_{ji}}\right)$, and is described with Eq. (3.1).

Equations (3.20) and (3.21) were solved by using the SIMPLE algorithm with the second-order scheme for the convection terms. The boundary conditions are uniform flow at the inlet of the CD, zero pressure at the outlet and no-slip at the walls. Symmetry in the geometry allows the half geometry model to be computed. Unstructured grid was adopted, and grid independence solution was achieved with about 63,000 tetrahedral cells.

COMPARISON OF 1D AND 3D CDS

Pressure drop in rigid-walled CD was calculated from Eq. (3.12) in the 1D biliary system model for the flow of non-Newtonian bile. This was compared with those from a CFD model for a 3D idealised rigid wall CD. The bile density was 1000 kg/m^3, and its flow rate was in the range of 0.12–9.42 ml/min, corresponding to the Reynolds number Re_{CD} $(= 4\rho Q/\pi\mu_\infty d_{CD})$ from 0.5 to 40 based on the CD diameter (d_{CD} = 5 mm). As in Chapter 2, the length of the CD was fixed at 100 mm, the CD wall and baffle thickness were 1 mm, the number of baffles n was ranged from 0 to 14 and the dimensionless baffle heights ξ were 0.3, 0.5 and 0.7.

The relationships between pressure drop and Reynolds number Re_{CD} are illustrated in Figure 3.4 at different numbers of baffles and dimensionless baffle heights. In Figure 3.4, the good agreement between the 1D model predictions and the 3D simulations is achieved in terms of the largest error of less than 10%, suggesting that the 1D biliary tract model is satisfactory for the non-Newtonian bile.

The value $n=0$ corresponds to laminar flow in a straight rigid CD without baffles. At the three dimensionless baffle heights used, all Δp_{CD} increases nonlinearly with Re_{CD}.

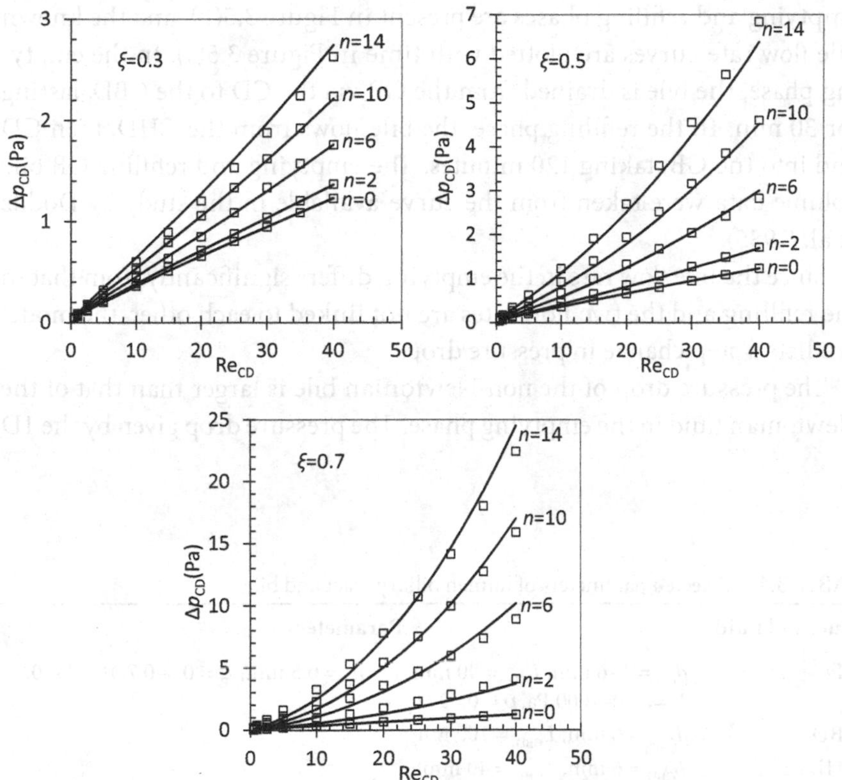

FIGURE 3.4 The pressure drop Δp_{CD} versus Reynolds number Re_{CD} for the non-Newtonian bile flow in the rigid CDs with different numbers of baffles n and dimensionless baffle heights ξ, symbols for 3D simulations, lines for 1D model.

In Figure 3.4(a), the dimensionless baffle height ($\xi = 0.3$) is the smallest and hence the flow clearance area is the largest. Therefore, the pressure drop for this case is the smallest. The converse is true in Figure 3.4(c). The baffles significantly increase the pressure drop; specially, the pressure drop rises nonlinearly with increasing Re_{CD} in each case. Increasing the number of baffles amplifies this effect noticeably.

PRESSURE DROP IN THE BILIARY SYSTEM

The pressure drops in one cycle of GB emptying and refilling were calculated with the 1D biliary tract model with rigid and elastic CD walls, respectively, for Newtonian and non-Newtonian bile. The parameters adopted in the pressure drop predictions are listed in Table 3.1.

The predicted pressure drops across the human biliary tract in the emptying and refilling phases are present in Figure 3.5(b), and the known bile flow rate curves are plotted with time in Figure 3.5(a). In the emptying phase, the bile is drained from the GB via the CD to the CBD, lasting for 30 min. In the refilling phase, the bile flows from the CHD, then CD and into the GB, taking 120 minutes. The emptying and refilling GB bile volume data were taken from the curve available in the study by Dodds et al. (1985).

Since the bile flow rate in the emptying differs significantly from that in the refilling and the two flow rates are not linked to each other, the model predicts a step change in pressure drop.

The pressure drop of the non-Newtonian bile is larger than that of the Newtonian fluid in the emptying phase. The pressure drop given by the 1D

TABLE 3.1 Selected parameters of human biliary tract and bile.

Duct and fluid	Parameters
CD	d_{CD} = 1–6 mm, L_{CD} = 40 mm, $h = h_b$ = 0.5 mm, ξ = 0.3–0.7, n = 0–10, E = 100–1000 Pa, υ = 0.49
CBD	d_{CBD} = 6 mm, L_{CBD} = 100 mm
CHD	L_{CBD} = 6 mm, L_{CHD} = 40 mm
Bile	ρ = 1000 kg/m^3 For the reference Newtonian fluid, $\mu = 1 \times 10^{-3}$ Pa·s For the non-Newtonian bile, $\mu_0 = 1 \times 10^{-2}$ Pa·s, $\mu_0 = 1 \times 10^{-3}$ Pa·s, m_1 = 160.5742 s, m_2 = 0.4843, the Carreau model

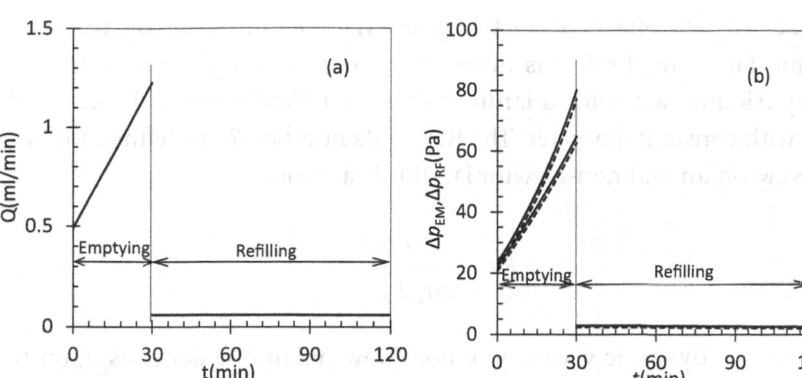

FIGURE 3.5 The bile flow rate (a) and pressure drop (b) curves with time in the human biliary tract in the emptying and refilling phases for the non-Newtonian (solid line) and Newtonian bile (dashed line), the CD parameters are given as follows: $d_{CD} = 1$ mm, $\xi = 0.5$, $n = 6$, $\alpha_{in} = 1$ and $E = 300$ Pa, the bile flow rate Q is plotted as a function of time in both the phases, which was derived from the GB bile volume-time curves by Dodds et al. (1985), see Chapter 2.

elastic model is higher than that by the 1D rigid one at the same flow rate owing to the narrower diameter caused from partially collapsed elastic CD wall in the emptying. Note that the pressure drops of the Newtonian and non-Newtonian fluids are less than 3 Pa in the refilling phase. This makes sense as the CD produces a negligible resistance for the bile to flow into a GB in the refilling.

THE DARCY FRICTION FACTOR OF CD

It has been demonstrated that the pressure drop in the CBD is less than 1.5%, and in the CHD, less than 0.15% of the pressure drop in the CD (see Chapter 2); therefore, the pressure drop mainly occurs in the CD. The pressure drop in the CD and effects of geometrical and rheological variables as well as flow rate on it should be examined carefully.

The Darcy friction factor, f, [Eq. (3.22)], is commonly used to compare the pressure drop in the duct flow (White 2011):

$$f = \frac{\Delta p_{CD}}{\frac{1}{2}\rho\left(\frac{L_{eq}}{d_{eq}}\right)\left(\frac{Q}{A_{eq}}\right)^2} \tag{3.22}$$

To clarify the effects of the CD geometry even more clearly, the Darcy friction factor in the CD is normalised by f_0, where f_0 (= $Re/64$) is the Darcy friction factor for a laminar flow of a Newtonian fluid in a rigid duct with constant diameter. The Reynolds number Re is defined for both the Newtonian and non-Newtonian fluids as follows:

$$Re = \frac{4\rho Q}{\pi \mu_\infty d_{eq}} \tag{3.23}$$

Since the dynamic viscosity of non-Newtonian bile depends upon the shear rate, it is not reasonable to define a Reynolds number for such a fluid. The Reynolds number defined with Eq. (3.23) just provides a comparison scale for the Newtonian and non-Newtonian bile. For the same geometry configuration, the same Reynolds number does not mean that the hydrodynamic similarity is held in both the fluid flows.

The influences of CD geometrical parameters and bile rheological property on the relationship between f/f_0 and Re are shown in Figure 3.6 for rigid CDs with the rigid wall model. The changes in the CD geometry are expressed in terms of n and ξ. Both figures show that the CD with baffles generally suffers from a higher pressure drop than the CD without baffles, and the non-Newtonian bile usually leads to a larger pressure drop than the Newtonian one since the μ_∞ is specified to be the dynamic viscosity of the Newtonian bile. It is most important that f/f_0 increases with decreasing Re from $Re = 20$ (from $Re = 7$ as $\xi = 0.7$).

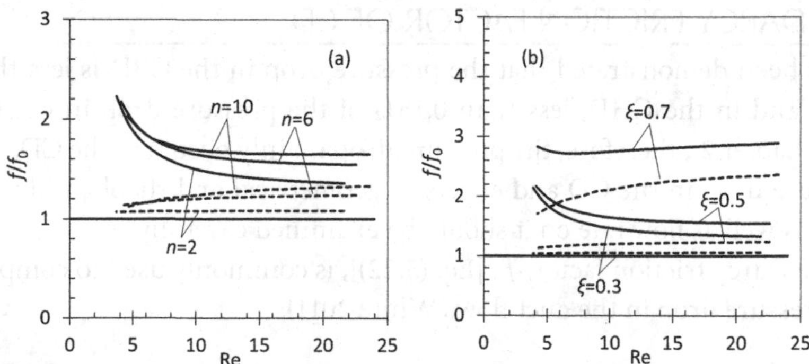

FIGURE 3.6 Rigid model f/f_0 against Re for the non-Newtonian (solid line) and Newtonian (dashed line) bile, d_{CD} = 1–6 mm, L_{CD} = 40 mm and Q = 1 ml/min, (a) for n = 2, 6, 10 and ξ = 0.5, and (b) for ξ = 0.3, 0.5, 0.7 and n = 6.

Since the bile flow rate Q is constant in the predictions, the shear rate is lowered by a decreased Re due to the increased d_{eq} based on Eq. (3.23). According to Carreau's model, Eq. (3.1), the dynamic viscosity μ is raised. Consequently, f/f_0 increases with decreasing Reynolds number. If $\xi = 0.7$, the equivalent diameter d_{eq} is so narrow that the larger shear rate remains. As a result, when $Re \leq 7$, a higher bile dynamic viscosity occurs; when $f/f_0 > 7$, a lower bile viscosity is present, and Re in terms of f/f_0 resembles to that of the Newtonian bile.

For the CD with elastic wall, f/f_0 of the Newtonian bile increases with increasing Re, but for the non-Newtonian bile it decreases asymptotically with increasing Re, as shown in Figure 3.7. The collapse effect narrows the tube diameter, in turn increases the shear rate. Consequently, the increase of f/f_0 with decreasing Re must occur at a much large d_{eq} or a small Re. It has been demonstrated that f/f_0 starts increasing from $Re = 5$ ($\xi = 0.7$) in the elastic duct. When the Reynolds number is more than this value, f/f_0 approaches the f/f_0 curve of the Newtonian bile.

Note that for both Newtonian and non-Newtonian flows, the dimensionless baffle height has more dominant effects on f/f_0 compared with the number of baffles.

The Darcy friction factor ratio f/f_0 given by using 3D flow simulations versus Reynolds number Re_{CD} is shown in Figure 3.8. In this case, the change in Reynolds number is only caused from the bile flow rate. The difference of f/f_0 between the non-Newtonian and the Newtonian bile

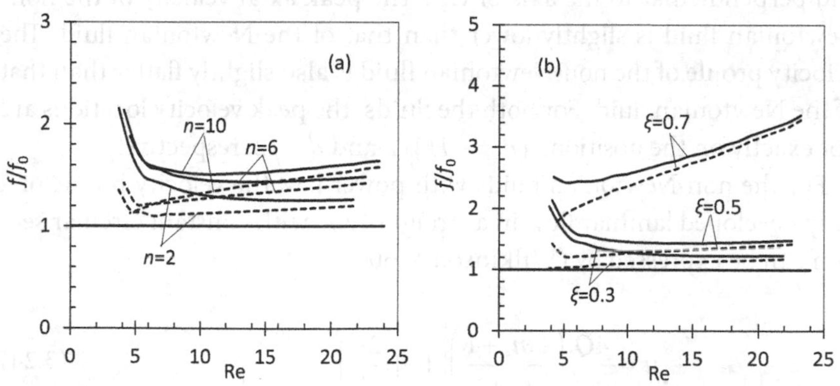

FIGURE 3.7 Elastic model f/f_0 against Re for the non-Newtonian (solid line) and Newtonian (dashed line) bile, d_{CD} = 1–6 mm, L_{CD} = 40 mm and Q = 1 ml/min, E = 300 Pa, (a) n = 2, 6, 10 and ξ = 0.5 and (b) for ξ = 0.3, 0.5, 0.7 and n = 6.

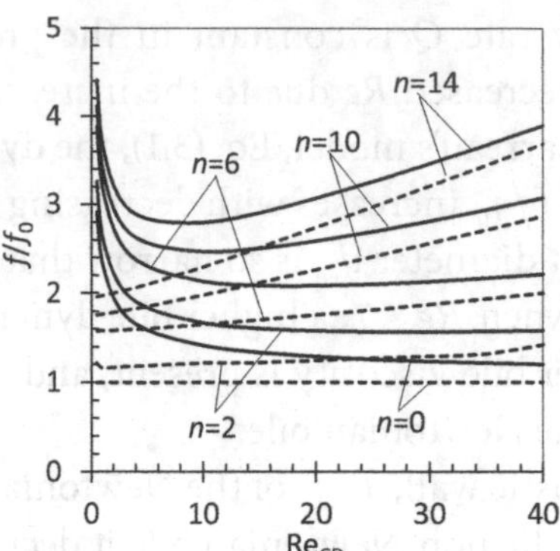

FIGURE 3.8 Darcy frication factor ratio versus Reynolds number for the non-Newtonian (solid line) and Newtonian (dashed line) bile, the factor is estimated by using 3D simulations in the CDs with rigid wall.

has been clarified. Compared to Figures 3.6 and 3.7, the 1D and 3D models seem to reveal the nearly identical trend of f/f_0 when Re_{CD} is decreased.

FLOW DETAILS IN 3D SIMULATIONS

Although the 1D model predicts the pressure drop well, it cannot give the detailed flow patterns in the CD with baffles. Figure 3.9 demonstrates the primary (axial) velocity profiles of the non-Newtonian and Newtonian fluids on the lines 1 and 2, which are in the symmetrical plane and perpendicular to the axis of CD. The peak axial velocity of the non-Newtonian fluid is slightly lower than that of the Newtonian fluid. The velocity profile of the non-Newtonian fluid is also slightly flatter than that of the Newtonian fluid. For both the fluids, the peak velocity locations are not exactly on the positions $(d_{CD}-H)/2$ and $d_{CD}/2$, respectively.

For the non-Newtonian fluids with power law, the velocity profile of a fully developed laminar flow in a straight tube with constant circular section can be expressed by (Wilkinson 1960)

$$u=\frac{4Q}{\pi d^2}\left(\frac{3m_3+1}{m_3+1}\right)\left[1-\left(\frac{2r}{d}\right)^{(m_3+1)/m_3}\right] \tag{3.24}$$

where $r\in\left[0,d/2\right]$ and $m_3\in\left(0,1\right]$. For the Newtonian fluid, $m_3=1$, Eq. (3.24) yields a well-known parabolic profile. For the non-Newtonian fluid described

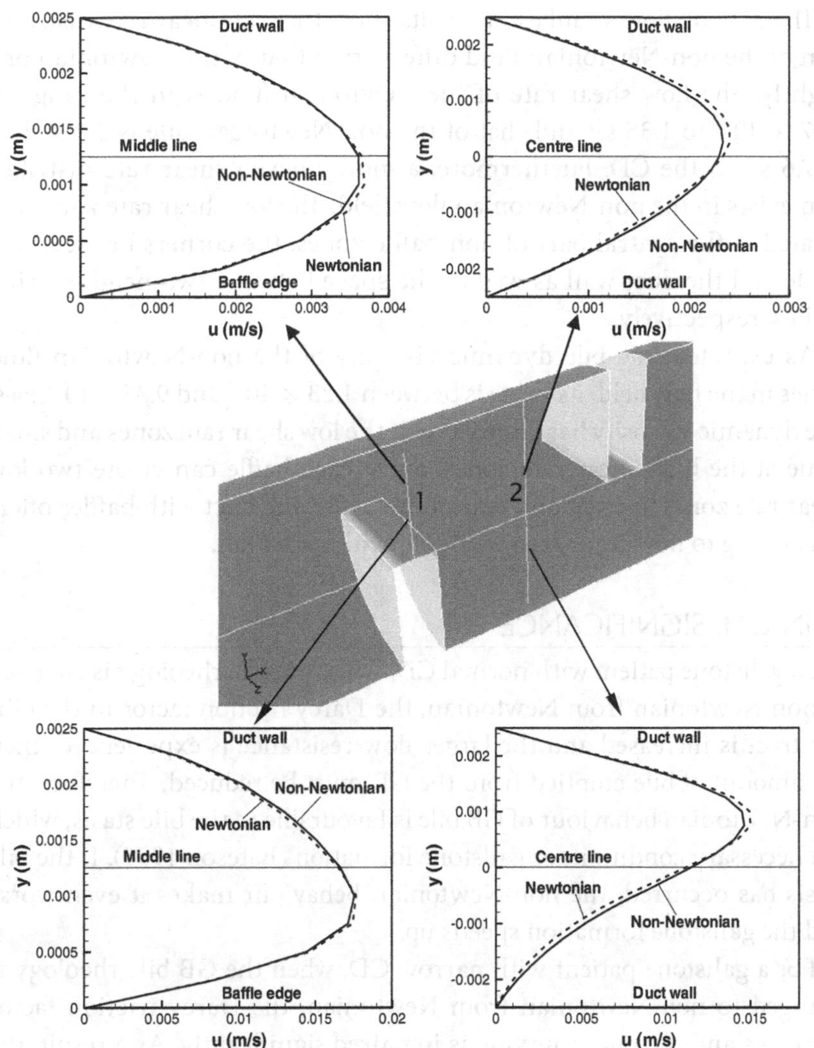

FIGURE 3.9 Axial velocity profile at two lines in a symmetrical plane of the CD for the non-Newtonian and Newtonian fluids at $Q = 1.18$ ($Re_{CD} = 5$) (top) and $Q = 5.89$ ml/min ($Re_{CD} = 25$) (bottom), for the CD $n = 10$, $d_{CD} = 5$ mm, $\xi = 0.5$, and the flow streams from the left to the right.

by the Carreau model, $0 < m_3 < 1$, Eq. (3.24) is applicable approximately. Nevertheless, the velocity profile for $0 < m_3 < 1$ is flatter than for $m_3 = 1$ from Eq. (3.24). This is the reason why the velocity profile of non-Newtonian fluid is flatter than the Newtonian one in a CD at the same Re or flow rate.

The 3D bile flow simulation results show that the shear rate distribution of the non-Newtonian fluid differs from that of the Newtonian one slightly. The flow shear rate of the Newtonian fluid is in the range of 3.57×10^{-2} to 1.36 s^{-1} and that of the non-Newtonian bile is 3.7×10^{-3} to 7.6 s^{-1} in the CD. Furthermore, a more uneven shear rate distribution exists in the non-Newtonian flow field. The low shear rate zones are located at the central part of non-baffle zones, the corners between the baffle and the duct wall as well as the space between two neighbouring baffles, respectively.

As expected, the bile dynamic viscosity of the non-Newtonian fluid varies in the flow field, its value is between 1.23×10^{-3} and 9.35×10^{-3} Pa·s. The dynamic viscosity has large value at the low shear rate zones and small value at the high shear rate zones. Since each baffle can create two low shear rate zones in front and rear of the baffle, the duct with baffles offers larger drag to flow compared with that without baffles.

CLINICAL SIGNIFICANCE

For a gallstone patient with normal CD, when the bile rheology is changed to non-Newtonian from Newtonian, the Darcy friction factor in the biliary tract is increased and the larger flow resistance is experienced, then the amount of bile emptied from the GB must be reduced. Therefore, the non-Newtonian behaviour of GB bile is favourable to the bile stasis, which is a necessary condition for gallstone formation (Bateson 1999). If the bile stasis has occurred, the non-Newtonian behaviour makes it even worse and the gallstone formation speeds up.

For a gallstone patient with narrow CD, when the GB bile rheology is changed to non-Newtonian from Newtonian, the Darcy friction factor increases and the bile emptying is impaired significantly. As a result, the bile stasis is promoted.

Even though the non-Newtonian behaviour occurrence proceeds before the bile stasis is unknown at the present time, the bile of two-thirds of gallstone patient is featured with non-Newtonian behaviour (Ooi 2004). This fact suggests that the gallstone disease likely has a link to the GB bile non-Newtonian behaviour. It has been illustrated that the increased CD resistance is observed prior to gallstone formation (Pitt et al. 1981). Meanwhile, the GB compliance remains unchanged (Pitt et al. 1982). However, the mechanism of increased CD resistance remains obscure (Pitt et al. 1982). Based on the current computational results of non-Newtonian bile, it would likely be speculated that the GB bile non-Newtonian behaviour might be one of the reasons for raising CD resistance.

REFERENCES

Bateson, M. C. 1999. Gallbladder disease. *British Medical Journal* 318:1745–1748.

Bober, W., and R. A. Kenyon. 1980. *Fluid Mechanics*. New York: John Wiley & Sons.

Bouchier, I. A., S. R. Cooperband, and B. M el-Kodsi. 1965. Mucous substances and viscosity of normal and pathological human bile. *Gastroenterology* 49:343–353.

Cancelli, C., and T. J. Pedley. 1985. A separated-flow model for collapsible-tube oscillations. *Journal of Fluid Mechanics* 157:375–404.

Catnach, S. M., P. D. Fairclough, R. C. Trembath, et al. 1992. Effect of oral erythromycin on gallbladder motility in normal subjects and subjects with gallstones. *Gastroenterology* 102:2071–2076.

Chen, J., and X. Y. Lu. 2004. Numerical investigation of the non-Newtonian blood flow in a bifurcation model with a non-planar branch. *Journal of Biomechanics* 37:1899–1911.

Chen, J., X. Y. Lu, and W. Wang. 2006. Non-Newtonian effects of blood flow on hemodynamics in distal vascular graft anastomoses. *Journal of Biomechanics* 39:1983–1995.

Coene, P. P., A. K. Groen, P. H. Davids, et al. 1994. Bile viscosity in patients with biliary drainage. *Scandinavian Journal of Gastroenterology* 29:757–763.

Cowie, A. G., and D. J. Sutor. 1975. Viscosity and osmolality of abnormal bile. *Digestion* 13:312–315.

Dodds, W. J., W. J. Groh, R. M. Darweesh, et al. 1985. Sonographic measurement of gallbladder volume. *American Journal of Roentgenology* 145:1009–1011.

Doty, J. E., H. A. Pitt, S. L. Kuchenbecker, et al. 1983. Role of gallbladder mucus in the pathogenesis of cholesterol gallstone. *American Journal of Surgery* 145:54–61.

Gijsen, F. J., F. N. van de Vosse, and J. D. Janssen. 1999. The influence of the non-Newtonian properties of blood on the flow in large arteries: Steady flow in a carotid bifurcation model. *Journal of Biomechanics* 32:601–608.

Gottschalk, M., and A. Lochner. 1990. Behavior of postoperative viscosity of bile fluid from T-drainage. *Gastroentroloisches Journal* 50:65–67.

Holzbach, R. T., M. Marsh, M. Olszewski, et al. 1973. Cholesterol solubility in bile. *Journal of Clinical Investigation* 52:1467–1479.

Jungst, D., A. Niemeyer, I. Muller, et al. 2001. Mucin and phospholipids determine viscosity of gallbladder bile in patients with gallstones. *World Journal of Gastroenterology* 7:203–207.

Kuchumov, A. G., V. Gilev, V. Popov, et al. 2014. Non-Newtonian flow of pathological bile in the biliary system: Experimental investigation and CFD simulations. *Korea-Australia Rheology Journal* 26:81–90.

Kuchumov, A. G. 2019. Biomechanical model of bile flow in the biliary system. *Russian Journal of Biomechanics* 23:224–248.

Li, W. G., X. Y. Luo, S. B. Chin, et al. 2008. Non-Newtonian bile flow in elastic cystic duct: One and three-dimensional modelling. *Annals of Biomedical Engineering* 36:1893–1908.

O'Callaghan, S., M. Walsh, and T. McGloughlin. 2006. Numerical modelling of Newtonian and non-Newtonian representation of blood in a distal end-to-side vascular bypass graft anastomosis. *Medical Engineering & Physics* 28:70–74.

Ooi, R. C. 2004. The flow of bile in human cystic duct. PhD diss., University of Sheffield, Sheffield, UK.

Pedley, T. J., B. S. Brook, and R. S. Seymour. 1996. Blood pressure and flow rate in the giraffe jugular vein. *Philosophical Transactions-Biological Sciences* 351:855–866.

Pitt, H. A., J. E. Doty, L. DenBesten, et al. 1982. Stasis before gallstone formation: Altered gallbladder compliance or cystic duct resistance? *American Journal of Surgery* 143:144–149.

Pitt, H. A., J. J. Roslyn, S. L. Kuchenbecker, et al. 1981. The role of cystic duct resistance in the pathogenesis of cholesterol gallstones. *Journal of Surgical Research* 30:508–514.

Saida, Y. 1992. Clinical and experimental evaluations of bile viscosity in the gallbladder. *Japanese Journal of Gastroenterological Surgery* 25:2129–2138.

Tera, H. 1963. Sedimentation of bile constituents. *Annals of Surgery* 157:467–472.

Thureborn, E. 1966. On the stratification of human bile and its importance for the solubility of cholesterol. *Gastroenterology* 50:775–780.

White, F. M. 2011. *Fluid Mechanics* (7th edition). New York: McGraw-Hill Companies Inc.

Wilkinson, W. L. 1960. *Non-Newtonian Fluids*. London: Pergamon Press.

CHAPTER 4

1D Dimensionless FSI and 3D FSI of CD

1D AND 3D FSI

The complicated 3D geometry and Newtonian or non-Newtonian bile flow in CDs have been presented by employing simple 1D models with the equivalent diameter and length, which were described in Chapters 2 and 3. However, the general numerical solutions were not tried.

Although the pressure drop given by the 1D models for the idealised 3D CD with rigid wall agrees very well with that of the fully 3D flow simulations by using FLUENT 6.2, the 1D models, which are featured with an equivalent diameter and length, cannot provide detailed flow patterns in the 3D CD, especially in FSI situations. Consequently, the agreement of the pressure drop of Newtonian and non-Newtonian bile flows in CDs with elastic wall and uniform baffles between 1D and 3D needs to be confirmed.

The FSI problem has been tackled numerically with FIDAP for a simple 2D CD with uniform baffles (Ooi et al. 2003). However, a comparison of pressure drop in a 3D rigid duct with that in elastic one has not been made by means of a 3D geometry and flow models, and the effect of FSI on the flow system behaviour remains unclear.

In this chapter, the Newtonian and non-Newtonian bile flows in the idealised 3D CDs in Chapters 2 and 3 are generalised, but also investigated numerically by using ADINA (ADINA R&D Inc., USA), which is finite element method (FEM) software with FSI capability. The results are compared with that of the rigid CDs by using FLUENT. The pervious FSI 1D models in Chapters 2 and 3 are validated. Good agreement was achieved between 1D and 3D for the compliant CD with uniform baffles.

DIMENSIONLESS 1D FSI ANALYSIS OF CD

The 1D FSI equations for either Newtonian or non-Newtonian bile can be written as

$$p_{\text{in}} - 2\int_0^x \frac{\tau_w}{r} dx + \frac{1}{2}\rho Q^2 \left(\frac{1}{A_{\text{in}}^2} - \frac{1}{A^2} \right) = p_e + \frac{\pi^{3/2} E h^3}{12\left(1-\upsilon^2\right) A^{3/2}} \left[\left(\frac{A}{A_{\text{eq}}} \right)^{10} - \left(\frac{A}{A_{\text{eq}}} \right)^{-3/2} \right] \tag{4.1}$$

$$\frac{Q}{\pi r^3} = \frac{1}{\tau_w^3} \int_0^{\tau_w} \gamma \tau^2 d\tau \tag{4.2}$$

The following dimensionless parameters are introduced

$$C_p = \frac{p_{\text{in}}}{\frac{1}{2}\rho\left(\frac{A}{A_{\text{eq}}}\right)^2}, C_{\text{DR}} = \frac{\frac{2L_{\text{eq}}\tau_{\text{weq}}}{r_{\text{eq}}}}{\frac{1}{2}\rho\left(\frac{A}{A_{\text{eq}}}\right)^2},$$
$$C_{\text{DF}} = \frac{\Delta p_{\text{CD}}}{\frac{1}{2}\rho\left(\frac{A}{A_{\text{eq}}}\right)^2}, M = \frac{\frac{E}{12\left(1-\upsilon^2\right)}}{\frac{1}{2}\rho\left(\frac{A}{A_{\text{eq}}}\right)^2} \tag{4.3}$$

$$Q' = \frac{Q}{\pi r_{\text{eq}}^3 \gamma_{\text{eq}}}, p_e' = \frac{p_e}{p_{\text{in}}}, \gamma' = \frac{\gamma}{\gamma_{\text{eq}}}, \tau' = \frac{\tau}{\tau_{\text{eq}}}, r' = \frac{r}{r_{\text{eq}}}, h' = \frac{h}{r_{\text{eq}}} \tag{4.4}$$

where C_p is the inlet pressure coefficient of CD, C_{DR} is the fluid friction loss coefficient for a rigid CD with r_{eq} and L_{eq}, C_{DF} is the fluid friction loss coefficient for an elastic CD with r_{eq} and L_{eq}, M is the FSI coefficient, Q' is the dimensionless bile flow rate, p_e' is the dimensionless external pressure of CD, γ' is the dimensionless shear rate, τ' is the dimensionless shear stress, γ_{weq} and τ_{weq} are the shear rate and shear stress at wall of a CD with r_{eq} and L_{eq}.

After these dimensionless coefficients are substituted into Eqs. (4.1) and (4.2), the dimensionless FSI equations are written as the following

$$C_p - C_{DR}\int_0^{x'} \frac{\tau'_w}{r'} dx' + \frac{1}{2}\rho Q^2\left(1-\frac{1}{\alpha^2}\right) = C_p p'_e + M\left(\frac{h'^3}{\alpha^{3/2}}\right)\left(\alpha^{10} - \alpha^{-3/2}\right) \quad (4.5)$$

$$\frac{Q'}{\pi r'^3} = \frac{1}{\tau'^3_w}\int_0^{\tau'_w} \gamma' \tau'^2 d\tau' \quad (4.6)$$

The fluid flow governing equation is written as

$$\frac{p}{p_{in}} = C_p - C_{DR}\int_0^{x'} \frac{\tau'_w}{r'} dx' + \frac{1}{2}\rho Q^2\left(1-\frac{1}{\alpha^2}\right) \quad (4.7)$$

where $x' = x/L_{eq}$. The unknown α can be obtained from Eqs. (4.5) and (4.6). Thus, the fluid pressure p can be estimated with Eq. (4.7).

When these equations are solved, the geometrical parameters of the elastic CD with uniform baffles are chosen to be as follows: d_{CD} = 1 mm, L_{CD} = 40 mm, ξ = 0.5, n = 6. The Young's modulus of the duct is in the range of 200–8000 Pa. The bile flow rates are follows: Q = 0.5, 0.7, 1.0 and 1.5 ml/min, and the bile is Newtonian with the density ρ = 1000 kg/m^3 and the dynamic viscosity μ = 1 mPa·s.

The fluid friction coefficient of the rigid CD, C_{DR}, and that of the elastic duct, C_{DF}, as well as the area ratio at the duct outlet, α_{out}, are illustrated in Figure 4.1, which are plotted as a function of the FSI coefficient, M. It can be observed that $C_{DF} > C_{DR}$. The transition points are indicated in the figure, where the friction loss coefficient of the rigid duct is equal to 99% of that of the elastic duct, i.e., $C_{DR} = 0.99C_{DF}$. Beyond these points, the flow pattern in the elastic duct is considered to be identical to that in the rigid one. A boundary through these transition points has been demonstrated in Figure 4.1. Obviously, the zone on the left side of the boundary shows a noticeable FSI, whereas the interaction can be ignored in the right side. The small M contributes to a low area ratio α_{out} and is responsible for a significant difference between C_{DR} and C_{DF}. This implies that small M causes a strong FSI.

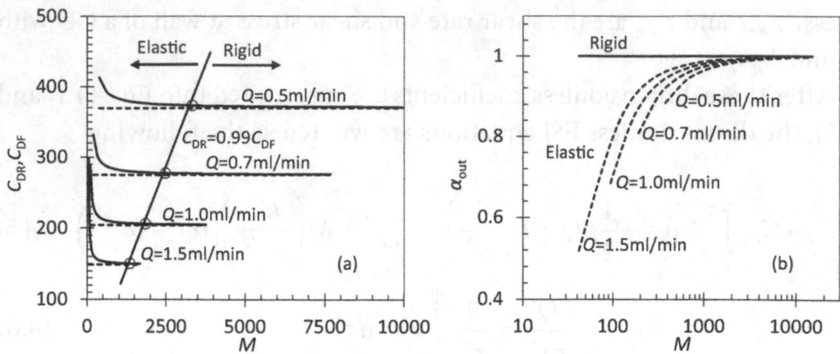

FIGURE 4.1 The fluid friction coefficients C_{DR}, C_{DF} and the area ratio at the CD outlet α_{out} against the FSI coefficient M, the solid lines for the rigid CD, the others for the elastic CD.

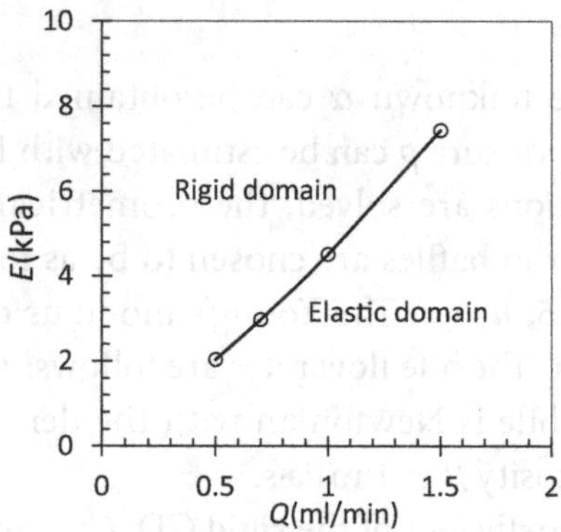

FIGURE 4.2 The Young's modulus versus the flow rate of bile at the transition points.

It is interesting to note that the Young's modulus is related to the bile flow rate at the transition points (Figure 4.2). In the zone below the boundary in the figure, FSI is dominant. Otherwise, in the zone above the boundary, FSI is negligible. An elastic CD can not only demonstrate rigid duct behaviour at a relative low bile flow rate, but also can have compliant feature at a large flow rate. Therefore, the FSI of a CD is determined by M completely, which indicates the relative magnitude of elasticity of CD to the bile flow dynamic head.

A BRIEF OF ADINA

The ADINA is a general-purpose system for the analysis of solid, fluid and FSI problems. Its codes are based on finite element (FE) for solid structures and FE or finite volume [flow-condition-based interpolation (FCBI) FE at high Reynolds number] for fluids. Structures can be modelled as 2D/3D solids, beams or shells with various Hooke's laws. The structure response can be linear or highly nonlinear. The contact problems are included. The fluid can be incompressible, slightly compressible or fully compressible. The flows formulated by the Navier-Stokes equations can be 2D planar, 2D axisymmetric or 3D. The fluid and structure can be coupled through their interface (FSI), porous media and thermal materials. ADINA has extensive material models, boundary conditions and user-friendly graphical system (Zhang et al. 2003).

A major aspect of FSI analysis is the coupling of the fluid and structure components. The coupling involves an exchange of information, such as stresses, velocities and/or displacements across the interface. Separate analysis methods for fluid and structure, a single analysis method and a single fluid-structure domain method are the three categories of the coupling schemes (Greenshields and Weller 2005).

The separate analysis method is the most popular one in the common coupling schemes currently. In the method, the individual fluid and structural problems are solved separately, then the link between fluid and structure is realised with data transfer across the interface. A fully coupled and convergent solution is achieved iteratively between fluid and structure at each time step. One drawback of the methods is considerably demanding on computational resources. ADINA employs this method to carry out the coupling between fluid and structure at the interface (Figure 4.3). In order to handle the moving interface between

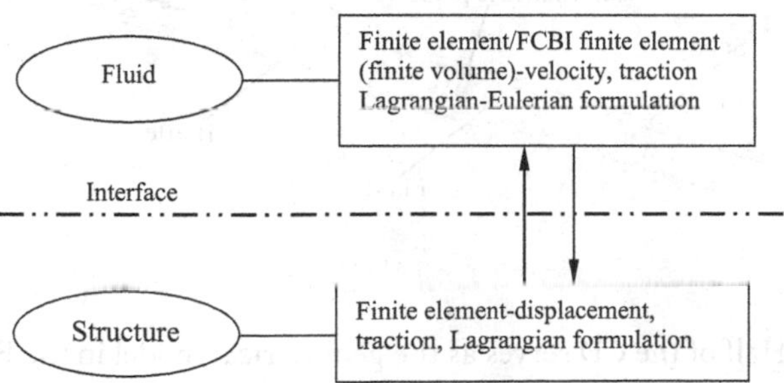

FIGURE 4.3 Coupling scheme of the separate analysis methods applied in ADINA.

fluids and solids, an arbitrary Lagrangian-Eulerian (ALE) formulation is applied to the fluid flow governing equations where the moving mesh velocity vector is included. The traction equilibrium and displacement compatibility must be satisfied at the interface of FSI (Zhang et al. 2003). The manuals issued by ADINA R&D Inc. should be referenced for more details on FSI.

COMPUTATIONAL MODELS

The Geometrical Model

As the 3D FSI simulations are time-consuming, here the focus will be on just one ideal CD. The duct has six uniformly staggered baffles, and its wall is rigid or elastic. The duct has been employed to validate the 1D fluid models proposed in Chapters 2 and 3. Due to symmetry in geometry, only half of the model is analysed (Figure 4.4). The inner diameter of the CD, d_{CD}, is 5 mm, the wall thickness, h, is 1 mm and baffle thickness, h_b, is 1 mm. Because the thickness/radius ratio is $2h/d_{CD} = 8/20 > 1/20$, which is the critical ratio for thin-shell structures, the CD wall has to be treated as 3D element (Ugural and Fenster 1987). A steady, 3D, laminar bile flow is applied at the duct inlet. The duct is subjected to an external pressure, p_e, and it is either deformed or collapsed under the transmural pressure, $p - p_e$. Here, p_e is assumed to be equal to the fluid pressure at the duct inlet, p_1.

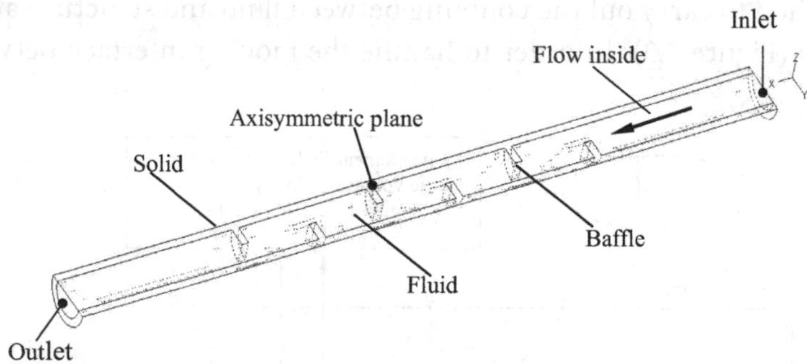

FIGURE 4.4 Half of the CD serves as the geometrical model in the FSI analysis, the CD has six uniformly staggered baffles, the right is the tube inlet and the left is the outlet.

Fluid and Solid Models

In the computations, both Newtonian and non-Newtonian fluids are considered. For the Newtonian bile, its density, ρ, and dynamic viscosity, μ, are 1000 kg/m^3 and 1×10^{-3} Pa·s, respectively. For the non-Newtonian bile, its density, ρ, is 1000 kg/m³ and the dynamic viscosity, μ, is decided by Carreau's model in Chapter 3. In order to meet the input requirement by ADINA, however, the two parameters in the model are adjusted as follows:

$$\mu = \mu_\infty + \left(\mu_0 - \mu_\infty\right)\left(1 + m_{1\text{ADINA}}\gamma^2\right)^{m_{2\text{ADINA}}} \tag{4.8}$$

where $\mu_\infty = 1\times10^{-3}$ Pa·s, $\mu_0 = 1\times10^{-2}$ Pa·s, $m_{1\text{ADINA}} = m_1^2 = 25784.0737\ \text{s}^{-2}$ and $m_{2\text{ADINA}} = (m_2 - 1)/2 = -0.2578$.

The CD wall and baffles are assumed to be a homogeneous, isotropic elastic solid material with the density, ρ_s, 1040 kg/m^3 and Poisson ratio, $\upsilon = 0.45$. The relationship between stress and strain is linear and the Young's modulus, E, is chosen to be 500 Pa.

Boundary Conditions

For the bile flow, the normal traction conditions are imposed at the CD inlet and outlet, respectively. At the inlet, the normal traction is zero. At the outlet, it is set to be one of the following values: −0.02, −0.05, −0.1, −0.3, −0.5, −1.0, −1.5, −2.0, −2.5, −3.0 and −3.5 Pa. The difference of the normal tractions at the inlet and the outlet is the pressure drop, which drives the flow. The bile flow rate for this given pressure difference can be calculated by ADINA. In the symmetrical plane, the velocity component in the z direction is chosen to be zero.

For the solid structure, the displacement in the x direction at the CD inlet and the outlet is chosen to be zero. The displacement in the z direction and the rotation about the x and y axes are zero in the symmetrical plane. In order to restrict the solid body, the two points on the outside surface of the CD, which have the lowest z coordinates, are specified at the inlet and outlet, respectively, and then their y displacement is fixed. The zero pressure load is applied on the outside surface of the duct, i.e. $p_e = 0$.

For the FSI problem, the interface between bile and inner side of the CD is set to be the FSI interface where no-slip velocity condition is applied.

Discretisation of Solid and Fluid Domains

The 4-node tetrahedral element is used to discretise the fluid and structure domain. It is a linear element with first-order accuracy in the spatial interpolations. The length of element edge is 0.3, 0.4 and 0.5 mm for the fluid domain. The length of element edge is 0.45 mm for the structure domain, and the number of elements is 52,000. The number of elements in the fluid domain is 180,000 when the mesh size is 0.3 mm. The effects of mesh size on the results of fluid dynamics are mentioned in the following section.

MODEL VALIDATION

First, the effect of mesh size on the pressure drop of the Newtonian bile flow in the rigid-walled CD is presented. Second, the pressure drops given by ADINA are compared with those calculated by using FLUENT 6.2 with the first- and second-order upwind schemes for the Newtonian bile flow in the CD of rigid wall, then the computational accuracy of ADINA is addressed. Third, the pressure drop and Darcy friction factor are discussed for both Newtonian and non-Newtonian bile flows in the CD with elastic and rigid walls. Finally, the flow details are illustrated.

Effect of Mesh Size

A series of pressure drops in terms of bile flow rate for the rigid CD is shown in Figure 4.5. The pressure drops are obtained by using ADINA8.4

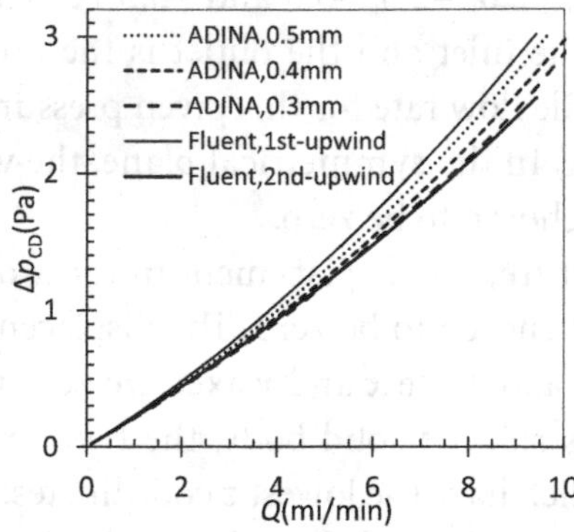

FIGURE 4.5 The pressure drop versus bile flow rate for the rigid CD, the pressure drops were obtained by using ADINA 8.4 with mesh sizes of 0.5, 0.4 and 0.3 mm for the Newtonian bile; for the convective terms in the Navier-Stokes equations, the FEM upwind scheme was used, the pressure drop given by using FLUENT 6.2 is as a reference with the mesh size of 0.5 mm, the first- and second-order upwind schemes were selected.

based on the three meshes with 0.5, 0.4 and 0.3 mm element edges. The bile is a Newtonian fluid. The FEM upwind scheme is applied to the convection terms in the Navier-Stokes equations in the computations of ADINA. These results are compared with those calculated by using FLUENT 6.2, in which the first- and second-order schemes are chosen and the mesh size is kept at 0.5 mm. It can be seen that the pressure drops given by ADINA in different mesh sizes are confined in the range of the pressure drops calculated by FLUENT 6.2. Moreover, as the mesh size gets small, the pressure drop decreases accordingly, especially at relatively high flow rate. When the mesh size is 0.3 mm, the corresponding pressure drop has been very close to that with the second-order scheme in FLUENT. Therefore, the mesh size of 0.3 mm is used in the fluid domain when the FSI computations are performed in ADINA.

The effect of mesh size of the solid domain on the parameters, such as the flow rate, Reynolds number, cross-sectional area ratio at the duct outlet and Darcy friction factor ratio is shown in Table 4.1, where the mesh sizes of the solid domain are 1.0, 0.75 and 0.45 mm, respectively, but the mesh size of the fluid domain is 0.3 mm and the pressure drop of the Newtonian bile across the CD is kept at 3.0 Pa in ADINA computations. It is clear that these parameters are not very sensitive to the mesh size of the solid domain. In other words, the requirement for the solid mesh size is not very high in order to estimate the parameters reliably. However, to ensure that the flow patterns are correct, the mesh sizes of 0.3 (fluid) and 0.45 mm (solid) are employed in ADINA FSI computations unless they are specified in special cases.

Effect of Upwind Scheme in ADINA

To investigate the computational accuracy of ADINA, the pressure drops estimated by using both FLUENT and ADINA are plotted in Figure 4.6 as a function of the bile flow rate in the rigid CD. In this case, the 4-node tetrahedral element was also used in FLUENT computations. The same mesh size (0.5 mm) was kept in both cases. The bile is a Newtonian fluid,

TABLE 4.1 Effect of mesh size in the solid domain.

Mesh size (mm)	Q (l/s)	Re_{CD}	α_{out}	f/f_o
1.00	9.7629	41.4349	0.9631	3.0163
0.75	9.7603	41.4240	0.96338	3.0172
0.45	9.7558	41.4051	0.9633	3.0185

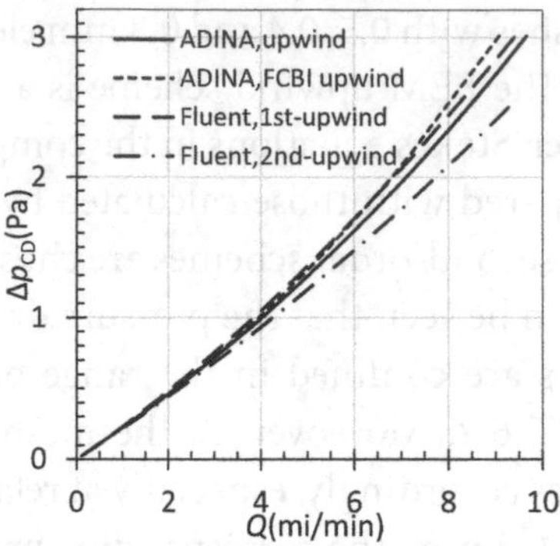

FIGURE 4.6 The pressure drop in terms of bile flow rate for the rigid CD, the pressure drops were calculated by using FLUENT 6.2 and ADINA 8.4, respectively, with the mesh size of 0.5 mm for the Newtonian bile, the first- and second-order upwind schemes were selected in FLUENT and two upwind schemes available were specified in ADINA 8.4.

and the first- and second-order schemes for the convection term in the Navier-Stokes equations are selected in FLUENT, respectively.

A noticeable difference can be observed in the pressure drop between the first- and second-order schemes. It should be noticed that the results in Chapters 2 and 3 are determined based on the second-order scheme in FLUENT. The result with the FCBI FEM upwind scheme in ADINA is close to that with the first-order scheme of FLUENT. The pressure drop due to ADINA FEM upwind scheme is slightly less than that obtained by the FCBI FEM upwind scheme. However, compared with FLUENT results, both schemes in ADINA seem to only have first-order accuracy. Since ADINA is equipped just with these two upwind schemes, it is impossible to choose a high-order upwind scheme to improve the accuracy and another possible way is to decrease mesh size to raise the accuracy. The increased number of elements will demand highly on the computational resources, especially for FSI problems. In the computations, the ADINA FEM upwind scheme is employed.

Effect of the Poisson Ratio

The Poisson ratio, υ, of the CD should be 0.5 like other human soft tissues. Although the Poisson ratio is kept at 0.45 in most of the numerical computations, $\upsilon = 0.3$ and 0.49 have also been attempted to examine the effect on the Darcy friction factor ratio, f/f_0, for the CD with a compliant

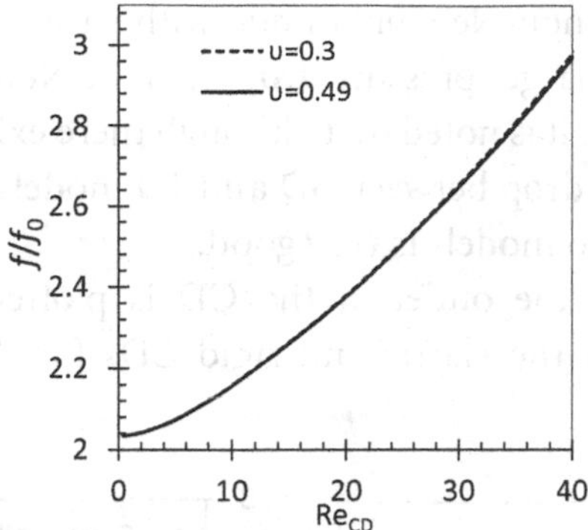

FIGURE 4.7 Darcy friction factor ratio against Reynolds number for two different Poisson ratios 0.3 and 0.49, Young's modulus of the CD wall is 500 Pa, the bile is a Newtonian fluid.

wall. Here, f and f_0 are the Darcy friction factor for the duct with baffles and without baffles, respectively, and f is defined as

$$f = \frac{\Delta p_{\mathrm{CD}}}{\frac{1}{2}\rho\left(\frac{4Q}{\pi d_{\mathrm{CD}}^2}\right)^2\left(\frac{L_{\mathrm{CD}}}{d_{\mathrm{CD}}}\right)} \tag{4.9}$$

Usually, f/f_0 is plotted as a function of Reynolds number Re_{CD} ($= 4\rho Q/\pi\mu d_{\mathrm{CD}}$ for Newtonian bile or $4\rho Q/\pi\mu_\infty d_{\mathrm{CD}}$ for non-Newtonian bile). The Darcy friction factor ratio, f/f_0, against Reynolds number, Re_{CD}, for two different Poisson ratios 0.3 and 0.49 has been illustrated in Figure 4.7. In these computations, the Young's modulus of the CD wall is 500 Pa, and the bile is a Newtonian fluid. The results are based on ADINA FSI simulations. It is concluded that the Poisson ratio has a negligible effect on f/f_0.

EFFECTS OF ELASTICITY AND NON-NEWTONIAN BILE

The effect of elasticity of the CD wall and non-Newtonian bile on the pressure drop across the CD can be observed in Figure 4.8, where the pressure drop is shown as a function of the flow rate of bile. The elastic CD can generate a larger pressure drop than the rigid duct for both Newtonian and non-Newtonian fluids.

Furthermore, the non-Newtonian bile with shear-thinning behaviour always experiences a larger pressure drop than the Newtonian bile in both elastic and rigid CDs. It is noted that although there exists a small discrepancy in the pressure drop between 3D and 1D models, the overall agreement between the two models is very good.

The area ratio at the outlet of the CD is plotted against the flow rate in Figure 4.9 in the elastic and rigid CDs for the Newtonian and

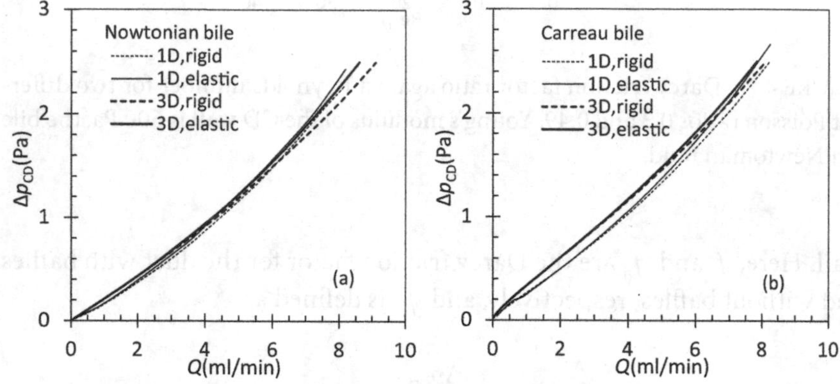

FIGURE 4.8 The pressure drop versus bile flow rate for the Newtonian (a) and non-Newtonian (b) bile flows in the CD with the elastic (E = 500 Pa, υ = 0.45) and rigid walls.

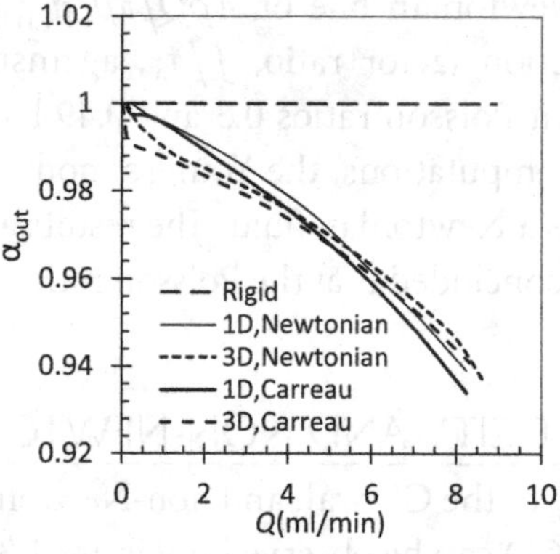

FIGURE 4.9 The area ratio at the outlet of the CD in terms of bile flow rate for the Newtonian and non-Newtonian bile flows in the CD with the elastic (E = 500 Pa, υ = 0.45) and rigid walls.

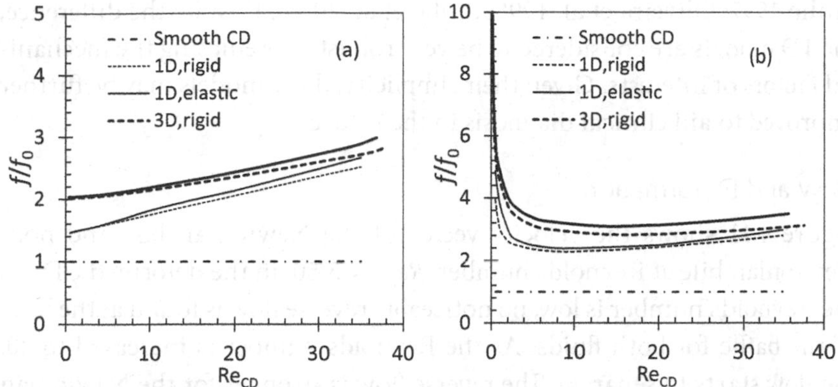

FIGURE 4.10 Darcy friction factor against Reynolds number for the Newtonian (a) and non-Newtonian (b) bile flows in the CD with elastic ($E = 500$ Pa, $\upsilon = 0.45$) and rigid walls.

non-Newtonian fluids. Since the baffles make the CD stiffer than the equivalent duct without baffles, there is a significant difference in the area ratio between 3D and 1D. It is also shown, both in 1D and 3D models, that the non-Newtonian bile tends to collapse the tube more than the Newtonian bile.

The Darcy friction factor ratio of the elastic and rigid CD is shown in Figure 4.10 for the Newtonian and non-Newtonian bile flows. The elastic duct and non-Newtonian bile also demonstrate a higher Darcy friction factor ratio compared with the rigid duct and Newtonian bile. Although some difference between 3D and 1D results can be observed, especially as the Reynolds number less than 25, the general trend of f/f_0 of the 1D models with Reynolds number is consistent with the behaviour of f/f_0 of the 3D model.

These results suggest that the simple 1D models in Chapters 2 and 3 can be used to predict the effects of elastic CD wall. The reasons for the disagreement between 1D and 3D results are that the minor loss in the 1D models is calculated on the basis of a constant coefficient, which is independent of the Reynolds number. It has been shown that the minor loss coefficient of Newtonian or non-Newtonian (shear-thinning power-law) fluid across a sudden axisymmetric expansion increased, rather than kept at a constant when the Reynolds number decreased from 30 (Oliveira and

Pinho 1997; Oliveira et al. 1998; Pinho et al. 2003). Despite the difference, the 1D models are considered to be very robust in predicting the mechanical factors of interests. Given their simplicity, these models may be further improved to aid clinical diagnosis in the future.

Flow and Deformation

Figure 4.11 shows the velocity vector of the Newtonian bile and non-Newtonian bile at Reynolds number $Re_{CD} \approx 3.50$ in the deformed CD. As the Reynolds number is low, no noticeable reverse flow is found at the back of the baffle for both fluids. As the Reynolds number is increased to 20, the flow starts to separate. The reverse flow is stronger for the Newtonian

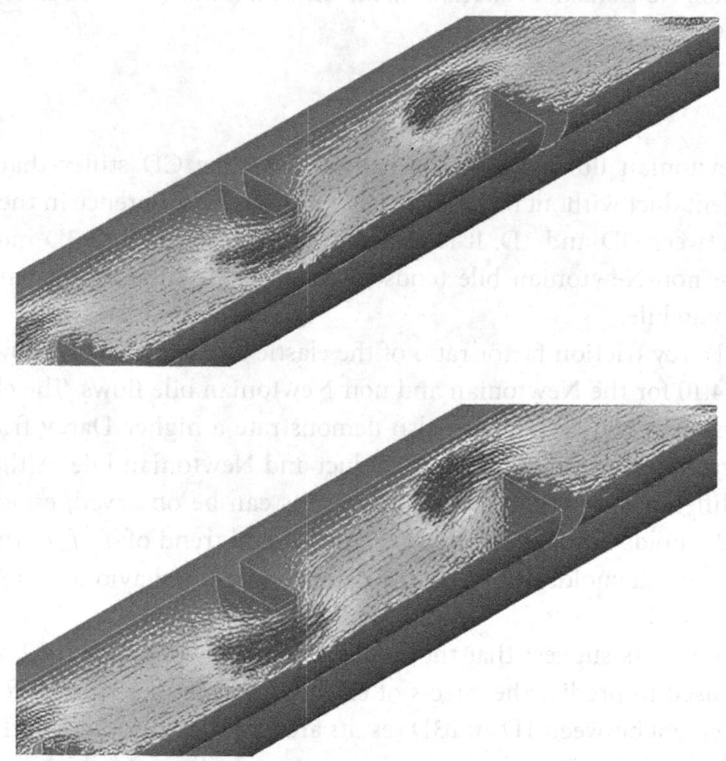

FIGURE 4.11 Velocity vector of the Newtonian bile (top) and non-Newtonian bile (bottom) at Reynolds number $Re_{CD} \approx 3.5$ in the deformed CD, even though 'dead water zone' can be found in the back of each baffle, no noticed reverse flow exists in the zone for both fluids.

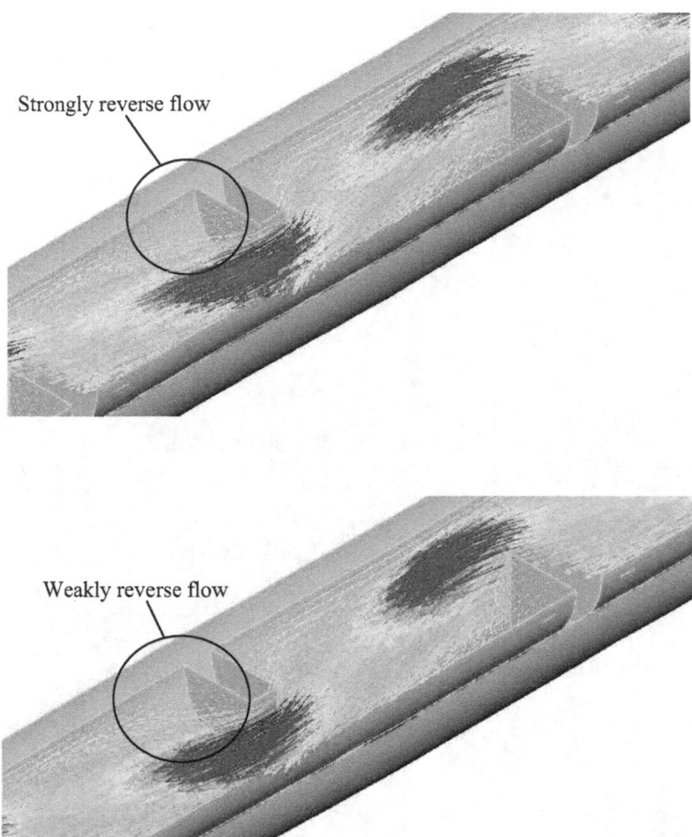

FIGURE 4.12 Velocity vector of the Newtonian bile (top) and non-Newtonian bile (bottom) at Reynolds number $Re_{CD} \approx 20$ in the deformed CD, the 'dead water zone' can be found in the back of each baffle; however, the reverse flow of Newtonian bile is more obvious than that of non-Newtonian bile.

bile compared with that of the non-Newtonian fluid (Figure 4.12). This is because the reversed flow tends to be suppressed by the higher viscosity of the non-Newtonian bile in the low shear-rate zone, i.e., the 'dead water' zone. This relatively high viscosity is also responsible for the larger pressure drop in the non-Newtonian bile as shown above.

The comparison of the configuration between the original and deformed CD is illustrated in Figures 4.13 and 4.14 at two Reynolds

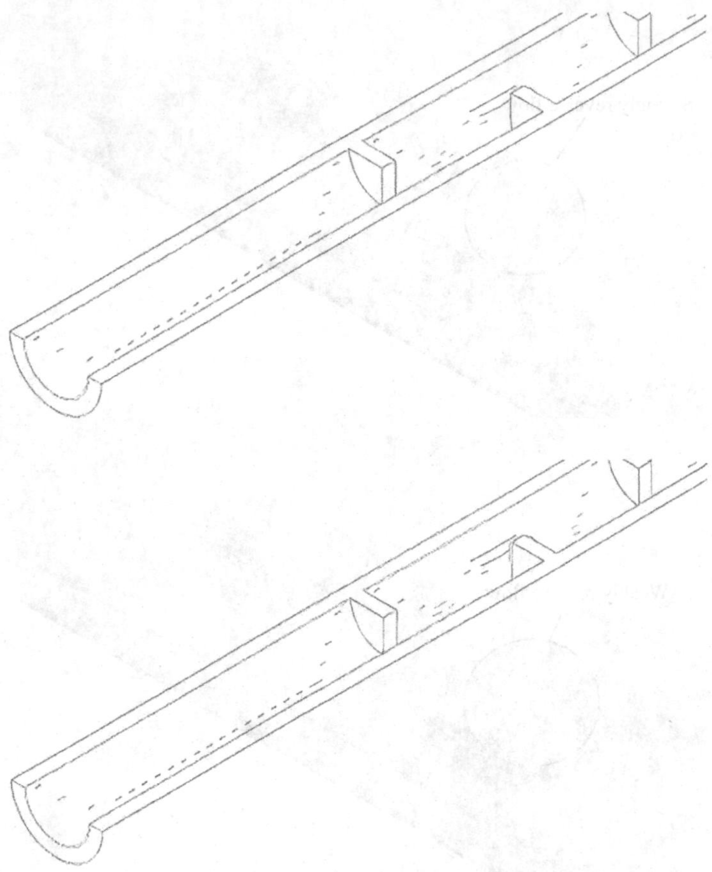

FIGURE 4.13 Comparison of the configuration between the original CD and the deformed duct for the Newtonian bile (top) and non-Newtonian bile (bottom) at Reynolds number $Re_{CD} \approx 3.5$, the displacement magnification factor is 30 in the plots; in this case, there is no noticeable difference between two configurations.

numbers $Re_{CD} \approx 3.5$ and 20 for both Newtonian and non-Newtonian fluids, respectively. The deformation of the CD becomes more significant as the Reynolds number increases.

There is a distinguished difference between the deformation of Newtonian and non-Newtonian bile at $Re_{CD} \approx 20$. When the CD is deformed, the gap between the baffle edge and the duct wall narrows, accordingly the minor loss increases and the pressure drop across the duct rises.

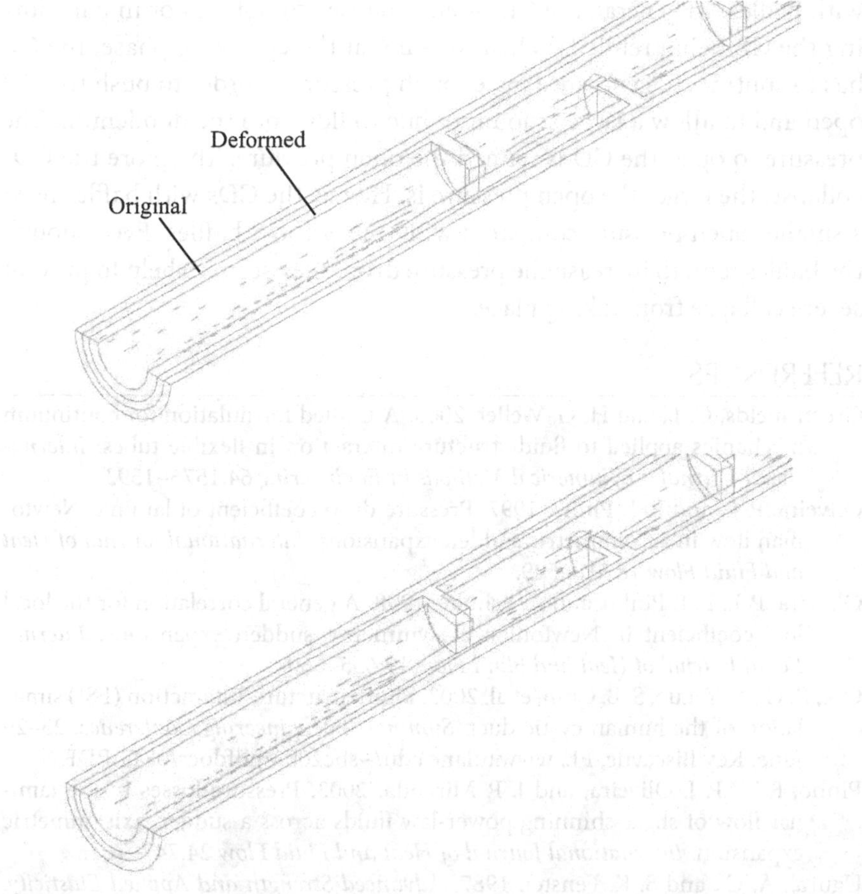

FIGURE 4.14 Comparison of the configuration between the original CD and the deformed duct for the Newtonian bile (top) and non-Newtonian bile (bottom) at Reynolds number $Re_{CD} \approx 20$, the displacement magnification factor is 30 in the plots; in this case, a significant difference between two configurations can be identified.

It should be noted that the baffle edge must attempt to contact with its opposite wall of the CD as the collapse is developed. Consequently, once the baffle contacts with the wall of the duct, there will be no bile flow in the duct at all. For the CD without baffles, the opposite two walls of the duct also may contact each other due to the collapse. However, this severe collapse needs a relatively large external pressure compared to the duct with baffles. In general, the CD often collapses completely or in part during the GB being refilled. When working at the emptying phase, the GB has to contract to generate a big enough pressure in order to push the CD open and to allow a large amount of bile to flow into the duodenum. The pressure to open the CD is named the open pressure. The more the CDs collapse, the larger the open pressure is. Hence, the CDs with baffles need a smaller open pressure compared with one without baffles. Even though the baffles tend to increase the pressure drop, they seem to help to prevent severe collapse from taking place.

REFERENCES

Greenshields, C. J., and H. G. Weller. 2005. A unified formulation for continuum mechanics applied to fluid-structure interaction in flexible tubes. *International Journal for Numerical Methods in Engineering* 64:1575–1593.

Oliveira, P. J., and F. T. Pinho. 1997. Pressure drop coefficient of laminar Newtonian flow in axisymmetric sudden expansions. *International Journal of Heat and Fluid Flow* 18:518–529.

Oliveira, P. J., F. T. Pinho, and A. Schulte. 1998. A general correlation for the local loss coefficient in Newtonian axisymmetric sudden expansions. *International Journal of Heat and Fluid Flow* 19:655–660.

Ooi, R. C., X. Y. Luo, S. B. Chin, et al. 2003. Fluid-structure-interaction (FSI) simulation of the human cystic duct. *Summer Bioengineering Conference*, 25–29 June, Key Biscayne, FL. www.tulane.edu/~sbc2003/pdfdocs/0811.PDF.

Pinho, F. T., P. J. Oliveira, and J. P. Miranda. 2003. Pressure losses in the laminar flow of shear-shinning power-law fluids across a sudden axisymmetric expansion. *International Journal of Heat and Fluid Flow* 24:747–761.

Ugural, A. C., and S. K. Fenster. 1987. *Advanced Strength and Applied Elasticity*. New York: Elsevier Science Publishing Co.

Zhang, H., X. Zhang, S. Ji, et al. 2003. Recent development of fluid-structure interaction capabilities in the ADINA system. *Computers and Structures* 81:1071–1085.

CHAPTER 5

Biomechanical Model for GB Pain

GB PAIN AND ITS DIAGNOSIS

GB pain, also known as ABP or functional biliary pain, is defined as a steady pain located in the epigastrium and right upper quadrant in the absence of gallstones or when no other structural abnormalities exist in the biliary tract (Delgado-Aros et al. 2003; Shaffer 2003) (see Figure 5.1). Clinically, GB pain is typically described as pain in the right upper part of the abdomen lasting for 30 min or more and provoked by a fatty meal, but not all patients experience these classical symptoms. Gallstones are the common cause, but only a small minority of the 10% of the population with stones experience pain.

This pain may occur in up to 7.6% in men and 20.7% in women (Corazziari et al. 1999) and has received more interest in recent years (Canfield et al. 1998; Corazziari et al. 1999; Shaffer 2003; Delgado-Aros et al. 2003; Rastogi et al. 2005). Patients with GB pain often pose diagnostic difficulties and undergo repeated ultrasound scans and oral cholecystography. Sonography or oral cholecystography combined with scintigraphy is commonly used to diagnose GB pain. Reproduction of pain within 5–10 min of an injection of CCK is also applied to select a group of patients who may benefit from cholecystectomy (Williamson 1988; Raptopoulos et al. 1985).

However, surgery is often conducted without any guarantee of relieving the GB pain symptoms. Only about 50% of patients with GB pain

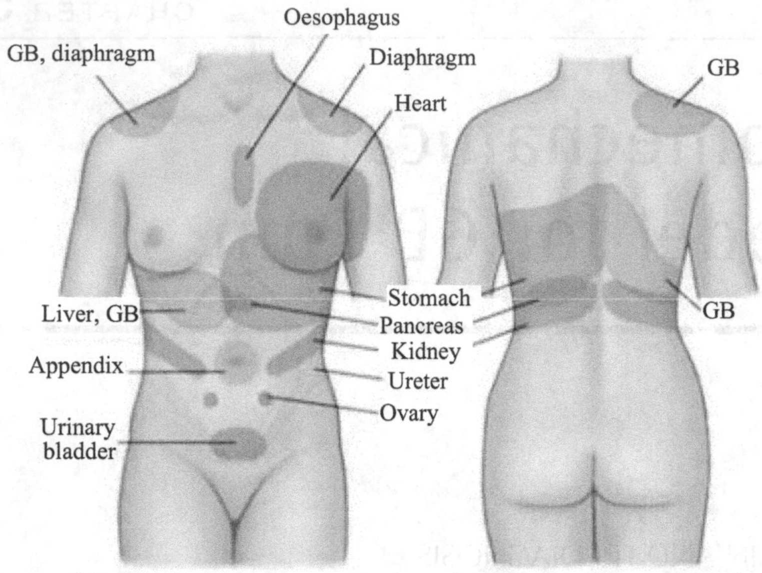

FIGURE 5.1 GB pain locations (the pictures are after Hale 2018; the copyright belongs to Dr Teka Hale, the use with his permission).

symptoms obtained symptomatic relief following surgery (Smythe et al. 1998). It is, therefore, important to have an approach for determining whether the pain actually happens in the GB.

Impaired motor function of the GB and sphincter of Oddi have long been suspected as a major factor contributing to GB pain. The presumed mechanism for the pain is obstruction leading to distension and inflammation. The obstruction might result from a lack of coordination between the GB and either the CD or the sphincter of Oddi due to increased flow resistance or tone (Corazziari et al. 1999). In other words, pain may be produced by contraction against resistance or stretch of the GB wall. When the GB is inflamed, artificial distension produces typical pain (Fuchs-Jensen 1995).

The pain provocation test has been utilised as a diagnostic tool to select patients with impaired GB motor function who may benefit from the cholecystectomy. In the test, CCK is injected intravenously to stimulate GB to contract and induce the biliary or GB pain. It is clinically accepted that when a GB ejection is less than 35% (Pickleman et al. 1985; Fink-Bennett et al. 1991; Kmiot et al. 1994; Michail et al. 2001; Misra et al. 1991; Ozden et al. 2003; Bingener et al. 2004; Riyad et al. 2007) or 40% (Yap et al.

1991; Ponce et al. 2004), then the GB motor function is considered to be impaired; otherwise, it is considered normal.

It has been found that the GB pain of some patients has been alleviated after their GBs were removed (Pickleman et al. 1985; Fink-Bennett et al. 1991; Misra et al. 1991; Kmiot et al. 1994; Michail et al. 2001; Ozden et al. 2003; Bingener et al. 2004; Brosseuk and Demetrick 2003; Yap et al. 1991; Ponce et al. 2004), but conflicting reports also exist (Sunderland and Carter 1988; Smythe et al. 1998, 2004; Rastogi et al. 2005). Thus, the impaired GB motor function is not the only factor responsible for the pain.

As a type of visceral pain, GB pain arises from the GB and biliary tract with obstruction of the CD or CBD, which elevates pressure within the human biliary system. Some researchers believe that the pain is directly related to intraluminal pressure of the biliary tract (Ness and Gebhart 1990). Gaensler (1951) examined the pain threshold of CBD for 40 patients before and after GB removal. The pain threshold varied from 14.7 to 59 mmHg (with a mean of 40 mmHg). Csendes et al. (1979) illustrated that the pain threshold is in the range of 15–60 mmHg (with a mean of 30 mmHg). Middelfart et al. (1998) showed that the pain threshold of 12 patients varied from 4 to 58 mmHg (with a mean of 23 mmHg). Note that when the GB contracts, the intraluminal pressure rises, then the GB geometrical size and shape should be altered. Consequently, the various stress distributions in the biliary tract and GB walls will differ from the pressure distribution. This means that the sensor of the pain in the biliary system may be related to other mechanical factors associated with the pressure and geometry, rather than the pressure alone.

Similar observations are made for pain in the oesophagus, duodenum, gastric antrum and rectum, which seems to respond to mechanoreceptors in the gastrointestinal tract wall (Drewes et al. 2003). These mechanoreceptors are found to depend on the luminal circumferential wall strain rather than pressure, tension and volume (Drewes et al. 2003; Barlow et al. 2001; Petersen et al. 2003; Gao et al. 2002; Gregersen et al. 2006).

The strain can be measured by using experiments (Jonderko and Bueno 1997) or estimated by stress with known tissue constitutive law. For GB with complicated geometrical shape, the measurement of strain may be very difficult; furthermore, the constitutive law for GB soft tissue is unavailable now. Consequently, it is not a good way to predict GB pain by applying the strain. However, the stress can be estimated simply by using force equilibrium equations for GBs. Therefore, it is assumed that

GB pain is related to the stress directly, which is responsible for activating the mechanoreceptors in the GB wall.

In this chapter, GB pain is studied from the bio mechanical point of view. A bio mechanical model is developed for the human biliary system in the emptying phase, based on a clinical test in which GB volume changes are measured in response to a standard stimulus and a recorded pain profile. The model can describe the bile emptying behaviour, the flow resistance in the human biliary tract, the peak total in-plane normal stress, including the passive and active stresses experienced by the GB during emptying, the EF and the peak pressure in the GB. The model is used to explore the potential link between GB pain and mechanical factors. It is found that the peak total normal stress may be used as an effective pain indicator for GB pain. When this model is applied to clinical data of volume changes due to CCK stimulation and pain from 37 patients, it shows a promising success rate of 78.4% in pain prediction. The model details were presented by Li et al. (2008); in this study, they are described briefly with some corrections.

ACQUISITION OF CLINICAL DATA

The GB stores and concentrates bile secreted by the liver between meals and then empties it into the duodenum by contracting in response to CCK secreted by the duodenum as food comes into the stomach. The motor function of the GB is its ability to contract and expel the bile. Failure to do this can result in stasis of the bile and encourage gallstone formation.

A CCK provocation test was carried on patients who had experienced repeated attacks of biliary pain in the absence of gallstones or any obvious causative findings (Cozzolino et al. 1963). After an overnight fast, the patients were given an intravenous infusion of saline (control) followed by an intravenous infusion of CCK (0.05 μg/kg body weight). Ultrasonography of the GB was used to monitor changes in shape, initial volume, emptying and wall thickness at 5 or 15-min intervals for 60 min. The patients did not know which drug was being given, and the test was only considered positive when the patients usual GB pain was reproduced following CCK infusion.

The pressure and volume variations with time are illustrated in Figure 5.2 schematically in a CCK provocation test. At point 1, the emptying is finished, the sphincter of Oddi is closed (see Figure 5.3), the GB is in a fasting state and its volume, pressure and stresses all reach their minimum levels (Lempke et al. 1963).

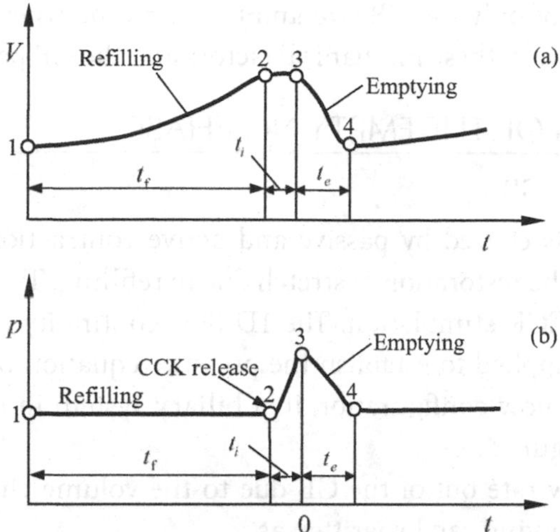

FIGURE 5.2 Diagrammatic representation of GB refilling and emptying. Refilling starts at point 1 and stops at point 2. Emptying begins at point 2 and lasts until point 4, when the next refilling starts. Note t_f is the refilling time, and t_e is the emptying time, $t_f \approx 6t_e$, but t_i is CCK response time.

Between points 1 and 2, a small but positive pressure difference between the liver and the GB exists, which allows the hepatic bile to be secreted slowly into the GB. During this refilling, although the GB volume is increasing, the pressure in the GB is basically constant as the muscle relaxes. At point 2, CCK is secreted and causes the GB to contract. The pressure in the GB rises rapidly up to point 3 in 3–5 min and exceeds the pressure in the CBD. In the meantime, the sphincter of Oddi relaxes and lowers the pressure in the CBD further. In consequence, the pressure in the GB is higher than the CBD, and the emptying phase takes place. For most of the subjects, the emptying lasts for about half an hour. The timescale of the refilling is usually the time lapse between two meals and is often longer than the emptying by six times.

The GB volume variation with time for a patient can be measured non-invasively using ultrasonography (Everson et al. 1980), which is readily available in clinical practice. The pressure inside the GB, however, can only be measured invasively, i.e., by placing a catheter into the GB. Here, it will develop a simple model to estimate pressure in the GB, flow resistance, shear and normal stress in the GB wall during the emptying based on non-invasive measurements of only the GB size and to determine whether there is any correlation between these mechanical factors and the GB pain experienced.

MODELLING OF THE EMPTYING PHASE

Pressure in the GB

GB emptying is caused by passive and active contractions. The passive contraction is the restoration of stretch due to refilling. The active contraction is due to CCK stimulation. The 1D flow continuity and momentum equations are applied to establish the pressure equation in the GB in the emptying. The flow configuration in a biliary system in the emptying is indicated in Figure 5.3.

The bile flow rate out of the GB due to the volume change of the GB during the emptying can be written as

$$Q_{GB} = -\frac{dV}{dt} \tag{5.1}$$

It is assumed that the bile flow is 1D and quasi-steady in the biliary tract. The bile flow rate, Q_{duct}, through the biliary duct under the difference of pressure between the GB and the duodenum (Figure 5.3) should be expressed by the flow momentum equation

$$Q_{duct} = \frac{p - p_d}{R} \tag{5.2}$$

where R is the flow resistance and p_d is the pressure in the duodenum, is taken to be about 6 mmHg (Dodds et al. 1989). As there is no bile flow

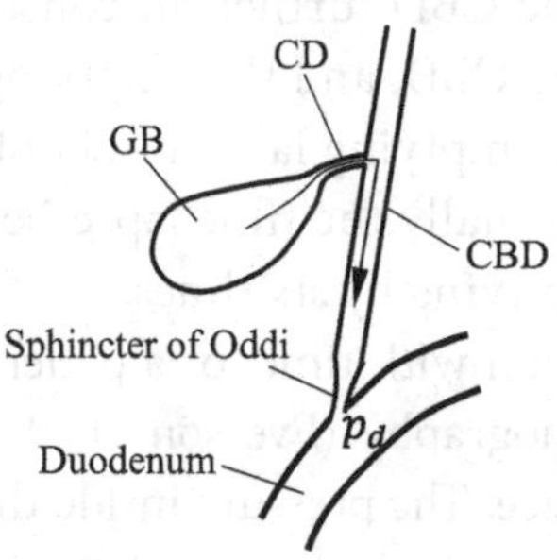

FIGURE 5.3 Bile flows into the duodenum from the GB through the CD and CBD due to the pressure difference $p - p_d$.

entering the biliary tract from the liver during the emptying, the flow rate out of the GB, Q_{GB}, is equal to the flow rate into the CD, Q_{duct}, i.e.,

$$-\frac{dV}{dt}=\frac{p-p_d}{R} \tag{5.3}$$

The GB volume change rate dV/dt is related to the pressure drop rate dp/dt with the GB compliance (Ryan and Cohen 1976; Schoetz et al. 1981)

$$\frac{dV}{dt}=C\frac{dp}{dt} \tag{5.4}$$

where C is the compliance of the GB. Combining Eqs. (5.3) and (5.4), we have

$$C\frac{dp}{dt}+\frac{p-p_d}{R}=0 \tag{5.5}$$

This is the same as the Windkessel model for the cardiovascular system (Fung 1984). The general solution of this first-ordinary linear differential equation is as follows when the zero time point is specified at the start of emptying

$$p=p_d+\left(p_e-p_d\right)e^{\left(\frac{t_e-t}{RC}\right)} \tag{5.6}$$

where p_e is devoted to the pressure at the end of emptying, $p_e = 11$ mmHg (Dodds et al. 1989), and t_e is the time taken for the complete emptying. The complete emptying is defined such that the GB volume $t_f \approx 6t_e$ V_e at the end of emptying is sufficiently small. Without loss of generality, it is assumed that $V_e = 0.3V_0$, where V_0 is the volume as the emptying starts. In other words, if GB motor function is impaired, then the GB cannot finish the complete emptying in its emptying time.

Usually, the GB compliance C differs from one patient to another. However, as a first approximation, $C = 2.731$ ml/mmHg is adopted herein based on the averaged value measured by Middelfart et al. (1998) for the human GBs.

Volume Change of the GB

From Eqs. (5.4) and (5.6), the GB volume is solved as

$$V=C(p_e-p_d)e^{\frac{(t_e-t)}{RC}}+\Omega \tag{5.7}$$

where Ω is a constant to be determined.

Using the clinical measurements of GB volume at the emptying start $(0, V_0)$, a time moment after emptying (t, V_t), and the end of emptying (t_e, V_e), Ω is obtained by

$$\Omega = V_e - C(p_e - p_d) \tag{5.8}$$

Substituting Ω into Eq. (5.7), R and t_e can be determined as follows

$$R = \frac{t}{C\ln[(V_0 - \Omega)/(V_1 - \Omega)]} \tag{5.9}$$

and

$$t_e = \frac{t\ln[(V_0 - \Omega)/(V_e - \Omega)]}{\ln[(V_0 - \Omega)/(V_t - \Omega)]} \tag{5.10}$$

These measurements also allow us to calculate the GB EF when the emptying time is 30 min as follows

$$\mathrm{EF} = \frac{V_0 - V_{30}}{V_0} \times 100\% \tag{5.11}$$

where V_{30} is the GB volume at $t = 30$ min for the emptying time, and note that a GB with a smaller EF means has a poorer ability in emptying.

Estimating the Passive Peak Stress

In order to estimate the peak stresses in GB wall tissue during emptying, it is assumed that the GB is an ellipsoid with a thin uniform wall thickness, h_{GB}. The ellipsoid has a major axis, D_3, and two minor axes, D_1 and D_2 ($D_1 \le D_2 \le D_3$), respectively, measured from its mid-plane surface (see Figure 5.4). Using the Cartesian coordinates as shown in that figure, the mid-plane surface is described by the following coordinate system

$$\begin{cases} x = 0.5D_1 \sin\theta\cos\varphi \\ y = 0.5D_2 \sin\theta\sin\varphi \\ z = 0.5D_3 \cos\theta \end{cases} \tag{5.12}$$

where θ and φ are the two independent angular variables for a point position on the surface, θ is in the meridian plane and measured from the

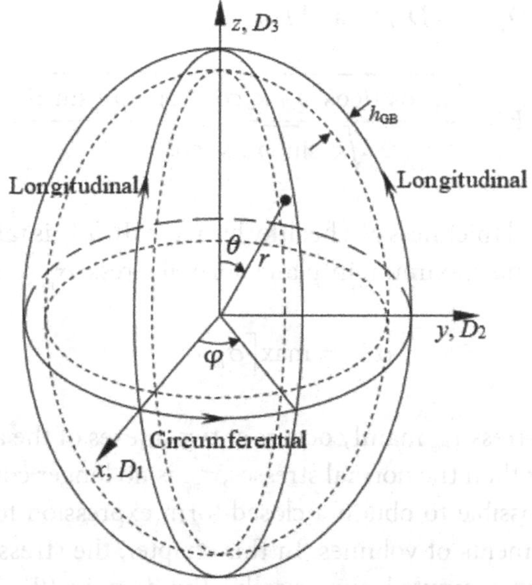

FIGURE 5.4 GB shape is assumed to be an ellipsoid during emptying, the major axis length is D_3, the longer minor axis length is D_2 and the shorter minor axis length is D_1 ($D_1 \le D_2 \le D_3$), the GB is with a uniform internal pressure. The stress due to the pressure at any point in mid-surface has three components: σ_θ (longitudinal), σ_φ (circumferential) and $\tau_{\theta\varphi}$ (in surface).

positive z axis, $\theta \in [0, \pi]$; whereas φ is in the latitudinal plane, which is perpendicular to the z axis and measured from the first quadrant of the x-z plane, $\varphi \in [0, 2\pi]$.

The stresses in the ellipsoid surface under a uniform inner fluid pressure load p can be calculated analytically (Novozhilov 1964):

$$\begin{cases} \sigma_\theta = \dfrac{pD_3\kappa_1\kappa_2}{4h_{\text{GB}}}\left[1-\dfrac{\kappa_1^2-\kappa_2^2}{\kappa_1^2\kappa_2^2}\cos 2\varphi\right]F \\ \sigma_\varphi == \dfrac{pD_3}{4\kappa_1\kappa_2 h_{\text{GB}}}\left[\begin{array}{l}\kappa_1^2\kappa_2^2+\left(\kappa_1^2+\kappa_2^2-2\kappa_1^2\kappa_2^2\right)\sin^2\theta \\ +\left(\kappa_1^2-\kappa_2^2\right)\cos^2\theta\cos 2\varphi\end{array}\right]\dfrac{1}{F} \\ \tau_{\theta\varphi} = \dfrac{pD_3}{4\kappa_1\kappa_2 h_{\text{GB}}}\left(\kappa_1^2-\kappa_2^2\right)\cos\theta\sin 2\varphi \end{cases} \tag{5.13}$$

where $\kappa_1 = D_1/D_3$, $\kappa_2 = D_2/D_3$ and F is

$$F = \frac{\sqrt{\kappa_1^2\cos^2\theta\cos^2\varphi + \kappa_2^2\cos^2\theta\sin^2\varphi + \sin^2\theta}}{\sqrt{\kappa_1^2\sin^2\varphi + \kappa_2^2\cos^2\varphi}} \tag{5.14}$$

The mean wall thickness of a healthy human GB, h_{GB}, is taken as 2.5 mm (Deitch 1981). The maximum in-plane normal stress, σ^{p}_{max}, is then

$$\sigma^{p}_{max} = \max\left[\sigma_\theta, \sigma_\varphi\right] \tag{5.15}$$

Since shear stress $\tau_{\theta\varphi}$ mainly occurs at two apexes of the axis D_3 and its value is smaller than the normal stresses, $\tau_{\theta\varphi}$ is no longer considered here.

It may be possible to obtain a closed-form expression for σ^{p}_{max} in different measurements of volumes. In this chapter, the stress distributions around GBs are computed numerically. The domain (θ, φ) was divided into 200 × 100 elements, then the values of σ_θ and σ_φ were calculated at each node of elements. This allows us to find the values and locations of the maximum stresses.

Contribution of the Active Normal Stress

During emptying, the GB also contracts due to CCK, which induces the active stress. In practice, the peak active stress σ^{a}_{max} depends on the CCK dose, though disagreements exist on how much the dependence is (Portincasa et al. 1994; Van De Heijning et al. 1999). In this chapter, for simplicity, it is proposed that all patients experience the same level of CCK stimulation, which induces the same peak active stress. Thus, a uniform response curve to CCK is used, estimated from various experiments (Takahashi et al. 1982; Traynor et al. 1984; Matsumoto et al. 1988), as shown in Figure 5.5. This curve can be interpolated by using

$$\sigma^{a} = \begin{cases} \sigma^{a}_{max}\sin\left(\dfrac{\pi t}{2t_{CCK}}\right) & t \le t_{CCK} \\ \sigma^{a}_{max}\left(1 - \dfrac{t - t_{CCK}}{t_{dc}}\right) & t > t_{CCK} \end{cases} \tag{5.16}$$

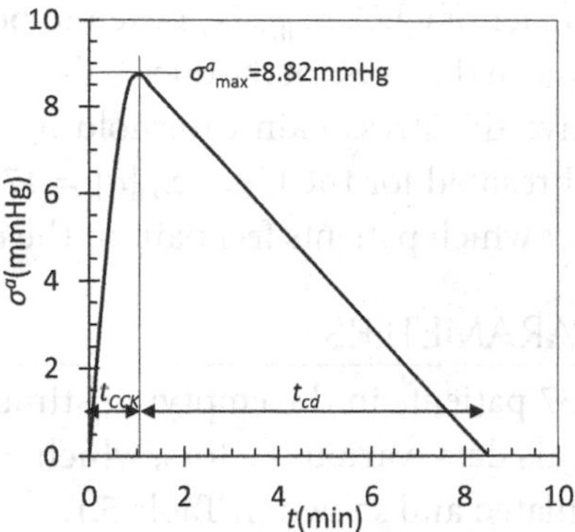

FIGURE 5.5 GB response curve to CCK (Milenov and Shahbazian 1995), t_i is the CCK response time, t_d is the CCK decaying time and σ^{a}_{max} is the peak active stress.

where σ^{a}_{max} is the maximum active stress which is 8.82 mmHg, t_{CCK} and t_{dc} are chosen to be t_{CCK} = 1 min and t_{dc} = 7.5 min (Milenov and Shahbazian 1995). The total maximum normal stress in the GB wall during the emptying is thus

$$\sigma_{max} = \sigma^{p}_{max} + \sigma^{a}_{max} \tag{5.17}$$

Pain Threshold

It is assumed that the total in-plane normal stress can activate the pain-sensitive mechanoreceptors in the GB wall. Hence, the pain prediction criterion is written as

$$\sigma_{max} > [\sigma] \tag{5.18}$$

where $[\sigma]$ is the threshold of the normal stress. Although there is not direct measurement of the threshold, $[\sigma]$, required in Eq. (5.18), it can be estimated from the measurements of Gaensler (1951), where the threshold of the pressure when pain is felt is around p_{CBD} = 35 mmHg in the CBDs of 33 patients.

Based on this pressure threshold, the circumferential stress in the CBD wall can be estimated by assuming the CBD is a straight circular pipe with uniform thickness

$$\sigma_{CBD} = \frac{p_{CBD} d_{CBD}}{2h_{CBD}} \tag{5.19}$$

where the mean diameter of CBD, d_{CBD}, is chosen to be 10 mm (Ferris and Vibert 1959) and the mean thickness of the duct wall, h_{CBD}, is 1 mm (Mahour et al. 1967). These give the stress pain threshold σ_{CBD} = 175 mmHg. By choosing the same threshold for the GB, i.e., $[\sigma]$ = 175 mmHg, Eq. (5.18) will be used to predict which patients feel pain in the emptying.

VARIATION OF PARAMETERS

The clinical data of 37 patients in the emptying stimulated by CCK were collected. Based on this data, various factors, which may be related to GB pain, have been calculated and shown in Table 5.1.

TABLE 5.1 Major parameters of GBs in the emptying.

GB	V_0 (ml)	t (min)	EF (%)	R	D_3 (mm), K_1, K_2	p_{max} (mmHg)	σ_{max} (mmHg)
1	16.6	15	4.2	392.6	54.1, 0.43, 0.46	15.2	77.61
2	33.0	20	5.5	217.6	59.7, 0.50, 0.59	19.4	136.82
3	25.5	22	9.7	134.1	72.2, 0.36, 0.36	17.5	102.82
4	36.8	27	12.2	90.7	64.9, 0.50, 0.52	20.4	127.82
5	13.3	20	13.8	131.6	74.1, 0.22, 0.28	14.4	210.82
6	21.1	15	14.8	94.4	68.8, 0.30, 0.41	16.4	212.82
7	23.0	10	16.6	80.0	57.3, 0.43, 0.55	18.9	139.82
8	33.5	20	20.5	53.8	66.7, 0.40, 0.54	19.6	213.82
9	23.1	10	21.3	60.1	61.1, 0.34, 0.53	17.1	213.82
10	21.0	15	23.6	64.6	73.0, 0.27, 0.30	15.3	300.82
11	36.3	21	24.5	42.5	81.7, 0.30, 0.43	20.3	340.82
12	22.0	11	26.2	49.7	71.5, 0.31, 0.37	16.6	173.82
13	20.1	15	28.0	48.5	63.2, 0.33, 0.47	16.2	202.82
14	21.5	28	30.0	43.2	63.5, 0.29, 0.55	16.5	295.82
15	25.5	15	32.1	34.3	72.0, 0.35, 0.42	18.4	161.82
16	12.6	10	32.3	55.0	50.6, 0.41, 0.45	14.2	84.82
17	21.7	10	37.8	32.9	55.9, 0.49, 0.49	16.6	87.82
18	18.0	10	39.5	35.7	57.6, 0.50, 0.51	15.4	80.82
19	59.9	15	40.4	19.3	92.3, 0.38, 0.39	26.3	183.82
20	24.1	21	47.0	23.6	58.0, 0.47, 0.50	17.2	97.82
21	15.6	15	51.2	27.9	74.5, 0.38, 0.41	15.0	117.82
22	33.1	15	60.5	14.0	41.0, 0.59, 0.73	19.5	243.82
23	33.9	10	61.5	13.5	77.5, 0.31, 0.44	19.7	294.82
24	28.3	10	61.7	59.0	63.0, 0.41, 0.52	18.3	161.82
25	42.6	10	63.0	11.4	75.4, 0.38, 0.50	22.0	249.82
26	30.1	15	63.8	13.7	75.8, 0.32, 0.41	18.7	212.82

GB	V_0 (ml)	t (min)	EF (%)	R	D_3 (mm), κ_1, κ_2	p_{max} (mmHg)	σ_{max} (mmHg)
27	23.0	15	70.2	14.1	64.8, 0.38, 0.42	16.9	100.82
28	26.2	15	71.1	12.6	68.0, 0.38, 0.41	17.7	101.82
29	24.1	15	75.5	12.1	56.7, 0.50, 0.51	17.2	94.82
30	26.7	11	79.3	10.4	68.6, 0.35, 0.45	17.8	186.82
31	21.6	9	89.8	6.9	74.2, 0.26, 0.39	16.5	300.82
32	18.2	15	97.1	5.0	55.4, 0.43, 0.47	15.7	175.82
33	25.7	7	100.0	1.7	63.0, 0.45, 0.51	17.6	128.82
34	19.0	20	100.0	9.4	67.0, 0.27, 0.44	15.9	278.82
35	10.0	11	100.0	6.5	45.8, 0.38, 0.52	13.6	111.82
36	23.7	15	100.0	6.7	76.1, 0.30, 0.34	17.1	149.82
37	26.2	12	100.0	11.1	53.8, 0.56, 0.57	17.7	108.82

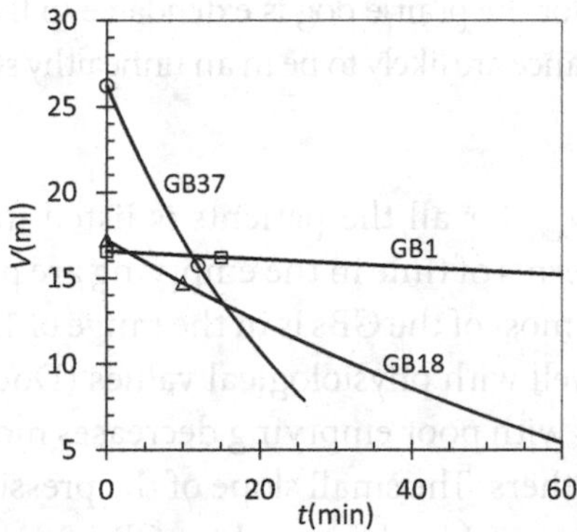

FIGURE 5.6 The GB volume variation with time in the emptying for three patients, the symbols are the clinical data and the solid curves are predicted with Eq. (5.7).

GB Volume and EF

All the initial GB volumes are in the range of 15–35 ml, except the patients 19 and 35, which have the initial volumes of 60 and 10 ml, respectively. The averaged initial GB volume is 25.3 ml. The GB volume change versus time is plotted in Figure 5.6 for three typical patients 1, 18 and 37, indicating poor, fair and super-emptying behaviours, respectively.

In Table 5.1, the GB emptying behaviour is clarified by using EF. If EF is less than 35%, then the GB emptying is impaired, otherwise, it is normal (Pickleman et al. 1985; Fink-Bennett et al. 1991; Kmiot et al. 1994; Michail et al. 2001; Misra et al. 1991; Ozden et al. 2003; Bingener et al.

2004). Clearly, the GBs 17–37 (about 57%) have normal emptying function and the rest of GBs (43%) have abnormal emptying characteristic.

Flow Resistance

The flow resistance varies from 1.7 to 392.6 mmHg/ml/min and shows a significant variation among patients. The GBs with good emptying function (with large EF) always have low flow resistance and those with poor emptying (with small EF) usually present high resistance. Basically, the resistance is in the range of 20–70 mmHg/ml/min. The average resistance is around 53.4 mmHg/ml/min, the resistance of patients 1, 2, 3 and 5, however, is all higher than 130 mmHg.

The resistance of the CD of the prairie dog was increased from 50 to 120 mmHg/ml/min when its GB changed from healthy status to the state with gallstones after being fed a cholesterol diet (Pitt et al. 1981; Doty et al. 1983a, 1983b). Thus, if the experimental finding for the prairie dog is extendable to the human, then those GBs with higher resistance are likely to be in an unhealthy state.

Pressure Profiles

The peak pressure p_{max} for all the patients is listed in Table 5.1, and the pressure profiles in terms of time in the emptying are present in Figure 5.7. The peak pressure of most of the GBs is in the range of 15–20 mmHg except GB 19. This agrees well with physiological values (Dodds et al. 1989). The pressure of GBs 1–14 with poor emptying decreases more slowly with time compared with the others. The small slope of the pressure curves is associated with poor emptying. In other words, a GB with poor emptying sustains higher pressure during a longer period.

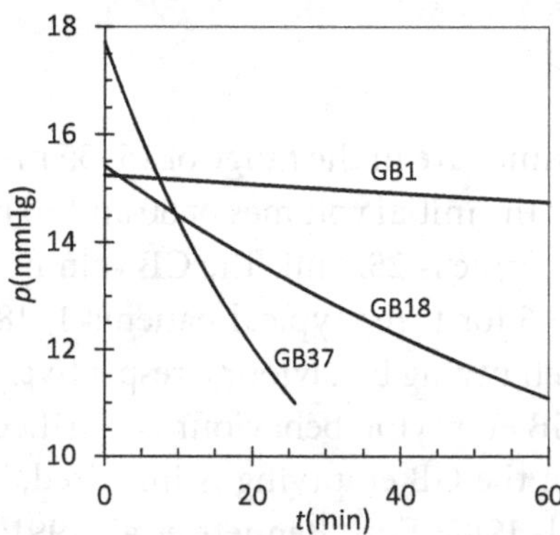

FIGURE 5.7 The pressure profiles with time in the emptying for GBs 1, 18 and 37, respectively; the curves are modelled by Eq. (5.6).

Peak Stress

The total peak normal stress, σ_{max} for all subjects are listed in Table 5.1. The peak in-plane normal stress is in the range of 80–340 mmHg. The mean peak normal stress is 175.9 mmHg.

Stress Patterns

The GB shapes in the patient group studied can be divided into two types roughly, i.e., type 1 for the GBs with $D_3 > D_2 > D_1$ ($\kappa_1 > \kappa_2$), type 2 for those with $D_3 > D_2 \approx D_1$ ($\kappa_1 \approx \kappa_2$). For example, GB12 is of type 1 and GB29 is type 2 and their corresponding stress patterns are shown in Figure 5.8. The peak normal and shear stresses level in type 1 usually are higher than

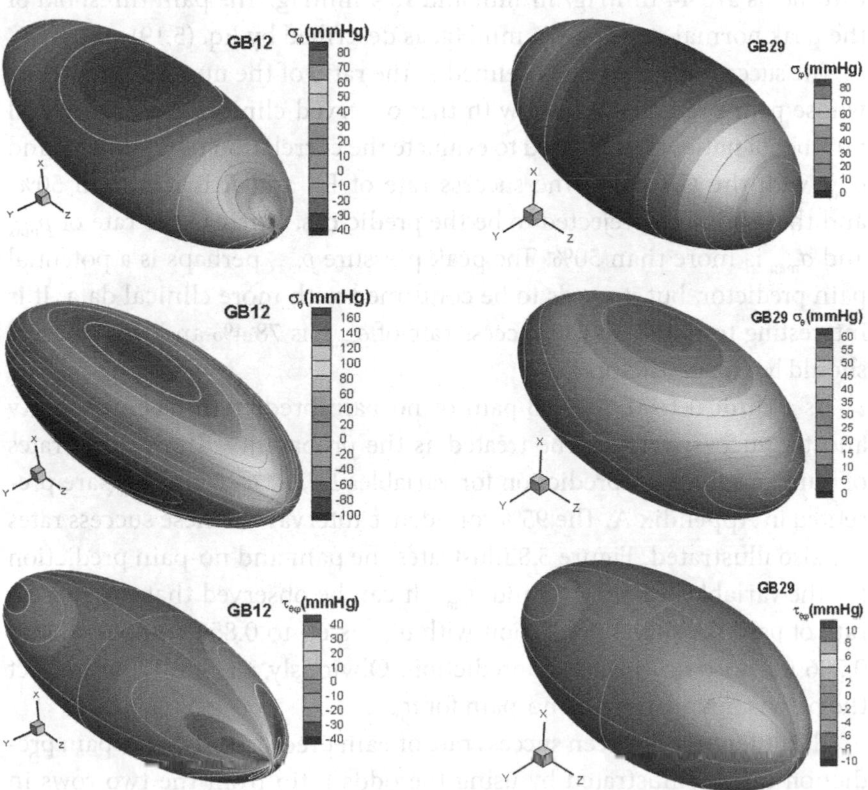

FIGURE 5.8 The stress contours in the GB wall at the start of emptying for GBs 12 (left) and 29 (right). The top frame is the in-plane normal stress σ_θ, the middle is the normal stress σ_φ, and the bottom is the shear stress $\tau_{\theta\varphi}$, the active stress is not involved in the contours.

that in type 2. The shear stress patterns for these two subjects are similar, but the normal stresses are very different and much more sensitive to the geometrical change.

CORRELATION WITH CLINICAL PAIN PROFILE

The GB pain probably is correlated with EF, flow resistance, peak pressure and peak normal stress. Consequently, the comparison of GB pain between prediction and clinical observation is summarised in Tables 5.2, 5.3, 5.4 and 5.5, respectively. The pain threshold in terms of EF is 35% (Pickleman et al. 1985; Fink-Bennett et al. 1991; Kmiot et al. 1994; Michail et al. 2001; Misra et al. 1991; Ozden et al. 2003; Dibaise and Oleyniko 2003; Bingener et al. 2004). The pain thresholds based on the flow resistance and peak pressure are determined by the condition of EF = 35%. These thresholds are 44 mmHg/ml/min and 15.4 mmHg. The pain threshold of the peak normal stress is 175 mmHg, as described by Eq. (5.19).

The success rate, which is defined as the ratio of the number of patients whose pain predicted agrees with that observed clinically over the total number of patients, is applied to evaluate the correlation of EF, R, p_{max} and σ_{max} with the GB pain. The success rate of EF and R is less than 50%, and they should be rejected to be the predictors, whereas the rate of p_{max} and σ_{max} is more than 50%. The peak pressure p_{max} perhaps is a potential pain predictor, but it needs to be confirmed with more clinical data. It is interesting to note that the success rate of σ_{tmax} is 78.4% and this variable should be the predictor.

As a clinical trial, the GB pain or no-pain prediction has probability and the success rate can be treated as the probability. The success rates of pain and no-pain prediction for variables EF, R, p_{max} and σ_{max} are presented in Appendix A. The 95% confidence intervals of these success rates are also illustrated. Figure 5.8 illustrates the pain and no-pain prediction for the variables EF, R, p_{max} and σ_{max}. It can be observed that the success rate of pain (positive) prediction with σ_{max} is up to 0.850 compared with 0.706 for no-pain (negative) prediction. Obviously, the ability to predict the pain is better than the no-pain for σ_{max}.

The difference between success rate of pain prediction and no-pain prediction can be illustrated by using the odds ratio from the two rows in the 2 × 2 contingency table (Table A.1 in Appendix A). An inference can be made for the odds ratio of pain and no-pain prediction, and the results

TABLE 5.2 GB pain prediction using bile EF.

GB	EF (%)	Pain prediction	Clinical observation	Agreement
1	4.2	Positive	Negative	No
2	5.5	Positive	Negative	No
3	9.7	Positive	Negative	No
4	12.2	Positive	Negative	No
5	13.8	Positive	Positive	Yes
6	14.8	Positive	Positive	Yes
7	16.6	Positive	Negative	No
8	20.5	Positive	Positive	*Yes*
9	21.3	Positive	Positive	Yes
10	23.6	Positive	Negative	No
11	24.5	Positive	Negative	No
12	26.2	Positive	Negative	No
13	28.0	Positive	Positive	*Yes*
14	30.0	Positive	Positive	Yes
15	32.1	Positive	Positive	Yes
16	32.3	Positive	Negative	No
17	37.8	Negative	Positive	No
18	39.5	Negative	Positive	No
19	40.4	Negative	Positive	No
20	47.0	Negative	Positive	No
21	51.2	Negative	Negative	Yes
22	60.5	Negative	Positive	No
23	61.5	Negative	Positive	No
24	61.7	Negative	Negative	Yes
25	63.0	Negative	Positive	No
26	63.8	Negative	Positive	No
27	70.2	Negative	Positive	No
28	71.1	Negative	Negative	*Yes*
29	75.5	Negative	Negative	Yes
30	79.3	Negative	Positive	No
31	89.8	Negative	Positive	No
32	97.1	Negative	Negative	Yes
33	100.0	Negative	Negative	Yes
34	100.0	Negative	Positive	No
35	100.0	Negative	Positive	No
36	100.0	Negative	Negative	Yes
37	100.0	Negative	Negative	Yes
Success rate			0.405	

TABLE 5.3 GB pain prediction by using flow resistance.

GB	*R* (mmHg/ml/min)	Pain predicted	Clinical observation	Agreement
33	1.7	Negative	Negative	Yes
32	5.0	Negative	Negative	Yes
35	6.5	Negative	Positive	No
36	6.7	Negative	Negative	Yes
31	6.9	Negative	Positive	No
34	9.4	Negative	Positive	No
30	10.4	Negative	Positive	No
37	11.1	Negative	Negative	Yes
25	11.4	Negative	Positive	No
29	12.1	Negative	Negative	Yes
28	12.6	Negative	Negative	Yes
23	13.5	Negative	Positive	No
26	13.7	Negative	Positive	No
22	14.0	Negative	Positive	No
27	14.1	Negative	Positive	No
19	19.3	Negative	Positive	No
20	23.6	Negative	Positive	No
21	27.9	Negative	Negative	Yes
17	32.9	Negative	Positive	No
15	34.3	Negative	Positive	No
18	35.7	Negative	Positive	No
11	42.5	Negative	Negative	Yes
14	43.2	Negative	Positive	No
13	48.5	Positive	Positive	Yes
12	49.7	Positive	Negative	No
8	53.8	Positive	Positive	Yes
16	55.0	Positive	Negative	No
24	59.0	Positive	Negative	No
9	60.1	Positive	Positive	Yes
10	64.6	Positive	Negative	No
7	80.0	Positive	Negative	No
4	90.7	Positive	Negative	No
6	94.4	Positive	Positive	Yes
5	131.6	Positive	Positive	Yes
3	134.1	Positive	Negative	No
2	217.6	Positive	Negative	No
1	392.6	Positive	Negative	No
Success rate			0.351	

TABLE 5.4 GB pain prediction by using peak pressure.

GB	p_{max} (mmHg)	Pain predicted	Clinical observation	Agreement
35	13.6	Negative	Positive	No
16	14.2	Negative	Negative	Yes
5	14.4	Negative	Positive	No
21	15.0	Negative	Negative	Yes
1	15.2	Negative	Negative	Yes
10	15.3	Negative	Negative	Yes
18	15.4	Positive	Positive	Yes
32	15.7	Positive	Negative	No
34	15.9	Positive	Positive	Yes
13	16.2	Positive	Positive	Yes
6	16.4	Positive	Positive	Yes
14	16.5	Positive	Positive	Yes
31	16.5	Positive	Positive	Yes
12	16.6	Positive	Negative	No
17	16.6	Positive	Positive	Yes
27	16.9	Positive	Positive	Yes
9	17.1	Positive	Positive	Yes
36	17.1	Positive	Negative	No
20	17.2	Positive	Positive	Yes
29	17.2	Positive	Negative	No
3	17.5	Positive	Negative	No
33	17.6	Positive	Negative	No
28	17.7	Positive	Negative	No
37	17.7	Positive	Negative	No
30	17.8	Positive	Positive	Yes
24	18.3	Positive	Negative	No
15	18.4	Positive	Positive	Yes
26	18.7	Positive	Positive	Yes
7	18.9	Positive	Negative	No
2	19.4	Positive	Negative	No
22	19.5	Positive	Positive	Yes
8	19.6	Positive	Positive	Yes
23	19.7	Positive	Positive	Yes
11	20.3	Positive	Negative	No
4	20.4	Positive	Negative	No
25	22.0	Positive	Positive	Yes
19	26.3	Positive	Positive	Yes
Success rate			0.568	

TABLE 5.5 GB pain prediction using the peak normal stress.

GB	σ_{max} (mmHg)	Pain predicted	Clinical observation	Agreement
18	80.82	Negative	Positive	*No*
16	84.82	Negative	Negative	*Yes*
17	87.82	Negative	Positive	No
29	94.82	Negative	Negative	Yes
20	97.82	Negative	Positive	No
27	100.82	Negative	Positive	*No*
28	101.82	Negative	Negative	Yes
3	102.82	Negative	Negative	Yes
37	108.82	Negative	Negative	Yes
35	111.82	Negative	Positive	No
21	117.82	Negative	Negative	*Yes*
1	119.82	Negative	Negative	Yes
4	127.82	Negative	Negative	Yes
33	128.82	Negative	Negative	Yes
2	136.82	Negative	Negative	Yes
7	139.82	Negative	Negative	Yes
36	149.82	Negative	Negative	Yes
15	161.82	Negative	Positive	No
24	161.82	Negative	Negative	Yes
12	173.82	Negative	Negative	Yes
32	175.82	Positive	Negative	No
19	183.82	Positive	Positive	*Yes*
30	186.82	Positive	Positive	Yes
13	202.82	Positive	Positive	Yes
5	210.82	Positive	Positive	Yes
6	212.82	Positive	Positive	Yes
26	212.82	Positive	Positive	Yes
8	213.82	Positive	Positive	Yes
9	213.82	Positive	Positive	Yes
22	243.82	Positive	Positive	Yes
25	249.82	Positive	Positive	Yes
34	278.82	Positive	Positive	Yes
23	294.82	Positive	Positive	Yes
14	295.82	Positive	Positive	Yes
10	300.82	Positive	Negative	*No*
31	300.82	Positive	Positive	Yes
11	340.82	Positive	Negative	No
Success rate			0.7838	

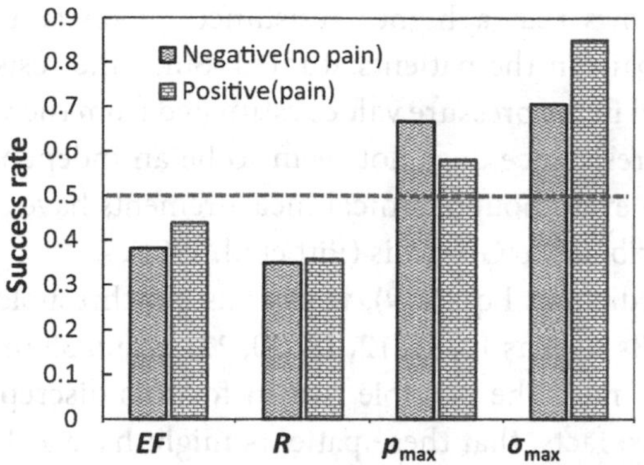

FIGURE 5.9 Pain and no-pain prediction for the variables EF, R, p_{max} and σ_{max}.

are summarised in Table A.3 for variable p_{max} and σ_{max}. The 95% confidence interval for odds ratio of success rate of pain and no pain prediction with σ_{max} is in (1.478, 1.426). Consequently, the success rate of pain prediction is higher by at least 3.5% than no pain prediction.

It is understood that the peak normal stress shows a good correlation with the pain observed in the patients. Despite bile EF and peak level of pressure being often applied in clinical practice, they along with flow resistance are not closely related to the GB pain.

The stresses in the GB wall depend not only on the pressure p, but also on its geometry, D_1, D_2 and D_3 as well as thickness h_{GB}, and their relative ratios (see Table 5.1). It has been suspected that the GB pressure in the emptying could be associated with GB pain. In fact, in the model, the effects of both the pressure and geometry change come into play in producing pain.

A poor emptying (a small EF) is not necessarily associated with the pain, GBs 1, 2, 3, 4, 7, 10, 11 and 12 have all showed the poor emptying, but do not experience any pain in terms of the model prediction or the clinical observation. The GB with super emptying (a large EF) can also experience pain, e.g., subject 31. Therefore, the EF is not considered to be a good index for pain prediction. Unfortunately, the impaired GB emptying still serves the clinical criterion for cholecystectomy (Shaffer 2003).

Table 5.3 shows that a higher resistance downstream is indirectly related to the pain in the patients studied. Since the resistance effect is already reflected in the pressure values estimated from the volume change. Thus, the flow resistance does not seem to be an independent index for pain prediction even though clinical measurements have shown that the resistance contributes to GB stasis (Pitt et al. 1981).

The pain prediction, Eq. (5.19), with constant threshold is correct for all but seven GBs such as 10, 11, 12, 18, 20, 27, 33 and 35 in Table 5.5 with a 78.4% success rate. The possible reason for this discrepancy could be attributed to two facts, that these patients might have a slightly lower or higher pain threshold than the level used, and the GB compliance involved in the model should be patient-specific.

Note that the pain sensitivity is subject to individual subjects. Using a standard value of 175 mmHg here is an approximation only. However, if this value varies between 150 and 200 mmHg, then the success rate of predication may be even higher.

In this chapter, the 'slow pump' concept for GBs during emptying has been utilised. Recent studies suggested that GB emptying or refilling was an alternative in which the GB serves as 'pump + bellows', not a 'slow pump' (Jazrawi et al. 2000; Lanzini and Northfield 1988). In the 'pump + bellows' case, the bile flow rate fluctuates during emptying (Figure 5.10). Presently, however, there are no available experimental data for the pressure or compliance fluctuation, frequency and magnitude.

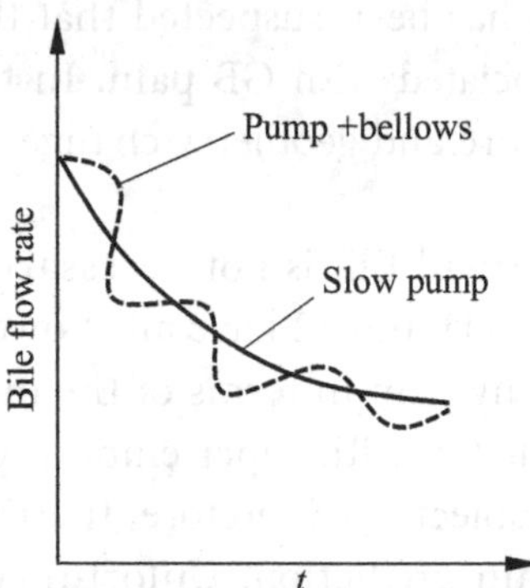

FIGURE 5.10 Comparison of the bile flow rate curve when GB emptying is considered as 'slow pump' or 'pump + bellows'.

Although the pain model has included both active and passive stresses, the active stress is not obtained from the smooth muscle bio mechanics; rather, it is the same typical form for all subjects. In practice, this also varies with individual subject. In the next chapter, patient-specific active stress will be examined.

REFERENCES

Barlow, J. D., H. Gregersen, and D. G. Thompson. 2001. Identification of the biomechanical factors associated with the perception of distension in the human oesophagus. *American Journal of Physiology* 282:G683–G689.

Bingener, J., M. L. Richards, W. H. Schwesinger, et al. 2004. Laparoscopic cholecystectomy for biliary dyskinesia. *Surgical Endoscopy* 18:802–806.

Brosseuk, D., and J. Demetrick. 2003. Laparoscopic cholecystectomy for symptoms of biliary colic in the absence of gallstones. *American Journal of Surgery* 186:1–3.

Canfield, A., S. P. Hetz, R. J. P. Schrive, et al. 1998. Biliary dyskinesia: A study of more than 200 patients and review of the literature. *Journal of Gastrointestinal Surgery* 2:443–448.

Corazziari, E., E. A. Shaffer, W. J. Hogan, et al. 1999. Functional disorders of the biliary tract and pancreas. *Gut* 45(Suppl II):II48–II54.

Cozzolino, H. J., F. Goldstein, R. R. Greening, et al. 1963. The cystic duct syndrome. *JAMA* 185:920–924.

Csendes, A., A. Kruse, P. Funch-Jensen, et al. 1979. Pressure measurements in the biliary and pancreatic duct systems in controls and in patients with gallstones, previous cholecystectomy, or common bile duct stones. *Gastroenterology* 77:1203–1210.

Deitch, E. A. 1981. Utility and accuracy of ultrasonically measured gallbladder wall as a diagnostic criteria in biliary tract disease. *Digestive Diseases and Sciences* 26:686–693.

Delgado-Aros, S., F. Cremonini, A. J. Bredenoord, et al. 2003. Systematic review and meta-analysis: Does gall-bladder ejection fraction on cholecystokinin cholescintigraphy predict outcome after cholecystectomy in suspected functional biliary pain? *Aliment Pharmacology* 18:167–174.

DiBaise, J. K., and D. Oleynikov. 2003. Does gallbladder ejection fraction predict outcome after cholecystectomy for suspected chronic acalculous gallbladder dysfunction? A systematic review. *American Journal of Gastroenterology* 98:2605–2611.

Dodds, W. J., W. J. Hogan, and J. E. Green. 1989. Motility of the biliary system. In *Handbook of Physiology*, ed. S. G. Schultz, 1055–1101. Bethesda: American Physiological Society.

Doty, J. E., H. A. Pitt, S. L. Kuchenbecker, et al. 1983a. Impaired gallbladder emptying before gallstone formation in the prairie dog. *Gastroenterology* 85:168–174.

Doty, J. E., H. A. Pitt, S. L. Kuchenbecker, et al. 1983b. Role of gallbladder mucus in the pathogenesis of cholesterol gallstone. *American Journal of Surgery* 145:54–61.

Drewes, A. M., J. Pedersen, W. Liu, et al. 2003. Controlled mechanical distension of human oesophagus: Sensory and biomechanical findings. *Scandinavian Journal of Gastroenterology* 38:27–35.

Everson, G. T., D. Z. Braverman, M. L. Johnson, et al. 1980. A critical evaluation of real-time ultrasonography for the study of gallbladder volume and contraction. *Gastroenterology* 79:40–48.

Ferris, D. O., and J. C. Vibert. 1959. The common bile duct: Significance of its diameter. *Annals of Surgery* 149:249–251

Fink-Bennett, D., P. DeRidder, W. Z. Kolozsi, et al. 1991. Cholecystokinin cholescintigraphy: Detection of abnormal gallbladder motor function in patients with chronic acalculous gallbladder disease. *Journal of Nuclear Medicine* 32:1695–1699.

Funch-Jensen, P. 1995. Sphincter of Oddi physiology. *Journal of Hepato-Biliary-Pancreatic Surgery* 2:249–254.

Fung, Y. C. 1984. *Biodynamics-Circulation*. New York: Springer-Verlag.

Gaensler, E. A. 1951. Quantitative determination of the visceral pain threshold in man. *Journal of Clinical Investigation* 30:406–420.

Gao, C., L. Arendt-Nielsen, W. Liu, et al. 2002. Sensory and biomechanical responses to ramp-controlled distension of the human duodenum. *American Journal of Physiology* 284:G461–G471.

Gregersen, H., T. Hausken, J. Yang, et al. 2006. Mechanosensory properties in the human gastric antrum evaluated using B-mode ultrasonography during volume-controlled antral distension. *American Journal of Physiology* 290:G876–G882.

Hale, T. 2018. Case presentation: On acute abdomen in pregnancy. www.slideshare.net/Haleta/2-abdominal-pain-in-pregnancy (accessed August 7, 2020).

Jazrawi, R. P., P. Pazzi, M. L. Petroni, et al. 2000. Role of the gallbladder in the pathogenesis of gallstone disease. In *Bile Acid in Hepatology Disease*, ed. T. C. Northfield, H. A. Ahmed, and R. P. Jazrawi, 182–191. Dordrecht, the Netherlands: Kluwer Academic Publishers.

Jonderko, K., and L. Bueno. 1997. Simultaneous assessment of gallbladder emptying in the dog by real-time ultrasonography and strain gauge. *Journal of Gastroenterology* 32:222–229.

Kmiot, W. A., E. P. Perry, I. A. Donovan, et al. 1994. Cholesterolosis in patients with chronic acalculous biliary pain. *British Journal of Surgery* 81:112–115.

Lanzini, A., and T. C. Northfield. 1988. Gallbladder motor function in man. In *Bile Acid in Hepatology Disease*, ed. T. C. Northfield, P. Zentler-Munro, and R. P. Jazrawi, 83–96. Dordrecht, the Netherlands: Kluwer Academic Publishers.

Lempke, R. E., R. D. King, and G. C. Kaise. 1963. Hydrodynamics of gallbladder filling. *Journal of American Medical Association* 186:152–155.

Li, W. G., X. Y. Luo, N. A. Hill, et al. 2008. Correlation of mechanical factors and gallbladder pain. *Computational & Mathematical Methods in Medicine* 9:27–45.

Mahour, G. H., K. G. Wakim, and D. O. Ferris. 1967. The common bile duct in man: Its diameter and circumference. *Annals of Surgery* 165:415–419.

Matsumoto, T., S. K. Sarna, R. E. Condon, et al. 1988. Canine gallbladder cyclic motor activity. *American Journal of Physiology* 255:G409–G416.

Michail, S., Preud'Homme, D., Christian, J., et al. 2001. Laparoscopic cholecystectomy: Effective treatment for chronic abdominal pain in children with acalculous biliary pain. *Journal of Pediatric Surgery* 36:1394–1396.

Middelfart, H. V., P. Jensen, L. Hojgaard, et al. 1998. Pain patterns after distension of the gallbladder in patients with acute cholecystitis. *Scandinavian Journal of Gastroenterology* 33:982–987.

Milenov, K., and A. Shahbazian. 1995. Cholinergic pathways in the effect of motilin on the canine ileum and gallbladder motility: In vivo and in vitro experiments. *Comparative Biochemistry Physiology* 112A:403–410.

Misra, D. C., G. B. Blossom, D. Fink-Bennett, et al. 1991. Results of surgical therapy for biliary dyskinesia. *Archives of Surgery* 126:957–960.

Ness, T. J., and G. F. Gebhart. 1990. Visceral pain: A review of experimental studies. *Pain* 41:167–234.

Novozhilov, V. V. 1964. *Thin Shell Theory*. Groningen: P Noordhoff Ltd.

Ozden, N., and J. K. DiBaise. 2003. Gallbladder ejection fraction and symptom outcome in patients with acalculous biliary-like pain. *Digestive Diseases and Sciences* 48:890–897.

Petersen, P., C. Gao, L. Arendt-Nielsen, et al. 2003. Pain intensity and biomechanical response during ramp-controlled distension of the human rectum. *Digestive Diseases and Sciences* 48:1310–1316.

Pickleman, J., R. L. Peiss, R. Henkin, et al. 1985. The role of sincalide cholescintigraphy in the evaluation of patients with acalculous gallbladder disease. *Archives of Surgery* 120:693–697.

Pitt, H. A., J. J. Roslyn, S. L. Kuchenbecker, et al. 1981. The role of cystic duct resistance in the pathogenesis of cholesterol gallstones. *Journal of Surgical Research* 30:508–514.

Ponce, J., V. Pons, R. Sopena, et al. 2004. Quantitative cholescintigraphy and bile abnormalities in patients with acalculous biliary pain. *European Journal of Nuclear Medicine and Molecular Imaging* 31:1160–1165.

Portincasa, P., A. Di Ciaula, G. Baldassarre, et al. 1994. Gallbladder motor function in gallstone patients: Sonographic and in vitro studies on the role of gallstones, and smooth muscle function and gallbladder wall inflammation. *Journal of Hepatology* 21:430–440.

Raptopoulos, V., C. C. Compton, P. Doherty, et al. 1985. Chronic acalculous gallbladder disease: Multiimaging evaluation with clinical-pathologic correlation. *American Journal of Roentgenology* 147:721–724.

Rastogi, A., A. Slivka, A. J. Moser, et al. 2005. Controversies concerning pathophysiology and management of acalculous biliary-type abdominal pain. *Digestive Diseases and Sciences* 50:1391–1401.

Riyad, K., C. R. Chalmers, A. Aldouri, et al. 2007. The role of ^{99m}technetium-labelled hepato imino diacetic acid (HIDA) scan in the management of biliary pain. *HPB* 9:219–224.

Ryan, J., and S. Cohen. 1976. Gallbladder pressure-volume response to gastrointestinal hormones. *American Journal of Physiology* 230:1461–1463.

Schoetz, D. J., W. W. LaMorte, W. E. Wise, et al. 1981. Mechanical properties of primate gallbladder: Description by a dynamic method. *American Journal of Physiology* 241: G376–G381.

Shaffer, E. 2003. Acalculous biliary pain: New concepts for an old entity. *Digestive and Liver Disease* 35:S20–S25.

Smythe, A., R. Ahmed, M. Fitzhenry, et al. 2004. Bethanechol provocation testing does not predict symptom relief after cholecystectomy for acalculous biliary pain. *Digestive and Liver Disease* 36:682–686.

Smythe, A., A. W. Majeed, M. Fitzhenry, et al. 1998. A requiem for the cholecystokinin provocation test? *Gut* 43:571–574.

Sunderland, G. T., and D. C. Carter. 1988. Clinical application of the cholecystokinin provocation test. *British Journal of Surgery* 75:444–449.

van de Heijning, B. J., P. C. van de Meeberg, P. Portincasa, et al. 1999. Effects of ursodeoxycholic acid therapy on in vitro gallbladder contractility in patients with cholesterol gallstones. *Digestive Diseases and Sciences* 44:190–196.

Takahashi, I., M. Nakaya, T. Suzuki, et al. 1982. Postprandial changes in contractile activity and bile concentration in gallbladder of the dog. *American Journal of Physiology* 243:G365–G371.

Traynor, O. J., R. R. Dozois, and E. P. Dimagno. 1984. Canine interdigestive and postprandial gallbladder motility emptying. *American Journal of Physiology* 246(4 Pt 1):G426–G432.

Williamson, R. C. 1988. Acalculous disease of gall bladder. *Gut* 29:860–872.

Yap, L., A. G. Wycherley, A. D. Morphett, et al. 1991. Acalculous biliary pain: Cholecystectomy alleviates symptoms in patients with abnormal cholescintigraphy. *Gastroenterology* 101:786–793.

CHAPTER 6

Passive and Active Stresses in the GB Wall

ACTIVE STRESS AND GLOBAL APPROACH

In Chapter 5, a biomechanical model was developed to predict the peak total in-plane stress in the GB wall and a strong correlation was established with GB pain being provoked by the CCK test based on 37 subjects. In the model, the active stress was treated in a very simple manner. As the mechanical stresses in the model were estimated entirely from non-invasive ultrasonographic routine measurements, this approach may be readily developed into a useful routine additional diagnostic tool if the concept of the stress-pain correlation can be validated with more numbers of patients. In the chapter, the earlier model will be developed further by investigating the patient-specific active stress induced by CCK.

On the one hand, it has been shown that the active stress of GB smooth muscle stimulated is time-dependent (Chijiiwa et al. 1994; Yamasaki et al. 1994) and patient-specific (Hould et al. 1988). On the other hand, extensive studies exist for the cardiac muscle. There are two approaches for modelling active stress in cardiac muscle in the systole (Shoucri 1991, 1998, 2000). The first is the structural method where the active stress developed can be determined from the complex structure of myofibre orientations by using sophisticated constitutive laws for the myocardium. This approach has been employed by Streeter et al. (1970) and by others that have provided comprehensive reviews (Hunter and Smaill 1988; Hunter et al. 1998; Avazmohammadi et al. 2019). The second is a global approach where active stress is considered as a continuous body force along three orthogonal coordinates rather than the force generated by a finite number of fibres (Shoucri 1991, 1998, 2000). Since this approach does not involve the complex structure of the myocardium, it provides a simple and effective tool for those who are only interested in the macro-mechanics of heart muscle contraction.

In this chapter, the global approach will be employed to analyse the active stress in GB smooth muscle. The predicted passive, active and total peak stresses were compared with the clinically recorded data of pain in the CCK provocation tests for 51 subjects/patients. It has been found that the peak total stress consistently correlates strongly with the CCK-induced pain. The approach details are referred to Li et al. (2011).

GB SMOOTH MUSCLE AND ASSUMPTIONS

Human GB isolated smooth muscle cells have a mean basal length 42 ± 1.2 μm in the patients with gallstones, 42 ± 1.1 μm in those with black pigment stones and 42 ± 1.3 μm in the healthy subjects (Chijiiwa et al. 1994; Yamasaki et al. 1994). However, these cells are distributed in the GB wall without preferred orientation. Furthermore, the muscle layer of the canine GB wall is in a 3D meshwork of smooth muscle bundles, which appear loosely and irregularly arranged on the mucosal aspect and consolidate to form a homogeneous plate-like layer on the serosal aspect (MacPherson et al. 1984).

Bearing these facts in mind, to develop the biomechanical model, the following assumptions are made: (1) the GB is considered as a thin-walled elastic ellipsoid membrane experiencing isotropic contraction; (2) the emptying and refilling phases of the GB are quasi-static without inertia forces and (3) GB smooth muscle is incompressible, isotropic, elastic and homogeneous.

CCK PROVOCATION TEST AND GB CYCLE

In this chapter, CCK provocation tests carried out in the hospital for 51 subjects with suspicious ABP described by Smythe et al. (1998) were utilised once again. After an overnight fast, patients were given an intravenous infusion of saline (control) followed by an intravenous infusion of CCK (0.05 μg/kg body weight). Ultrasonography of the GB was used to monitor changes in shape, initial volume, change of volume and wall thickness at 5- or 15-min intervals for 60 min. The test was only considered positive when usual GB pain in the patients was reproduced following CCK infusion (Smythe et al. 1998, 2004). The pressure and volume versus time curves as well as pressure-volume curve in a CCK provocation test are depicted in Figure 6.1.

When the refilling phase starts at point D, the sphincter of Oddi is closed and the GB is in fasting, and the volume and pressure are at the

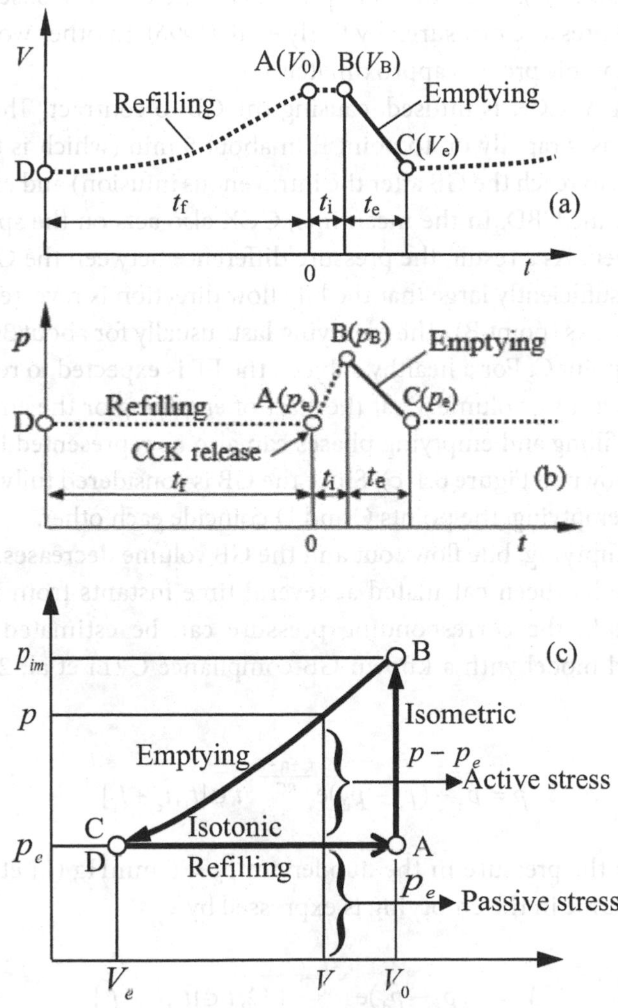

FIGURE 6.1 GB volume and pressure curves with time in refilling and CCK-induced emptying phases as well as pressure-volume curve: (a) volume curve, (b) pressure curve and (c) pressure-volume curve.

minimum basal levels (Lempke et al. 1963). Between points D and A (the end of the refilling), a small but positive pressure difference between the liver and the GB exists (Herring and Simpson 1907) to allow the hepatic bile to be secreted slowly into the GB. The refilling time t_f (from D to A) scale is usually the time lapse between two meals, which is often much longer than the emptying time t_e (from B to C). During t_f, the GB relaxes and the internal pressure p_e is considered as constant, i.e., $p_e = 11$ mmHg (Dodds et al. 1989). Such an assumption is in line with the observation of human GB pressure measured by Borly et al. (1996). In other words, refilling is an isotonic process approximately.

At point A, CCK is infused, causing the GB to contract. The pressure in the GB rises rapidly up to point B in about 5 min (which is the time t_i CCK needs to reach the GB after the intravenous infusion) and exceeds the pressure in the CBD. In the meantime, CCK also acts on the sphincter of Oddi to open. As a result, the pressure difference between the GB and the CBD is so sufficiently large that the bile flow direction is reversed, and the emptying starts (point B). The emptying lasts usually for about 30 min and finishes at point C. For a healthy subject, the EF is expected to reach about 70% of the fasting volume V_0 at the start of emptying or the end of refilling. The refilling and emptying phases can also be represented by the *p-V* diagram shown in Figure 6.1(c). Since the GB is considered fully relaxed at the end of emptying, the points C and D coincide each other.

In the emptying, bile flows out and the GB volume decreases. Since the GB volume has been calculated at several time instants from its images after point B, the corresponding pressure can be estimated by using Windkessel model with a known GB compliance C (Li et al. 2008), that gives

$$p = p_d + (p_e - p_d)e^{\frac{t_e + t_i - t}{RC}}, t \in [t_i, t_e + t_i] \tag{6.1}$$

where p_d is the pressure in the duodenum $p_d = 6$ mmHg (Li et al. 2008), and the volume in the emptying is expressed by

$$V = C(p_e - p_d)e^{\frac{t_e + t_i - t}{RC}} + \Omega, t \in [t_i, t_e + t_i] \tag{6.2}$$

where $\Omega = V_e - C(p_e - p_d)$, t_e is the emptying time for EF to reach 70%, R is the flow resistance and C is taken to be 2.731 ml/mmHg from the experimental data by Middelfart et al. (1998). For known experimental data (t_i, V_0), (t, V_t) and ($t_i + t_e$, V_e), where $V_e = 0.3V_0$, R and t_e are determined with the following expressions

$$R = \frac{t - t_i}{C\ln[(V_0 - \Omega)/(V_t - \Omega)]}, t \in [t_i, t_e + t_i] \tag{6.3}$$

and

$$t_e = \frac{(t - t_i)\ln[(V_0 - \Omega)/(V_e - \Omega)]}{\ln[(V_0 - \Omega)/(V_t - \Omega)]}, t \in [t_i, t_e + t_i] \tag{6.4}$$

where V_t is the volume measured at a specific time t in the emptying.

STRESSES IN THE GB WALL

The longitudinal cross section of a GB is illustrated in Figure 6.2 when the GB is subject to an internal bile pressure p, which gives rise to a passive stress σ^p and an active stress σ^a. Note that active stress is due to CCK stimulus only, while passive stress is caused by volume changes at a constant pressure.

In fact, active stress can be triggered by various stimuli, including electric field stimulation (Shimada 1993; Parkman et al. 1997), ionic action potential (Renzetti et al. 1990; Zhang et al. 1993), CCK (Ryan 1985; Brotschi et al. 1989; Lee et al. 1989; Shaffer et al. 1992; Ishizuka et al. 1993; Morales et al. 2005) and ACh (Ryan 1985; Brotschi et al. 1989; Lambert and Ryan 1990; Washabau et al. 1991; Parkman et al. 2001).

The basic mechanism of these stimuli is to establish an effective Ca^{2+} flux across GB smooth muscle cell membranes to activate myosin light-chain

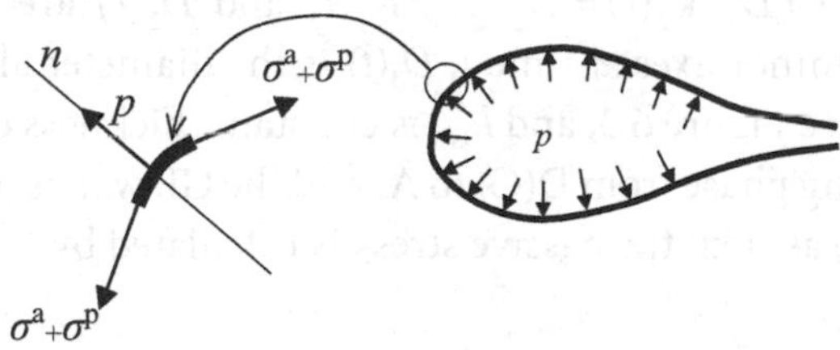

FIGURE 6.2 The GB is subject to an increased internal pressure during CCK infusion, which generates the active stress, σ^a, in the muscle and the passive stresses, σ^p, the other tissue in the GB wall.

phosphorylation; subsequently, latch cross bridges start to form in the cells and generate contraction (Washabau et al. 1991, 1994).

If the GB maintains a membrane ellipsoidal shape in the refilling and emptying phases, which is assumed in the study by Li et al. (2008), then the stress components (σ_θ, σ_φ, $\tau_{\theta\varphi}$) in the spherical coordinate system (r, φ, θ) under a net internal pressure p are expressed as (Novozhilov 1964)

$$\begin{cases} \sigma_\theta = pF_\theta\left(h_{\mathrm{GB}}, D_3, k_1, k_2, \varphi\right) \times F_n\left(k_1, k_2, \theta, \varphi\right) \\ \sigma_\varphi = pF_\varphi\left(h_{\mathrm{GB}}, D_3, k_1, k_2, \theta, \varphi\right) \times \dfrac{1}{F_n\left(k_1, k_2, \theta, \varphi\right)} \\ \tau_{\theta\varphi} = pF_\tau\left(h_{\mathrm{GB}}, D_3, k_1, k_2, \theta, \varphi\right) \end{cases} \tag{6.5}$$

where p is decided by Eq. (6.1) in the emptying phase from B to C, the four spatial quantities F_θ, F_φ, F_τ and F_n are functions describing the geometric properties of a GB are given by

$$\begin{cases} F_\theta\left(h_{\mathrm{GB}}, D_3, \kappa_1, \kappa_2, \varphi\right) = \dfrac{D_3\kappa_1\kappa_2}{4h_{\mathrm{GB}}}\left(1 - \dfrac{\kappa_1^2 - \kappa_2^2}{\kappa_1^2\kappa_2^2}\cos 2\varphi\right) \\ F_\varphi\left(h_{\mathrm{GB}}, D_3, \kappa_1, \kappa_2, \theta, \varphi\right) = \dfrac{D_3}{4\kappa_1\kappa_2 h_{\mathrm{GB}}}\left[\begin{array}{l} \kappa_1^2\kappa_2^2 + \left(\kappa_1^2 + \kappa_2^2 - 2\kappa_1^2\kappa_2^2\right)\sin^2\theta \\ +\left(\kappa_1^2 - \kappa_2^2\right)\cos^2\theta\cos 2\varphi \end{array}\right] \\ F_\tau\left(h_{\mathrm{GB}}, D_3, \kappa_1, \kappa_2, \theta, \varphi\right) = \dfrac{D_3}{4\kappa_1\kappa_2 h_{\mathrm{GB}}}\left(\kappa_1^2 - \kappa_2^2\right)\cos\theta\sin 2\varphi \\ F_n\left(\kappa_1, \kappa_2, \theta, \varphi\right) = \dfrac{\sqrt{\kappa_1^2\cos^2\theta\cos^2\varphi + \kappa_2^2\cos^2\theta\sin^2\varphi + \sin^2\theta}}{\sqrt{\kappa_1^2\sin^2\varphi + \kappa_2^2\cos^2\varphi}} \end{cases} \tag{6.6}$$

where $\kappa_1(t) = D_1 / D_3$, $\kappa_2(t) = D_2 / D_3$, $D_1(t)$ and $D_2(t)$ are the diameters along the two minor axes at time t, $D_3(t)$ is the diameter along the major axis at time t, see Figure 6.3, and h_{GB} is constant thickness of the GB wall.

In the refilling phase from D(C) to A, and the GB wall is under the basal pressure $p = p_e$, as such the passive stress is calculated by

$$\begin{cases} \sigma_\theta^{\mathrm{p}} = p_e F_\theta\left(h_{\mathrm{GB}}, D_3, \kappa_1, \kappa_2, \varphi\right) \times F_n\left(\kappa_1, \kappa_2, \theta, \varphi\right) \\ \sigma_\varphi^{\mathrm{p}} = p_e F_\varphi\left(h_{\mathrm{GB}}, D_3, \kappa_1, \kappa_2, \theta, \varphi\right) \times \dfrac{1}{F_n\left(\kappa_1, \kappa_2, \theta, \varphi\right)} \\ \tau_{\theta\varphi}^{\mathrm{p}} = p_e F_\tau\left(h_{\mathrm{GB}}, D_3, \kappa_1, \kappa_2, \theta, \varphi\right) \end{cases} \tag{6.7}$$

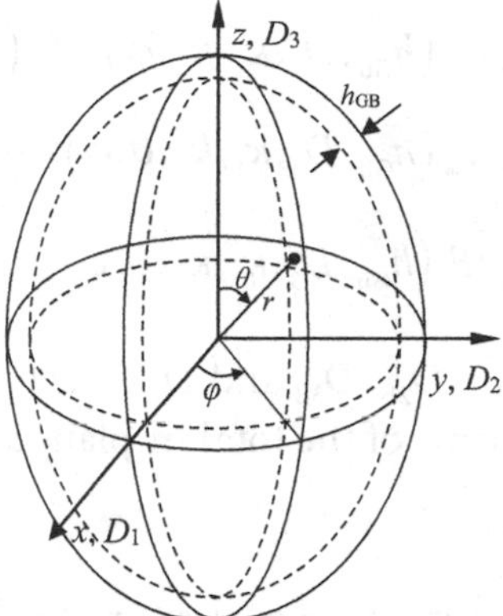

FIGURE 6.3 The coordinate system for the ellipsoid model of GB.

where $\kappa_1(t)$, $\kappa_2(t)$ and $D_3(t)$ are the same as those in the emptying.

In the isometric phase from A to B, the $\kappa_1(t)$, $\kappa_2(t)$ and $D_3(t)$ remain unchanged. At point A, the CCK is applied; it reaches the GB in about 5 min (point B), and the GB starts to contract; subsequently, the internal pressure increases. Since this GB response is almost instantaneous (<15 s), the process should be considered isometric. In the isometric contraction, the active force of smooth muscle is an exponential function of time (Meiss 1975); thus, the GB pressure and volume can be expressed as

$$\begin{cases} p = p_e e^{q} t \\ V_B \approx V_0 \end{cases} \tag{6.8}$$

where $p_e < p < p_B$, $q = \ln(p_B / p_e) / t_i$, V_B and p_B are the GB volume and pressure at point B, and q is a constant. From Eq. (6.1), p_B is determined by

$$p_B = p_d + (p_e - p_d) e^{\frac{t_e}{RC}} \tag{6.9}$$

The pressure difference $p_B - p_e$ in the emptying does work on the bile and delivers the bile into the duodenum. The amount of this work is represented by the area enclosed by boundaries AB, BC and CA.

Furthermore, the pressure difference results in the peak active stress at point B and the stress components are given by

$$\begin{cases} \sigma_\theta^{a} = (p_B - p_e) F_\theta\left(h_{GB}, D_3, \kappa_1, \kappa_2, \varphi\right) \times F_n\left(\kappa_1, \kappa_2, \theta, \varphi\right) \\ \sigma_\varphi^{a} = (p_B - p_e) F_\phi\left(h_{GB}, D_3, \kappa_1, \kappa_2, \theta, \varphi\right) \times \dfrac{1}{F_n\left(\kappa_1, \kappa_2, \theta, \varphi\right)} \\ \tau_{\theta\varphi}^{a} = (p_B - p_e) F_\tau\left(h_{GB}, D_3, \kappa_1, \kappa_2, \theta, \varphi\right) \end{cases} \tag{6.10}$$

where $\kappa_{1B} = D_{1B}, D_{3B}$, $\kappa_{2B} = D_{2B}, D_{3B}$ and $D_{1B} = D_1(t_i)$, D_{2B}, $D_{2B} = D_2(t_i)$, and $D_{3B} = D_3(t_i)$. The components of the total in-plane stress in the GB wall are then

$$\sigma_\theta = \sigma_\theta^{p} + \sigma_\theta^{a}, \sigma_\varphi = \sigma_\varphi^{p} + \sigma_\varphi^{a}, \tau_{\theta\varphi} = \tau_{\theta\varphi}^{p} + \tau_{\theta\varphi}^{a} \tag{6.11}$$

and the peak stress, which is GB shape-dependent, is

$$\sigma_{max}^{a} = \max[\sigma_\theta^{a}, \sigma_\varphi^{a}], \sigma_{max}^{p} = \max[\sigma_\theta^{p}, \sigma_\varphi^{p}], \sigma_{max} = \max[\sigma_\theta, \sigma_\varphi] \tag{6.12}$$

Note that Eq. (6.11) predicts a similar total peak stress to the model derived by Li et al. (2008) as shown in Chapter 5; however, significant differences exist in the modelling of both passive and active stresses here. In Chapter 5, the same magnitude of the active stress was used for all subjects, while both the passive and active stresses are patient dependent herein. The active stress induced by the CCK can be used to provide further insights in the smooth muscle cross-bridge cycling kinetics during isometric contraction.

The Threshold for GB Pain

Like Chapter 5, here the stress threshold for GB pain is based on the experimental data on CBD inflation by saline (Gaensler 1951), which showed that the average pressure threshold for pain was p_{CBD} = 35 mmHg in the CBDs for 33 patients. Thus, the circumferential stress in the duct wall can be estimated as

$$[\sigma] = \frac{p_{CBD} d_{CBD}}{2h_{CBD}} = 175\,\text{mmHg} \tag{6.13}$$

where the mean diameter of CBD, d_{CBD} = 10 mm (Ferris and Vibert 1959), and the mean thickness of the duct wall h_{CBD} is 1 mm (Mahour et al. 1967). Based on this threshold, GB pain is predicted whenever

$$\sigma_{max} \geq [\sigma] \tag{6.14}$$

GB VOLUME, PRESSURE AND STRESS PROFILES

The GB sample size for CCK provocation test has been extended from 37 subjects in Chapter 5 to 51 patients. The volume and pressure of the subjects are derived as functions of time based on the volume measured from the images and Eqs. (6.1) and (6.2). The volume and pressure profiles are shown in Figure 6.4 for three typically selected GBs 1, 9 and 37. The initial time is when CCK is applied, and it was assumed that CCK reached the GB within 5 min for all subjects. In the isometric contraction, the volume of these GBs remains constant and the pressure increases until the emptying starts. The corresponding *p*-*V* diagrams for these subjects are illustrated in Figure 6.5 along with the computed peak passive and active stresses plotted as a function of time.

The maximum values of the peak stresses do not correspond to the maximum pressure, e.g., GB 9 has a lower peak pressure compared with GB 37, but the former has higher passive and active stresses than the latter due to the GB geometry.

As the GB size decreases with time in the emptying phase, the passive stress reduces owing to GB shape change and decrease in D_3 [see Eq. (6.6) and Figure 6.5].

The active stress also peaks at the isometric contraction just before the emptying starts and decays with time afterwards. The variation of passive and active stresses in the GBs with time resembles to that of myocardium (Claus et al. 2003).

The contours of three stress components of GB 9 are shown in Figure 6.6. It shows that the active stress shares the same pattern as the passive stress. The reason for this effect is that the GB maintains the identical ellipsoidal shape when the GB volume in the refilling is the same to the volume in the emptying in the model.

PREDICTION OF GB PAIN

The maximum peak stresses in the emptying of 51 subjects are listed in Table 6.1. For reference, the initial GB geometry, maximum pressure, EF estimated at t = 30 min after CCK infusion are also included. GB pain profiles were predicted by using the stress criterion defined in Eq. (6.14).

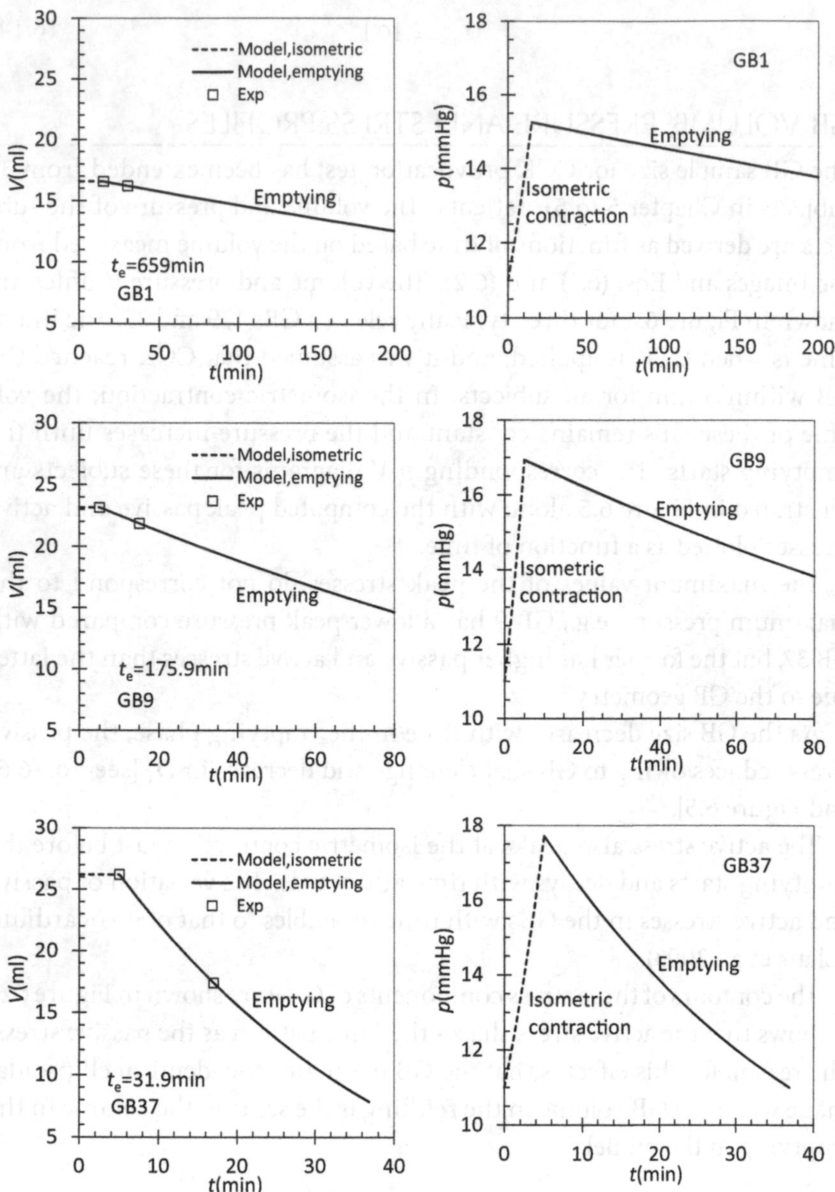

FIGURE 6.4 The volume (left) and pressure (right) profiles versus time in the emptying for three typically selected GBs 1, 9 and 37, the circles stand for the experimental data point in the isometric contraction, the squares represent the experimental data in the emptying, GB 1 shows very poor emptying and the lowest pressure, but GB 37 has very good emptying and the highest pressure.

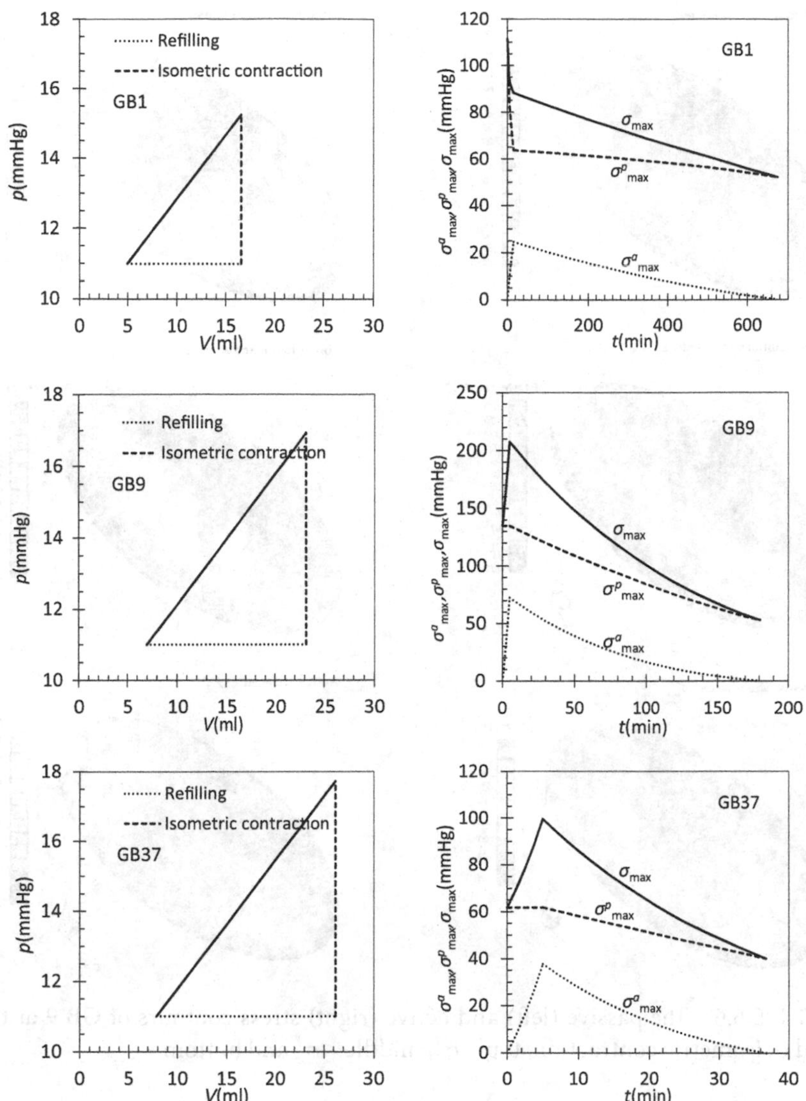

FIGURE 6.5 The pressure-volume diagram (left) and peak active (σ^a_{max}), passive (σ^p_{max}) and total (σ_{max}) stresses (right) in the isometric and emptying phases, the total stress in the GB 9 wall is over 200 mmHg, those peak stresses are usually in the θ direction.

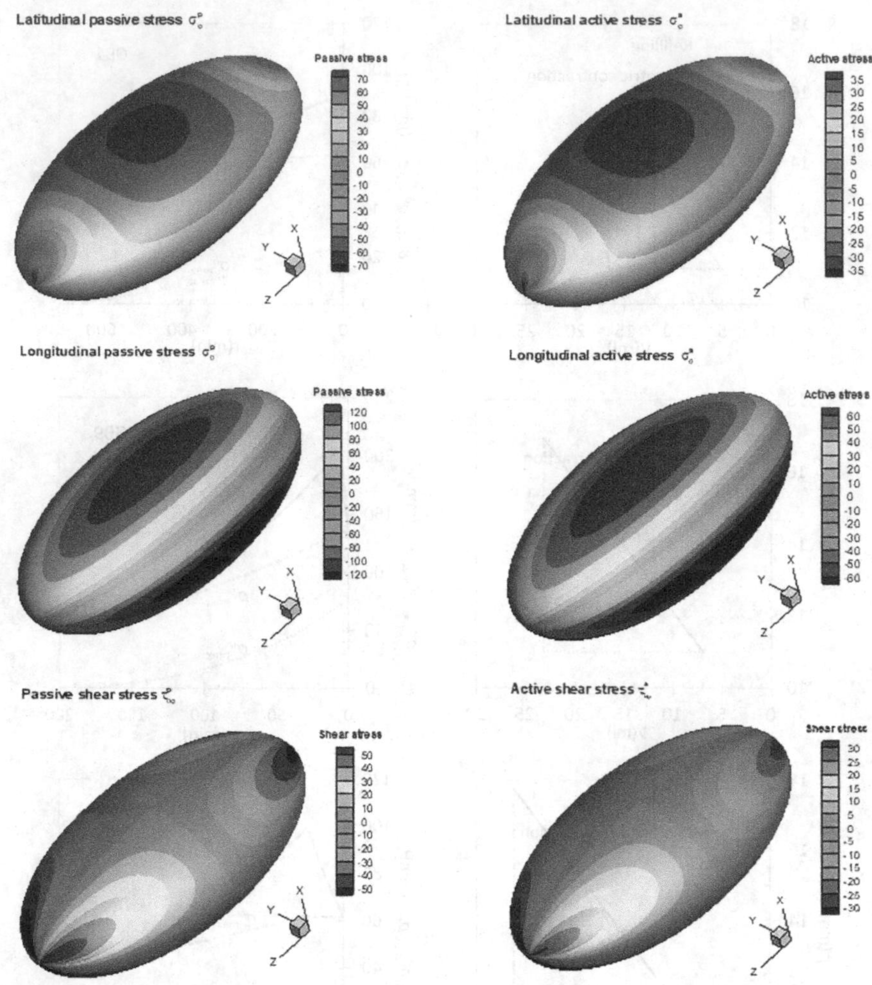

FIGURE 6.6 The passive (left) and active (right) stress contours of GB 9 at the end of isometric contraction, top – σ_{θ}^{p}, middle – σ_{φ}^{p} and bottom – $\tau_{\theta\varphi}^{p}$.

TABLE 6.1 Parameters and predictions for 51 acalculous GBs.

GB	k_1, k_2, D_3 (mm)	EF (%)	p_{max} (mmHg)	σ_{max} (mmHg)	σ^{p}_{max} (mmHg)	σ^{a}_{max} (mmHg)	CCK test	VAS
1	0.26, 0.31, 72.4	4.5(+)	15.2	111	111	19	pain (−)	N/A
2	0.50, 0.59, 59.7	5.4(+)	19.4(+)	128	73	56	pain (−)	N/A
3	0.36, 0.36, 72.2	11.4(+)	16.4	94	57	37	pain (−)	N/A
4	0.57, 0.61, 57.5	15.5(+)	20.4(+)	119	64	55	pain (−)	N/A
5	0.22, 0.28, 74.1	10.7(+)	14.4	202(+)	154	48	pain (+)	34
6	0.30, 0.41, 68.8	10.0(+)	16.4	204(+)	137	67(+)	pain (+)	54
7	0.43, 0.55, 57.3	14.0(+)	16.9(+)	131	123	46	no pain	N/A
8	0.40, 0.54, 66.7	21.9(+)	19.6(+)	205(+)	115	90	pain (+)	71
9	0.35, 0.55, 61.1	16.1(+)	16.7(+)	205(+)	135	70	pain (+)	N/A
10	0.27, 0.45, 69.1	5.4(+)	16.4	292(+)	196	96	pain (−)	N/A
11	0.28, 0.41, 82.0	15.1(+)	20.3(+)	332(+)	200	152	pain (−)	N/A
12	0.28, 0.39, 68.3	21.3(+)	16.6(+)	165	147	56	pain (−)	N/A
13	0.33, 0.47, 63.2	39.7	16.2	194(+)	134	62	pain (+)	86
14	0.29, 0.55, 63.5	20.6(+)	16.5	287(+)	191	96	pain (+)	29
15	0.35, 0.42, 72.0	80.8	18.4(+)	153	91	61	pain (+)	72
16	0.29, 0.34, 50.7	32.3(+)	16.6(+)	76	76	13	no pain	N/A
17	0.40, 0.46, 57.5	32.4(+)	16.6(+)	79	64	27	pain (+)	32
18	0.29, 0.32, 71.6	93.7	15.9	72	64	22	pain (+)	34
19	0.38, 0.39, 92.3	49.4	26.3(+)	175(+)	73	102	pain (+)	50
20	0.41, 0.53, 56.7	48.7	17.2(+)	89	57	32	pain (+)	52
21	0.38, 0.40, 74.5	66.3	19.5(+)	109	63	47	pain (−)	N/A
22	0.23, 0.44, 62.8	54.4	15.0	235(+)	235	20	pain (+)	89
23	0.31, 0.44, 77.5	44.7	19.7(+)	286(+)	160	126	pain (+)	22
24	0.41, 0.52, 63.0	27.4(+)	18.3(+)	153	92	61	pain (−)	37
25	0.38, 0.50, 75.4	27.9(+)	21.9(+)	241(+)	121	120	pain (+)	N/A
26	0.32, 0.41, 75.8	19.1(+)	18.7(+)	204(+)	120	84	pain (+)	N/A
27	0.38, 0.42, 64.8	70.2	16.9(+)	92	67	32	pain (+)	76
28	0.38, 0.41, 68.0	71.5	17.7(+)	93	57	35	pain (−)	67
29	0.50, 0.52, 56.1	37.8	17.1(+)	86	55	31	pain (−)	N/A
30	0.32, 0.39, 71.2	91.8	17.8(+)	178(+)	110	68	pain (+)	N/A
31	0.26, 0.39, 74.2	100	16.5	292(+)	194	98	pain (+)	68
32	0.31, 0.37, 71.5	10.1(+)	16.6(+)	167	110	56	pain (−)	48
33	0.41, 0.48, 63.0	95	17.6(+)	120	78	42	pain (−)	N/A
34	0.27, 0.44, 67.0	82.6	15.9	270(+)	187	83	pain (+)	N/A
35	0.38, 0.53, 45.8	100	13.6	103	86	19	pain (+)	54
36	0.30, 0.34, 76.1	16.3(+)	17.1(+)	141	91	50	pain (−)	44
37	0.56, 0.57, 53.8	77	17.7(+)	100	62	38	pain (−)	78
38	0.37, 0.47, 69.5	10.7(+)	18.9(+)	182(+)	106	76	pain (+)	17

(Continued)

TABLE 6.1 Parameters and predictions for 51 acalculous GBs.

GB	k_1, k_2, D_3 (mm)	EF (%)	p_{max} (mmHg)	σ_{max} (mmHg)	σ^p_{max} (mmHg)	σ^a_{max} (mmHg)	CCK test	VAS
39	0.51, 0.53, 57.3	8.0(+)	19.0(+)	103	60	43	pain (+)	N/A
40	0.25, 0.27, 73.1	22.2(+)	14.8	101	101	24	pain (+)	76
41	0.44, 0.47, 55.4	36.4	16.2	236(+)	201	35	pain (+)	N/A
42	0.46, 0.46, 82.1	2.8(+)	26.7(+)	182(+)	75	107	pain (−)	N/A
43	0.23, 0.37, 88.6	14.2(+)	19.1(+)	473(+)	272	201	pain (+)	N/A
44	0.24, 0.29, 81.5	15.2(+)	16.0	223(+)	188	70	pain (−)	N/A
44	0.41, 0.48, 57.5	9.1(+)	16.0	96	66	30	pain (−)	46
46	0.45, 0.52, 60.0	6.5(+)	17.6(+)	107	67	40	pain (−)	73
47	0.32, 0.39, 70.5	34.4(+)	17.1(+)	166	107	59	pain (−)	N/A
48	0.35, 0.57, 61.2	0.1(+)	17.1(+)	223(+)	143	80	pain (+)	62
49	0.29, 0.43, 57.3	1.25(+)	14.2	181(+)	141	40	pain (+)	37
50	0.47, 0.44, 71.9	8.75(+)	21.2(+)	147	62	85	pain (−)	N/A
51	0.38, 0.64, 55.1	1.62(+)	16.5	192(+)	128	64	pain (+)	N/A
Threshold			0.35	16.6	175			N/A
Success rate			0.412	0.412	0.765			

[EF] = 0.35 is commonly used as clinical threshold, and [p] = 16.6 mmHg is an average value of the pressures estimated of 51GBs at EF = 0.35. The success rate for each of [EF], [p] and [σ] threshold is defined as the number of subjects whose pain was correctly predicted over the total number of subjects, + indicates GB painful, – does not GB painful.

The pain score visual analogue scale (VAS) (Langley and Sheppeard 1985) based on McGill pain questionnaire (Melzack 1975) was applied to specify the GB pain level in CCK provocation tests. These pain scores (where available) are listed in Table 6.1. The physical pain threshold is assessed by the patients themselves by ticking the 'pain/not pain' box in the questionnaire. As these scores are highly subjective, the pain prediction with the stress is on a clear-cut basis of pain/not pain rather than on the severity of the pain.

The pain predictions shown in Table 6.1 are those obtained at [σ] = 175 mmHg. This gives a success rate of 75%, compared to the overall success rate of 78.4% for 37 samples in Chapter 5. The success rates of EF and pressure are poor with [EF] = 0.35 and [p] = 16.6 mmHg as the GB pain indexes, respectively.

Since the stress pain threshold was based on an average value of pain thresholds of 33 patients when their CBDs are subject to pressure increase,

TABLE 6.2 Statistics outcome of positive and negative predictions against physical symptom of pain based on the stress threshold $[\sigma] = 175$ mmHg.

Parameter	**Against physical pain induced by CCK**	
	Positive (pain)	**Negative (no pain)**
Success	20 (0.74)	19 (0.792)
Failure	7 (0.26)	5 (0.208)
Sample size	27	24
Confidence interval (95%)	(0.547, 0.871)	(0.587, 0.911)
Ratio of odds (positive/negative)	0.752	
Asymptotic standard error	−0.404	
Confidence interval (95%) of the ratio	(1.661, 0.340)	

it is important that a small difference in this value does not have a notable impact on the overall success rate. If the threshold is shifted by ±10% from the default value of 175 mmHg, i.e., $[\sigma] = 192.5$ and 157.5 mmHg, respectively, the success rate is changed to 71.2% and 76.9 %. Thus, the maximum peak stress presents the strongest correlation with the physical symptom of pain in the patients.

The counts of success and failure against the CCK-induced pain are shown in the 2×2 contingency table (Table 6.2) along with the statistics using the logistic transformation (see Appendix A). In the table, the success rate of positive prediction (0.74) is slightly lower than that of negative prediction (0.792), with the ratio of odds (positive/negative) of 0.752. This is different from the previous observation based on the 37 samples in Chapter 5 where a slightly higher positive prediction (0.850) was shown.

The biomechanical model in Chapter 5 has been extended to account properly for both active and passive stresses by considering isotonic refilling and isometric contraction in CCK provocation tests. The model is applied to a larger sample size (51), and the results strongly support the hypothesis that the peak total (active + passive) stress inside the GB wall is associated with CCK-induced GB pain.

CIRCUMFERENTIAL STRETCH AS GB PAIN INDEX

It has been shown experimentally that the circumferential strain is responsible for the pain in human oesophagus (Barlow et al. 2002), duodenum (Gao et al. 2003) and rectum (Petersen et al. 2003). It is believed that mechanoreceptors in the wall of those organs is the circumferential strain/stretch rather than the total or passive tension/stress. For the human duodenum in active state, the stretch threshold for pain in the circumferential direction is found to be 2.1 ± 0.2 (Gao et al. 2003). Similar observation is made in human gastroenterological pain (Drewes et al. 2003). It is likely that the pain receptors in the nerves respond to the muscle strain.

The geometrical shape of the GB is more complex than these organs, it is very hard to derive the strain field directly in the human GBs. Here, just the mean circumferential stretch is going to be estimated approximately based on the ellipsoid model. Based on the ultrasonographical images, the local stretch cannot be estimated because the marked material points are not observed, instead the mean stretch can be evaluated.

Since the GB is a thin-walled elastic ellipsoid membrane in *Stresses in the GB Wall* section, the GB shape has been known at every time instant by interpolating the ultrasound pictures of GB at two time instants with an exponential function of time in the emptying.

An ellipsoid GB with axes D_1, D_2 and D_3 at a time instant is illustrated in Figure 6.7. The arc lengths of the ellipsoid in the planes x-y (D_1-D_2), y-z (D_2-D_3) and x-z (D_1-D_3) are determined in the emptying. The arc of ellipsoid in the plane x-y is circumferential, but those in the planes y-z and x-z

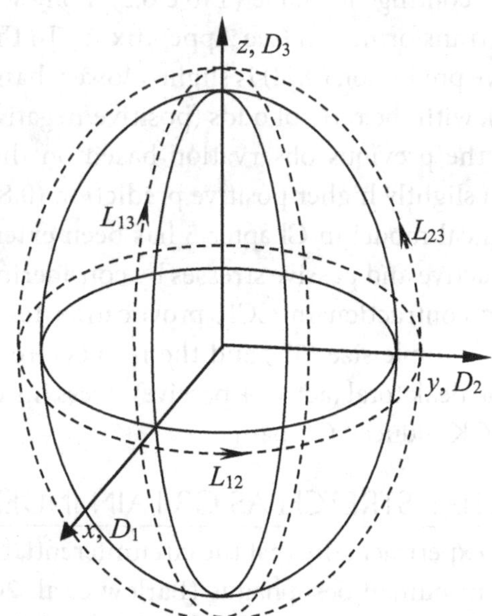

FIGURE 6.7 GB ellipsoid model with three arc lengths L_{12}, L_{23} and L_{13}, respectively, in the planes x-y (D_1-D_2), y-x (D_2-D_3) and x-z (D_1-D_3).

are meridian. For a revolutionary ellipsoid, the lengths of the arcs in the planes *y-z* and *x-z* are identical; otherwise, they are different.

To determine the stretch, the instant arc lengths are normalised by using the lengths in the reference state. Here, the GB at 70% EF serves as the reference state. Accordingly, the mean stretch can be expressed by

$$\begin{cases} \lambda_{12} = L_{12} / L_{12\text{ref}} \\ \lambda_{23} = L_{23} / L_{23\text{ref}} \\ \lambda_{13} = L_{13} / L_{13\text{ref}} \end{cases} \tag{6.15}$$

These three arc lengths of three ellipses respectively in planes *x-y* (D_1-D_2), *y-z* (D_2-D_3) and *x-z* (D_1-D_3) can be estimated approximately in the following expresses proposed by Muir in 1883 (Almkvist and Berndt 1988):

$$\begin{cases} L_{12} = 2\pi\left(\dfrac{D_1^{1.5} + D_2^{1.5}}{2^{2.5}}\right)^{1/1.5} \\ L_{23} = 2\pi\left(\dfrac{D_2^{1.5} + D_3^{1.5}}{2^{2.5}}\right)^{1/1.5} \\ L_{13} = 2\pi\left(\dfrac{D_1^{1.5} + D_3^{1.5}}{2^{2.5}}\right)^{1/1.5} \end{cases} \tag{6.16}$$

The stretches λ_{12}, λ_{23} and λ_{13} have the following relations to the strain components ε_{12}, ε_{23} and ε_{13} proposed by Gao et al. (2003) for the human duodenum:

$$\begin{cases} \varepsilon_{12} = \left(L_{12} - L_{12\text{ref}}\right) / L_{12\text{ref}} = \lambda_{12} - 1 \\ \varepsilon_{23} = \left(L_{23} - L_{23\text{ref}}\right) / L_{23\text{ref}} = \lambda_{23} - 1 \\ \varepsilon_{13} = \left(L_{13} - L_{13\text{ref}}\right) / L_{13\text{ref}} = \lambda_{13} - 1 \end{cases} \tag{6.17}$$

The stretch or strain is related to the stress in the GB wall, and the stretch or strain is determined just by the GB shape and dimension. In this context, the peak stretch may be an index of GB pain. It is assumed that the peak stretch causes the GB pain. Once the peak stretch is more than the threshold, i.e., $\lambda_{max} \geq [\lambda]$, the pain will be onset. The peak stretch is the maximum in λ_{12}, λ_{23} and λ_{13}, namely

$$\lambda_{max} = \max\left[\lambda_{12}, \lambda_{23}, \lambda_{13}\right] \tag{6.18}$$

Figure 6.8 shows the arc lengths of GBs 1, 9 and 37 in terms of emptying time. These lengths reduce with emptying progress. The circumferential length (L_{12}) always is the smallest one. The difference in the other arc lengths L_{23} and L_{13} does not seem significant.

Figure 6.9 presents the stretches λ_{12}, λ_{23} and λ_{13} of GBs 1, 9 and 37 as a function of time. These stretches decline with increasing time of emptying, but they are in different slopes. A peak stretch, which was built up in the refilling, occurs at the start of emptying. Usually, the CCK effect is valid in 30 min, indicating the emptying hardly goes on beyond 30 min. In this case, GB 1 is subject to serious stretch because of its poor emptying function, but GB 37 seems less stretched due to its best emptying behaviour.

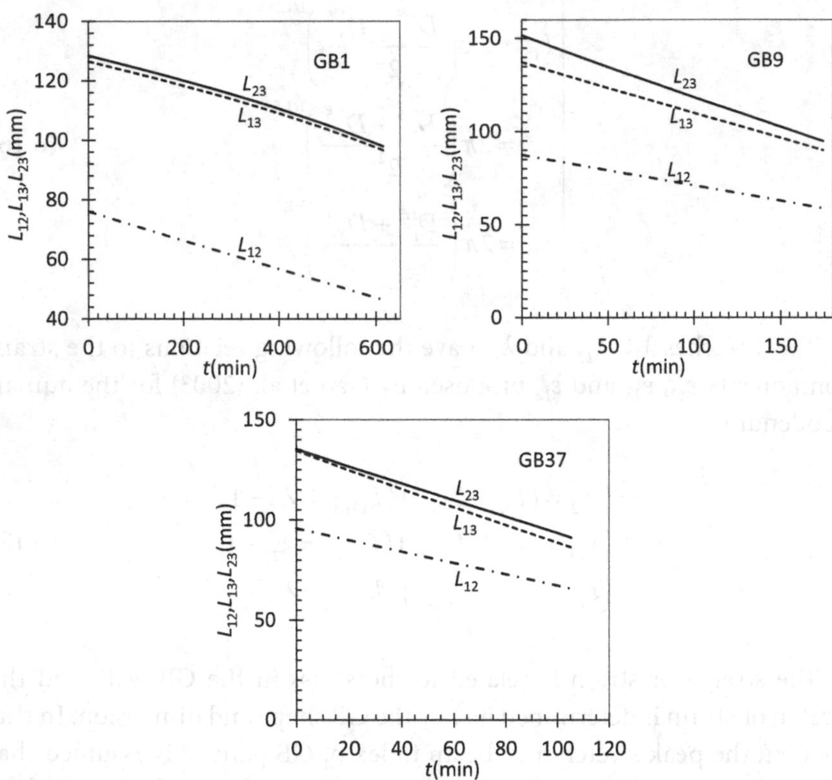

FIGURE 6.8 The arc lengths L_{12}, L_{23} and L_{13} of GBs 1, 9 and 37 versus emptying time.

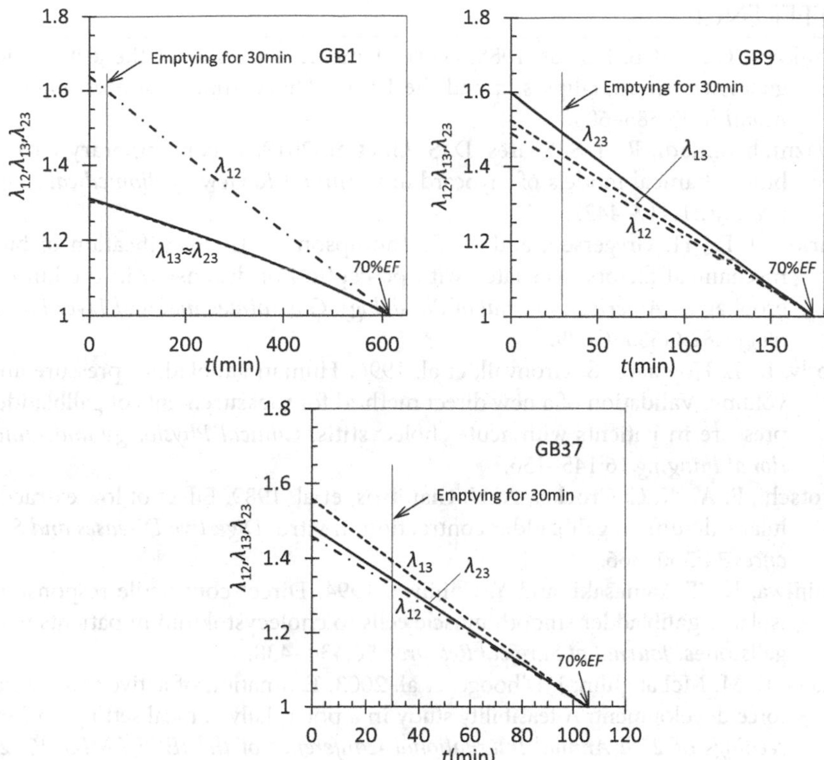

FIGURE 6.9 Stretch λ_{12}, λ_{23} and λ_{13} of GBs 1, 9 and 37 versus time in the emptying.

A suitable threshold of stretch should be able to achieve a best success rate compared with the CCK pain test outcomes. The threshold of stretch can be decided by using trial-and-error method. Here, the threshold is $[\lambda] = 1.55$ compared with the stretch threshold 2.1 ± 0.2 for the human duodenum (Gao et al. 2003). With that threshold, a 0.569 success rate was achieved, which is better than the success rate of 0.412 based on EF and peak pressure. For the human GB, the peak stretch does not seem to be a good index for GB pain.

Furthermore, strain of stretch cannot be measured non-invasively and estimated accurately. The simple and straightforward stress prediction, however, is only derived from non-invasive volume measurements; thus, this approach has clinical significance potentially.

REFERENCES

Almkvist, G., and B. Berndt. 1988. Gauss, Landen, Ramanujan, the arithmetic-geometric mean, ellipses, π and the Ladies Diary. *American Mathematical Monthly* 95:585–608.

Avazmohammadi, R., J. S. Soares, D. S. Li, et al. 2019. A contemporary look at biomechanical models of myocardium. *Annual Review of Biomedical Engineering* 21:417–442.

Barlow, J. D., H. Gregersen, and D. G. Thompson. 2002. Identification of biomechanical factors associated with perception of distension in the human esophagus. *American Journal of Physiology-Gastrointestinal and Liver Physiology* 282:G683–G689.

Borly, L., L. Hojgaard, S. Gronvall, et al. 1996. Human gallbladder pressure and volume: Validation of a new direct method for measurements of gallbladder pressure in patients with acute cholecystitis. *Clinical Physiology and Functional Imaging* 16:145–156.

Brotschi, E. A., K. C. Crocker, A. N. Gianitsos, et al. 1989. Effect of low extracellular calcium on gallbladder contraction in vitro. *Digestive Diseases and Sciences* 34:360–366.

Chijiiwa, K., T. Yamasaki, and Y. Chijiiwa. 1994. Direct contractile response of isolated gallbladder smooth muscle cells to cholecystokinin in patients with gallstones. *Journal of Surgical Research* 56:434–438.

Claus, P., M. McLaughlin, J. D'hooge, et al. 2003. Estimation of active myocardial force development: A feasibility study in a potentially clinical setting. In *Proceedings of 25th Annual International Conference of the IEEE EMBS*, 17–21 September, Cancun, Mexico.

Dodds, W. J., W. J. Hogan, and J. E. Green. 1989. Motility of the biliary system: In *Handbook of Physiology*. Bethesda: American Physiological Society.

Drewes, A. M., H. Gregersen, and L. Arendt-Nielsen. 2003. Experimental pain in gastroenterology: A reappraisal of human studies. *Scandinavian Journal of Gastroenterology* 38:1115–1130.

Ferris, D. O., and J. C. Vibert. 1959. The common bile duct: Significance of its diameter. *Annals of Surgery* 149:249–251

Gaensler, E. A. 1951. Quantitative determination of the visceral pain threshold in man. *Journal of Clinical Investigation* 30:406–420.

Gao, C., L. Arendt-Nielsen, W. Liu, et al. 2003. Sensory and biomechanical responses to ramp-controlled distension of the human duodenum. *American Journal of Physiology-Gastrointestinal and Liver Physiology* 284:G461–G471.

Herring, P., and S. Simpson. 1907. The pressure of bile secretion and the mechanism of bile absorption in obstruction of the bile duct. *Proceedings of Royal Society of London-Series B* 79:517–532.

Hould, F. S., G. M. Fried, A. G. Fazekas, et al. 1988. Progesterone receptors regulate gallbladder motility. *Journal of Surgical Research* 45:505–512.

Hunter, P., A. D. McCulloch, and H. E. ter Keurs. 1998. Modelling the mechanical properties of cardiac muscle. *Progress in Biophysics and Molecular Biology* 69:289–331.

Hunter, P., and B. Smaill. 1988. The analysis of cardiac function: A continuum approach. *Progress in Biophysics and Molecular Biology* 52:101–164.

Ishizuka, J., M. Murakami, G. A. Nichols, et al. 1993. Age-related changes in gallbladder contractility and cytoplasmic Ca^{2+} concentration in the guinea pig. *American Journal of Physiology-Gastrointestinal and Liver Physiology* 264:G624–G629.

Lambert, R., and J. P. Ryan. 1990. Response to calcium of skinned gallbladder smooth muscle from newborn and adult guinea pigs. *Pediatric Research* 28:336–338.

Langley, G. B., and H. Sheppeard. 1985. The visual analogue scale: Its use in pain measurement. *Rheumatology International* 5:145–148.

Lee, K. Y., P. Biancani, and J. Behar. 1989. Calcium sources utilized by cholecystokinin and acetylcholine in the cat gallbladder muscle. *American Journal of Physiology-Gastrointestinal and Liver Physiology* 256:G785–G788.

Lempke, R. E., R. D. King, and G. C. Kaiser. 1963. Hydrodynamics of gallbladder filling. *Journal of American Medical Association* 186:152–155.

Li, W. G., X. Y. Luo, N. A. Hill, et al. 2008. Correlation of mechanical factors and gallbladder pain. *Computational and Mathematical Methods in Medicine* 9:27–45.

Li, W. G., X. Y. Luo, N. A. Hill, et al. 2011. A mechanical model for CCK-induced acalculous gallstone pain. *Annals of Biomedical Engineering* 39:786–800.

MacPherson, B. R., G. W. Scott, J. P. Chansouria, et al. 1984. The muscle layer of the canine gallbladder and cystic duct. *Acta Anatomica* 120:117–122.

Mahour, G. H., K. G. Wakim, and D. O. Ferris. 1967. The common bile duct in man: Its diameter and circumference. *Annals of Surgery* 165:415–419.

Meiss, R. 1975. Graded activation in rabbit mesotubarium smooth muscle. *American Journal of Physiology* 229:455–465.

Melzack, R. 1975. The McGill pain questionnaire: Major properties and scoring methods. *Pain* 1:277–299.

Middelfart, H. V., P. Jensen, L. Hojgaard, et al. 1998. Pain patterns after distension of the gallbladder in patients with acute cholecystitis. *Scandinavian Journal of Gastroenterology* 33:982–987.

Morales, S., P. J. Camello, G. M. Mawe, et al. 2005. Characterization of intracellular Ca^{2+} stores in gallbladder smooth muscle. *American Journal of Physiology-Gastrointestinal and Liver Physiology* 288:G507–G513.

Novozhilov, V. V. 1964. *Thin Shell Theory*. Groningen: P Noordhoff Ltd.

Parkman, H., R. Garbarino, and J. P. Ryan. 2001. Myosin light chain phosphorylation correlates with contractile force in guinea pig gallbladder muscle. *Digestive Diseases and Sciences* 46:176–181.

Parkman, H., A. Pagano, J. S. Martin, et al. 1997. Electric field stimulation-induced guinea pig gallbladder contractions (role of calcium channels in acetylcholine release). *Digestive Diseases and Sciences* 42:1919–1925.

Petersen, P., C. Gao, L. Arendt-Nielsen, et al. 2003. Pain intensity and biomechanical responses during ramp-controlled distension of the human rectum. *Digestive Diseases and Sciences* 48:1310–1316.

Renzetti, L. M., M. B. Wang, and J. P. Ryan. 1990. Contribution of intracellular calcium to gallbladder smooth muscle contraction. *American Journal of Physiology-Gastrointestinal and Liver Physiology* 259:G1–G5.

Ryan, J. 1985. Calcium and gallbladder smooth muscle contraction in the guinea pig: Effect of pregnancy. *Gastroenterology* 89:1279–1285.

Shaffer, E., A. Bomzon, H. Lax, et al. 1992. The source of calcium for CCK-induced contraction of the guinea-pig gall bladder. *Regulatory Peptides* 37:15–26.

Shimada, T. 1993. Voltage-dependent calcium channel current in isolated gallbladder smooth muscle cells of guinea pig. *American Journal of Physiology-Gastrointestinal and Liver Physiology* 264:G1066–G1076.

Shoucri, R. 1991. Theoretical study of pressure-volume relation in left ventricle. *American Journal of Physiology-Heart and Circulatory Physiology* 260:H282–H291.

Shoucri, R. 1998. Studying the mechanics of left ventricular contraction. *IEEE Engineering in Medicine and Biology Magazine* 17:95–101.

Shoucri, R. 2000. Active and passive stresses in the myocardium. *American Journal of Physiology-Heart and Circulatory Physiology* 279:H2519–H2528.

Smythe, A., R. Ahmed, M. Fitzhery, et al. 2004. Bethanechol provocation testing does not predict symptom relief after cholecystectomy for acalculous biliary pain. *Digestive and Liver Disease* 36:682–686.

Smythe, A., A. W. Majeed, M. Fitzhery, et al. 1998. A requiem for the cholecystokinin provocation test? *Gut* 43:571–574.

Streeter, D. D., R. Vaishnav, D. J. Patel, et al. 1970. Stress distribution in the canine left ventricle during diastole and systole. *Biophysical Journal* 10:345–363.

Washabau, R. J., M. B. Wang, C. Dorst, et al. 1991. Effect of muscle length on isometric stress and myosin light chain phosphorylation in gallbladder smooth muscle. *American Journal of Physiology-Gastrointestinal and Liver Physiology* 260:G920–G904.

Washabau, R. J., M. B. Wang, C. Dorst, et al. 1994. Role of myosin light-chain phosphorylation in guinea pig gallbladder smooth muscle contraction. *American Journal of Physiology-Gastrointestinal and Liver Physiology* 266: G469–G474.

Yamasaki, T., K. Chijiiwa, and Y. Chijiiwa. 1994. Direct contractile effect of motilin on isolated smooth muscle cells from human gallbladder. *Journal of Surgical Research* 56:89–93.

Zhang, L., A. D. Bonev, M. T. Nelson, et al. 1993. Ionic basis of the action potential of guinea pig gallbladder smooth muscle cells. *American Journal of Physiology-Cell Physiology* 265:C1552–C1561.

CHAPTER 7

Cross Bridges of GB Smooth Muscle Contraction

CCK ACTIVATION PATHWAY AND CROSS BRIDGES

A positive correlation between GB pain and peak normal stress in the GB wall was identified in Chapter 5, and further work was carried out to evaluate both the passive and active stresses in GBs and the active stress has played an important role in GB pain prediction in Chapter 6.

The active stress is related to smooth muscle in the human GB. At the micro-structure level, active stress of smooth muscle can be modelled effectively using the three-element approach, including a Ca^{2+}-driven cross-bridge contracting element, two passive springs respectively in series and in parallel (Zulliger et al. 2004; Stålhand et al. 2008; Kroon 2010). The approach leads to new constitutive formulations of smooth muscle to form a framework for soft tissue modelling. However, the difficulty in using such models is that they often require more than five coefficients of material properties that must be determined experimentally, making it impossible to apply them to patient-specific studies.

In this chapter, the active stress estimated from the biomechanical model in Chapter 6 and the patient-specific ultrasonographic data of 51 subjects were used to determine the rate constants and Ca^{2+} levels in the GB smooth muscle cross-bridge model.

The GB is an organ that stores bile. A meal or hormone CCK initiates GB contraction and sphincter of Oddi relaxation simultaneously, then a bile emptying starts to occur, allowing the bile to flow into the duodenum. CCK causes GB contraction by stimulating cholinergic neurons and the smooth muscle, which requires intracellular calcium (Yu et al. 1994). At a high CCK concentration of 10^{-6} to 10^{-4} mM/l for human GB intact smooth muscle cells or 10^{-4} to 10^{-2} mM/l for muscle strips, the intracellular free calcium (Ca^{2+}) above a certain concentration triggers a series of events, leading to force generation in the GB smooth muscle (Yu et al. 1994, 1998).

Additional experiments (Yu et al. 1993, 1994, 1998) showed that CCK initially bound to $G_i\alpha_3$ (a G protein) in the cell membrane. In response to activation of G-protein-coupled receptors, the phosphatidylinositol-specific phospholipase C beta 3 (PLC-β3) hydrolyses phosphatidylinositol biphosphate (PIP_2) into inositol 1,4,5-triphosphate (IP_3) and diacylglycerol (DAG). At a high dose of CCK, IP_3 stimulates sarcoplasmic reticulum (SR) to release more Ca^{2+}, which then binds to calmodulin (CaM), activating myosin light-chain kinase (MLCK) to allow myosin to be phosphorylated (P-myosin). In this case, CaM inhibits the protein kinase C (PKC) pathway (Harnett et al. 2005). The P-myosin attaches to actin filaments and creates cross bridges to generate tension (Yu et al. 1998). The CCK activation pathway is illustrated in Figure 7.1.

For GB smooth muscle that is muscle without sarcomere, the contraction induced by CCK requires a certain level of extracellular Ca^{2+} concentration and utilises increased intracellular Ca^{2+} (Brotschi et al. 1989; Shaffer et al. 1992). In other words, GB smooth muscle contracts via the high CCK dose pathway as shown in Figure 7.1.

It was demonstrated that Ca^{2+} CaM-dependent phosphorylation of the 20,000-Da myosin light chain by MLCK and dephosphorylation by myosin light-chain phosphatase (MLCP) were responsible for contraction of arterial, airway, vascular and uterine smooth muscle (Hai and Murphy 1988a). It was postulated that a latch bridge was formed by dephosphorylation of an attached phosphorylation and cross-bridge and latch-bridge formation and detachment were driven by a higher molecular mass actin. The structure of model of four-state is shown in Figure 7.2(a) (Hai and Murphy 1988a). The model was assumed to have two constraints to get a unique set of rate constants. First, the rate constants of phosphorylation of MLCK and dephosphorylation of MLCP are equal, i.e., $K_1 = K_6$ and $K_2 = K_5$. Second, the attachment-to-detachment ratio is chosen as 4:1, i.e.,

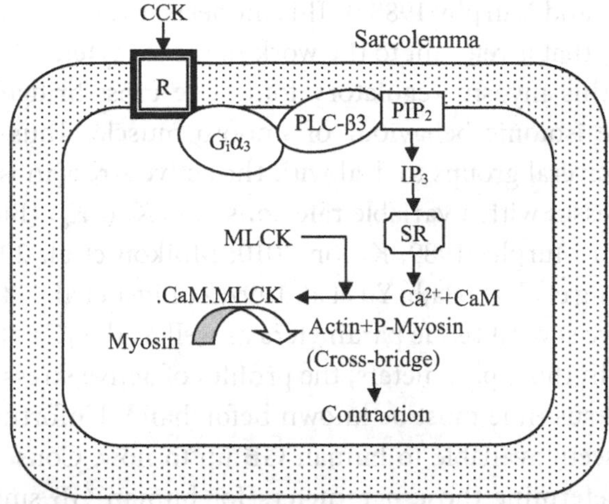

FIGURE 7.1 The pathways of CCK activation to GB smooth muscle, there are two pathways, CCK passing through R (receptor) activates $G_i\alpha_3$ proteins at membrane, then $G_i\alpha_3$ hydrolyses PIP_2 into IP_3; for the high dose of CCK, however, IP_3 stimulates SR to release stored Ca^{2+}; subsequently, Ca^{2+} binds to CaM and activates MLCK to cause P-myosin, active P-myosin attaches actin filament to form cross bridges for force generation.

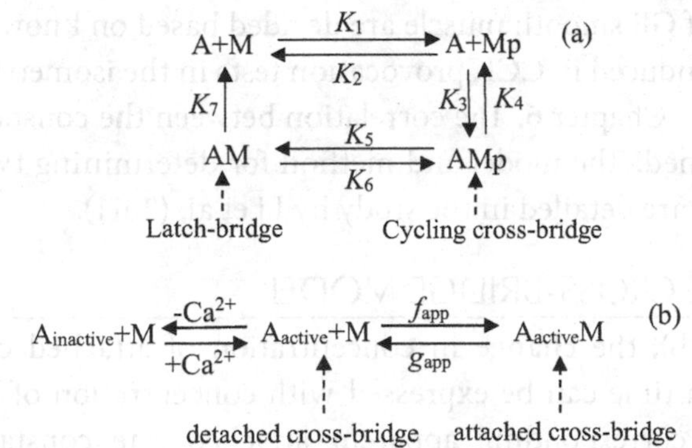

FIGURE 7.2 Four-state thick filament-regulated latch cross-bridge model (a) and the equivalent Huxley's two-state thin filament-regulated cross bridge (b), where A = actin (thin filament), M = detached dephosphorylated cross bridge, Mp = detached phosphorylated cross bridge, AM = attached dephosphorylated cross bridge (latch-bridge), and $A_{inactive}$ and A_{active} = actin in absence and presence of Ca^{2+}, respectively.

$K_3/K_4 = 4$ (Hai and Murphy 1988a). This model is myosin regulation-based and is the one that is relevant to the work in this chapter.

However, this myosin regulatory four-state cross-bridge model fails to handle the isotonic behaviour of smooth muscle. Thus, it has been extended by several groups to deal with the active stress in isometric and isotonic processes with a variable rate constants K_1 (=K_6) (Bursztyn et al. 2007; Hai and Murphy 1989; Kroon 2010; Mbikou et al. 2011; Murtada et al. 2010; Yang et al. 2003a,b; Yu et al. 1997; Zulliger et al. 2004). All those models require several tissue parameters as well as the [Ca^{2+}] profile with time. To obtain those parameters, the profiles of active stress, Mp + AMp and [Ca^{2+}] versus time must be known beforehand. Unfortunately, those data do not seem available to human GB up to now. Consequently, it is unlikely to determine those parameters for human GB smooth muscle based on the four-state cross-bridge model presently.

Actually, to simplify the four-state cross-bridge model and reduce the number of required experimental parameters, Hai and Murphy (1988b) converted the four-state cross-bridge model into a two-state cross-bridge model (Huxley 1957) by lumping together dephosphorylation and phosphorylation cross bridges (Hai and Murphy 1988b; Yu et al. 1997) [see Figure 7.2(b)]. This will allow us to have apparent rate constants in the model just by using a known active stress profile.

In this chapter, two apparent rate constants that can characterise the cross bridge of GB smooth muscle are decided based on known GB active stress profile induced in CCK provocation tests in the isometric and emptying phases in Chapter 6. The correlation between the constants and GB pain is examined. The model and method for determining two apparent rate constants are detailed in the study by Li et al. (2011).

TWO-STATE CROSS-BRIDGE MODEL

In Figure 7.2(b), the change in concentration of attached cross bridge [$A_{active}M$] with time can be expressed with concentration of [A_{active}] and [M] and the corresponding apparent attached rate constant f_{app} and detached rate constant g_{app} as the following (Hai and Murphy 1988a):

$$\frac{\mathrm{d}w(t^*)}{\mathrm{d}t^*} = [1 - w(t^*)]f_{app} - w(t^*)g_{app} \tag{7.1}$$

where $w(t^*)$ is the concentration of [$A_{active}M$] or fraction of attached cross bridges, t^* indicates the time of chemical reaction, [] denotes the concentration of species, A is actin (thin filament) and M is detached dephosphorylated cross bridge. It was shown rate constants f_{app} and g_{app} can be calculated with the following equation by using seven rate constants K_1 through K_7 in Figure 7.2(a) (Hai and Murphy 1988b)

$$\begin{cases} f_{app} = \dfrac{k_1 k_3 (k_4 k_6 + k_4 k_7 + k_5 k_7)}{(k_1 + k_2)(k_4 k_6 + k_4 k_7 + k_5 k_7) + k_3 k_5 k_7} \\ g_{app} = \dfrac{k_4 k_6 + k_4 k_7 + k_5 k_7}{k_5 + k_6 + k_7} \end{cases} \tag{7.2}$$

To get K_1 through K_7 for a specific smooth muscle, the profiles of active stress and [Mp], [AMp] must be available beforehand, where [Mp] is detached phosphorylated concentration and [AMp] is attached phosphorylated cross-bridge concentration. For human GB smooth muscle, however, there are currently no such. Thus, we are unlikely able to have recourse to Eq. (7.2).

Equation (7.1) represents chemical reactions in the cross bridges of smooth muscle cells. To clarify the timescale of the reactions, an *in vivo* experiment was tried for a number of patients with GB. In the experiment, a specific amount of CCK was injected into the GB and the internal pressure inside the GB was monitored with a miniature pressure transducer simultaneously. Consequently, a typical diagram of GB pressure response under a CCK stimulant is illustrated in Figure 7.3. It is indicated that just in 14 s the GB pressure

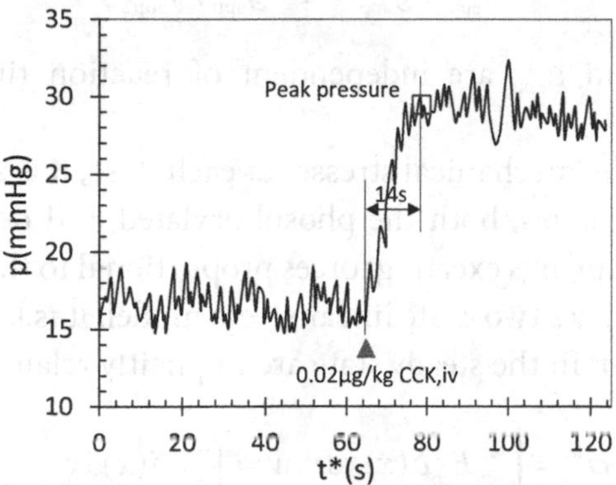

FIGURE 7.3 *In vivo* pressure inside a GB versus time before and after a certain CCK stimulant is applied, the GB develops a steady high presser in 14 s based on the basal pressure, iv = intravenous infusion.

reaches the highest steady level from the basal one. This suggests the timescale of the reaction described by Eq. (7.1) is around 14 s for GB smooth muscle.

In CCK provocation tests, the 51 patients with ABP were experienced an overnight fast and then were given an intravenous infusion of saline (control) followed by a continuously intravenous infusion of CCK (0.05 μg/kg body weight) for 10 min to diagnose attacks of biliary pain. After 10 min, a high enough pressure established in the GB pushes the bile out, causing the emptying phase. This means the timescale of CCK test is 10 min. Clearly, the CCK infusion timescale is much longer than the CCK-induced reaction timescale for cross-bridge formation. Thus, cascade reactions for cross-bridge formation take place in GB smooth muscle during 10-min CCK infusion or active stress developing period.

Since the CCK-induced reaction timescale is so short compared with the CCK infusion timescale that the chemical reaction in Eq. (7.1) can be regarded to take place and achieve a steady state at an instance in every continuous dose, the quasi-steady reaction is adopted herein. In other words, the active stress developing phase at each dose of CCK consists of enormous quasi-steady chemical reactions. At each dose of CCK, Eq. (7.1) can be reduced to

$$(1-w)f_{\text{app}} - wg_{\text{app}} = 0 \tag{7.3}$$

Solving Eq. (7.3) for w, it reads

$$w = \frac{f_{\text{app}}}{f_{\text{app}} + g_{\text{app}}} = \frac{1}{1 + g_{\text{app}} / f_{\text{app}}} \tag{7.4}$$

where w, f_{app} and g_{app} are independent of reaction time t^*, but CCK dose-dependent.

To calculate the mechanical stresses at each dose, it is assumed that the attached cross bridges, both the phosphorylated and dephosphorylated, behave as linear springs exerting forces proportional to their displacement x. From the Huxley's two-state linear stress model, it is known that w and the active stress σ^{a} in the steady state are implicitly related by

$$\sigma^{\text{a}} = \int_{-\infty}^{\infty} E_{\text{cb}} \delta(x) x \mathrm{d}x, w = \int_{-\infty}^{\infty} \delta(x) \mathrm{d}x \tag{7.5}$$

where E_{cb} is spring constant and $\delta(x)$ is the distribution function of total number of attachments along cross bridges. The cross bridges detach when the displacement is either negative or greater than a critical value (defined as $x = 1$, x is the normalised displacement). In the steady state of the reaction, $\delta(x)$ is uniform along the filaments of cross bridge (Wong 1972, 1973), $\delta(x) = \delta_0$, is independent of x, i.e.,

$$\delta(x) = \begin{cases} \delta_0, 0 \le x \le 1 \\ 0, \text{otherwise} \end{cases} \quad (7.6)$$

Then Eq. (7.5) leads to

$$\sigma^{\mathrm{a}} = \frac{E_c b \delta_0}{2}, w = \delta_0 \quad (7.7)$$

In the CCK infusion period, the active stress is dose-dependent and increased with time; subsequently, it is a function of infusion time instant t. Accordingly, w, f_{app} and g_{app} should be a function of such time instant. Based on Eq. (7.7), the fraction of attached cross bridges $w(t)$ is proportional to the active stress

$$w(t) \propto \sigma^{\mathrm{a}}(t) \quad (7.8)$$

Equation (7.8) is scaled by the peak active stress $\sigma^{\mathrm{a}}_{\max}$ and maximum fraction of attachment $w_{\max}$, so that

$$w(t) = \left(\frac{w_{\max}}{\sigma^{\mathrm{a}}_{\max}}\right)\sigma^{\mathrm{a}}(t) \quad (7.9)$$

The maximum active stress, $\sigma^{\mathrm{a}}_{\max}$, is a constant for any individual GB smooth muscle. The peak fraction of attached cross bridges, $w_{\max}$, is also a constant. It is known that for cardiac muscle; $w_{\max} = 0.3–0.53$ (Robertson et al. 1981; Solaro and Rarick 1998), but for smooth muscle, $w_{\max} = 0.5–0.78$ due to the latch-bridge effect (Hai 1991; Hai and Murphy 1988a; Kamm and Stull 1985a; Ratz et al. 1989).

Substituting Eq. (7.9) into Eq. (7.4), the cross-bridge kinetics to $\sigma^{\mathrm{a}}(t)$ is related by

$$\left[1 - \left(\frac{w_{\max}}{\sigma^{\mathrm{a}}_{\max}}\right)\sigma^{\mathrm{a}}(t)\right] f_{\mathrm{app}}(t) - \left(\frac{w_{\max}}{\sigma^{\mathrm{a}}_{\max}}\right)\sigma^{\mathrm{a}}(t) g_{\mathrm{app}}(t) = 0 \quad (7.10)$$

Equation (7.10) can also be written as

$$f_{\text{app}}(t)=\left(\frac{w_{\max}}{\sigma^{\text{a}}_{\max}}\right)\left[\frac{\sigma^{\text{a}}(t)g_{\text{app}}(t)}{1-(w_{\max}/\sigma^{\text{a}}_{\max})\sigma^{\text{a}}(t)}\right] \tag{7.11}$$

Estimation of the Active Stress Due to CCK

$\sigma^{a}(t)$ is the peak active stress in the GB in the isometric contraction (CCK infusion period) [see Figure 6.1(c)]. This is estimated from the biomechanical model in Chapter 6. In the isometric phase, the active stress is determined by the transmural pressure and the geometry of the GB

$$\sigma^{\text{a}}(t)=[p(t)-p_{\text{e}}]\max[F_{\theta}F_{n},F_{\varphi}/F_{n}] \tag{7.12}$$

where $\max[F_{\theta}F_{n},F_{\varphi}/F_{n}]$ means to take the maximum from $F_{\theta}F_{n}$ and $F_{\varphi}F_{n}$, the F_{θ}, F_{φ} and F_n are geometrical functions given by

$$\begin{cases} F_{\theta}\left(h_{\text{GB}},D_3,\kappa_1,\kappa_2,\varphi\right)=\dfrac{D_3\kappa_1\kappa_2}{4h_{\text{GB}}}\left(1-\dfrac{\kappa_1^2-\kappa_2^2}{\kappa_1^2\kappa_2^2}\cos 2\varphi\right) \\ F_{\varphi}\left(h_{\text{GB}},D_3,\kappa_1,\kappa_2,\theta,\varphi\right)=\dfrac{D_3}{4\kappa_1\kappa_2h_{\text{GB}}}\begin{bmatrix}\kappa_1^2\kappa_2^2+\left(\kappa_1^2+\kappa_2^2-2\kappa_1^2\kappa_2^2\right)\sin^2\theta \\ +\left(\kappa_1^2-\kappa_2^2\right)\cos^2\theta\cos 2\varphi\end{bmatrix} \\ F_{n}\left(\kappa_1,\kappa_2,\theta,\varphi\right)=\dfrac{\sqrt{\kappa_1^2\cos^2\theta\cos^2\varphi+\kappa_2^2\cos^2\theta\sin^2\varphi+\sin^2\theta}}{\sqrt{\kappa_1^2\sin^2\varphi+\kappa_2^2\cos^2\varphi}}, \end{cases} \tag{7.13}$$

where D_1, D_2 and D_3 are the three major axis lengths of the ellipsoid fitting the GB shape and h_{GB} is the GB wall thickness. Note that $\kappa_1 = D_1/D_3$ and $\kappa_2 = D_1/D_3$ are constants and estimated from the GB ultrasonographical images of a patient in the isometric period; $p_e = 11$ mmHg is the refilling basal pressure in a GB in Chapter 6; θ and φ are the angular coordinates of a spherical coordinate system on the ellipsoid GB wall (Figure 6.3). The peak active stress occurs at the end of the isometric contraction

$$\sigma^{\text{a}}_{\max}=(p_{\max}-p_{\text{e}})\max[F_{\theta}F_{n},F_{\varphi}/F_{n}] \tag{7.14}$$

where $p_{\max}$ is the GB pressure at the end of isometric contraction.

The isometric contraction was induced by infusion of CCK to patients over the period of 10 min (Smythe et al. 1998); thus, the isometric contraction is a quasi-steady process and the GB shape remains ellipsoidal; hence, the shape functions F_θ, F_n and F_φ are independent of time, while the pressure in the GB obeys (Meiss 1975)

$$p(t) = p_e e^{qt}, q = \frac{\ln\left(\frac{p_{max}}{p_e}\right)}{t_i} \tag{7.15}$$

where t_i is the GB isometric contraction time in the course of CCK injection (10 min) and q is a constant. Then $\sigma^a(t)/\sigma^a_{max}$ is expressed as

$$\frac{\sigma^a(t)}{\sigma^a_{max}} = \frac{(e^{qt}-1)}{(e^{qt_i}-1)} \tag{7.16}$$

Taking Eq. (7.16) into Eq. (7.11), f_{app} reads

$$f_{app}(t) = \frac{w_{max}(e^{qt}-1)g_{app}(t)}{(e^{qt_i}-1) - w_{max}(e^{qt}-1)} \tag{7.17}$$

The rate constants in Eq. (7.17) seem to be the function of time because they are functions of [Ca^{2+}] which increase as CCK is infused, but this time dependence is much slower than the transient state described by Eq. (7.1). Equation (7.17) also implies that the final value of f^{max}_{app} (when $t = t_i$) is proportional to g_{app} only at a given value of w_{max} although f_{app} profile is dependent on the value of q. The reason for this effect is that a faster cycling process of cross bridge requires larger attachment and detachment rate constants.

Estimation of g_{app} for the Human GB

g_{app} is nearly independent of [Ca^{2+}] and approximately remains unchanged in the cycling of cross bridges (Brenner 1988). Therefore, g_{app} should be independent of time. However, there are no direct experimental data on g_{app} for human GB smooth muscle; consequently, g_{app} must be estimated from the limited data available for smooth muscle of guinea pigs or the others.

The following method is proposed to estimate g_{app} for human GB smooth muscle. First, the rate constants, K_1–K_7 are determined based on the four-state cross-bridge model for swine artery, bovine tracheal and guinea pig GB smooth muscles, where the active stress and Mp and AMp concentrations were measured. The following chemical reactions of four-state cross-bridge model for smooth muscle proposed by Hai and Murphy (1988a) were solved numerically by using the fourth-order Runge-Kutta method

$$\begin{cases} \dfrac{d[M]}{dt^*} = -K_1[M] + K_2[Mp] + K_7[AM] \\ \dfrac{d[Mp]}{dt^*} = K_4[AMp] + K_1[M] - (K_2 + K_3)[Mp] \\ \dfrac{d[AMp]}{dt^*} = K_3[Mp] + K_6[AM] - (K_4 + K_5)[AMp] \\ \dfrac{d[AM]}{dt^*} = K_5[AMp] - (K_7 + K_6)[AM], \end{cases} \tag{7.18}$$

where an initial condition $[M] = 1$, $[Mp] = [AMp] = [AM] = 0$ is applied.

The K_1–K_7 are optimised to best fit the experimental data obtained for swine carotid artery smooth muscle (Singer and Murphy 1987), tracheal smooth muscle (Kamm and Stull 1985b) and guinea pig GB smooth muscle (Washabau et al. 1994) by using the *lsqnonlin* function in MATLAB. The rate constants K_1 through K_7 for the swine carotid artery smooth muscle and tracheal smooth muscle are given by Hai and Murphy (1988a). The rate constants K_1 through K_7 for the guinea pig GB smooth muscle are determined here which give the best experimental fit for the time-dependent active stress and the concentration [Mp] + [AMp], as shown in Figure 7.4.

Second, after these rate constants available, f_{app} and g_{app} will be calculated with Eq. (7.2). Since there was an unlimited supply of Ca^{2+} in the experiments, the rate constants determined should be the maximum values, f_{app}^{max} and g_{app}^{max}. All the estimated rate constants are listed in Table 7.1. The maximum concentration of $[A_{active}M]$ w_{max} $(=1/(1+g_{app}^{max}/f_{app}^{max}))$ is about 0.77 for both the swine carotid artery and

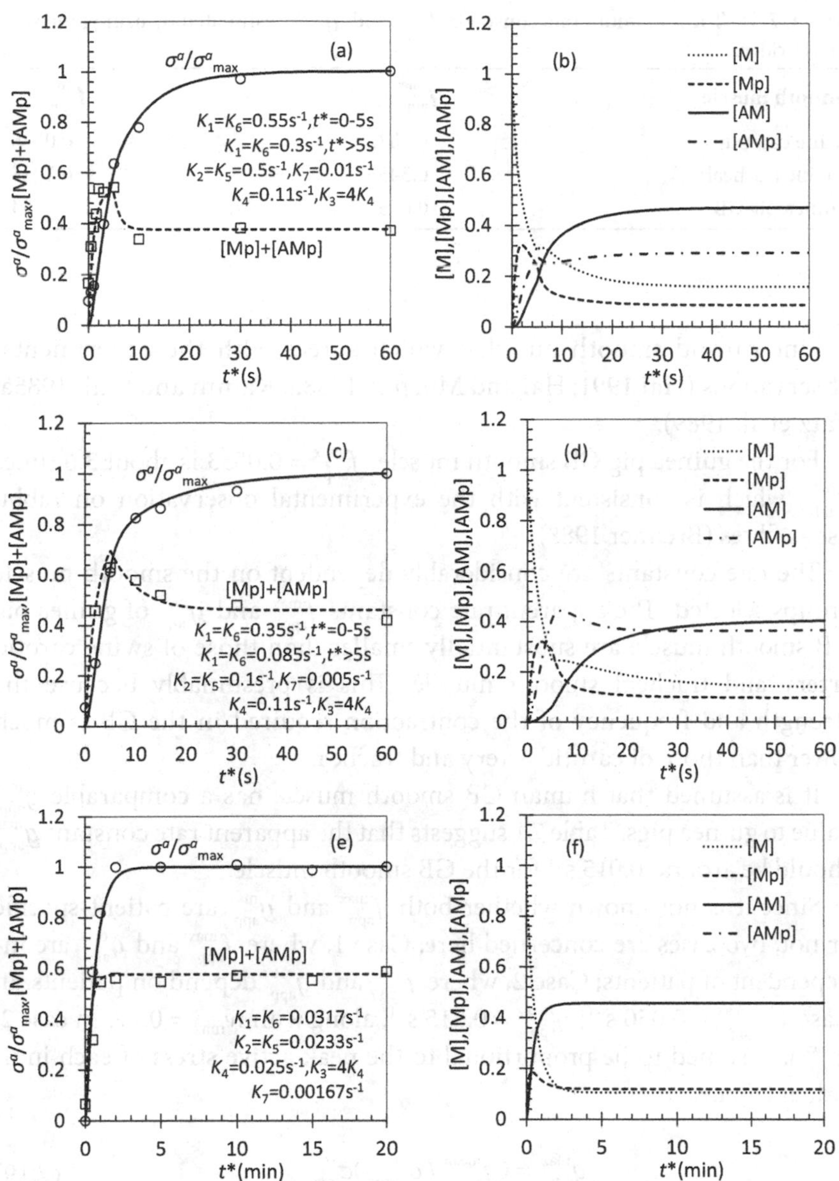

FIGURE 7.4 Dimensionless active stress (left) and concentrations (right) plotted as a function of time, (a) and (b) for swine carotid artery smooth muscle (Singer and Murphy 1987), (c) and (d) for tracheal smooth muscle (Kamm and Stull 1985b), (e) and (f) for guinea pig GB smooth muscle (Washabau et al. 1994). The symbols for experimental data and curves for predictions of four-state cross-bridge model, the active stress σ^a was normalised by its peak value σ^a_{max}.

TABLE 7.1 The maximum rate constants f_{app}^{max} and g_{app}^{max} estimated by using four-state cross bridge.

Smooth muscle	g_{app}^{max} (s^{-1})	f_{app}^{max} (s^{-1})
Swine carotid	0.2032	0.0575
Bovine tracheal	0.3455	0.1021
Guinea pig GB	0.0553	0.0154

bovine carotid smooth muscles, which agrees with the experimental observations (Hai 1991; Hai and Murphy 1988a; Kamm and Stull 1985a; Ratz et al. 1989).

For the guinea pig GB smooth muscle, $f_{app}^{max} = 0.0553$ is about 3.6 times g_{app}^{max}, which is consistent with the experimental observation on rabbit psoas fibres (Brenner 1988).

The rate constants are considerably dependent on the smooth muscle groups selected. The apparent rate constants f_{app}^{max} and g_{app}^{max} of guinea pig GB smooth muscle are substantially smaller than those of swine carotid artery and tracheal smooth muscle. This is presumably because the strength and frequency of the contraction required in the GB is much lower than those of carotid artery and trachea.

It is assumed that human GB smooth muscle has a comparable g_{app}^{max} value to guinea pigs. Table 7.1 suggests that the apparent rate constant g_{app}^{max} should be around 0.015 s^{-1} for the GB smooth muscle.

Since it is not known whether both f_{app}^{max} and g_{app}^{max} are patient-specific or not, two cases are concerned here: Case 1, where f_{app}^{max} and g_{app}^{max} are independent of patients; Case 2, where f_{app}^{max} and g_{app}^{max} depend on patients. In Case 1, $f_{app}^{max} = 0.050$ s^{-1}, $g_{app}^{max} = 0.015$ s^{-1}, along with $w_{max} = 0.77$. In case 2, g_{app}^{max} is assumed to be proportional to the peak active stress of each individual GB, i.e.,

$$g_{app}^{max} = (g_{app}^{mean} / \sigma_{mean}^{a})\sigma_{max}^{a} \tag{7.19}$$

where σ_{mean}^{a} is the mean peak active stress level based on the 51 GB samples, $\sigma_{mean}^{a} = 59$ mmHg and g_{app}^{mean} is the mean apparent detachment rate constant for 51 GBs, $g_{app}^{mean} = 0.015$ s^{-1} (Table 7.1), suggesting $g_{app}^{mean}/\sigma_{mean}^{a} = 2.54$ (mmHg·s)$^{-1}$.

ACTIVE STRESS AND INCREASED CA^{2+} LEVEL

The CCK-induced GB contraction utilises Ca^{2+} from the intracellular stores (Lee et al. 1989; Renzetti et al. 1990; Shaffer et al. 1992). The relationship between the active stress and $[Ca^{2+}]$ is time-dependent and implicit through the constants K_1 and K_7 of the cross-bridge model (Hai and Murphy 1988a). If the steady state is considered as discussed in *Two-State Cross-Bridge Model* section, then the steady active stress should be proportional to the total increase of $[Ca^{2+}]$ supplied

$$\Delta[Ca^{2+}] = c_0\sigma^a(t) \tag{7.20}$$

where c_0 is a constant. In fact, this relationship between the steady active stress and $\Delta[Ca^{2+}]$ was found experimentally (Ishizuka et al. 1993).

c_0 is determined by using the ratio of the mean increased calcium level $\Delta[Ca^{2+}]^{mean}$ to the mean active stress σ^a_{mean}, i.e., $c_0 = \Delta[Ca^{2+}]^{mean} / \sigma^a_{mean}$. σ^a_{mean} can be calculated from the average value of the active stress of 51 GB samples (σ^a_{mean} = 59 mmHg), but $\Delta[Ca^{2+}]^{mean}$ needs to be estimated from the experimental data for guinea pigs once more.

For the adult guinea pig GB, the mean intracellular free Ca^{2+} level in normal GB smooth muscle is about 82.5 M/l (Shen et al. 2007), and the mean intracellular free Ca^{2+} level in healthy guinea pig heart muscle is about 151 nM/l (Thompson et al. 2000). Therefore, the ratio of the mean free Ca^{2+} of smooth and heart muscles for guinea pig GB is about 0.546. For human myocardium, the resting intracellular free Ca^{2+} level is around 250 nM/l (Gwathmey and Hajjar 1990; Beuckelmann et al. 1992), thus the free $[Ca^{2+}]^{mean}$ in human GB smooth muscles could be around 250 × 0.546 = 136.6 nM/l. It is also known that the peak value of Ca^{2+} for the guinea pig GB smooth muscles is normally 17% higher than the resting level under the CCK-8S stimulus of 50 μM/l (Si et al. 2009). Applying the same percentage to human GB smooth muscles, the mean calcium level is $\Delta[Ca^{2+}]^{mean}$ = 136.6 × 0.17 = 23.2 nM/l.

PEAK APPARENT RATE CONSTANTS

In Case 2, the peak apparent rate constants f^{max}_{app} and g^{max}_{app} for the 51 GBs after the end of the CCK infusion are calculated by using Eqs. (7.17) and (7.19), and listed in Table 7.2 for all the patients. The other parameters, such as EF, q, σ^a_{max}, f^{max}_{app}, g^{max}_{app} and $\Delta[Ca^{2+}]_{max}$. In Case 1, since

TABLE 7.2 Peak active stress, maximum rate constants and maximum increased [Ca^{2+}] level.

GB	EF (%)	q (min^{-1})	σ^{a}_{max} (mmHg)	f^{max}_{app} (s^{-1})	g^{max}_{app} (s^{-1})	$\Delta[Ca^{2+}]_{max}$ (nM/L)	CCK Test	VAS
1	4.5	0.0327	19	0.0163	0.0049	7.47	pain(-)	N/A
2	5.4	0.0570	56	0.0329	0.0141	22.02	pain(-)	N/A
3	11.4	0.0498	37	0.0312	0.0093	14.55	pain(-)	N/A
4	15.5	0.0619	55	0.0469	0.0140	21.63	pain(-)	N/A
5	10.7	0.0270	48	0.0406	0.0121	18.87	pain(+)	34
6	10.0	0.0399	67(+)	0.0572(+)	0.0171	26.35(+)	pain(+)	54
7	14.0	0.0429	46	0.0390	0.0116	18.09	pain(-)	N/A
8	21.9	0.0577	90(+)	0.0763(+)	0.0228	35.39(+)	pain(+)	71
9	16.1	0.0418	70(+)	0.0595(+)	0.0178	27.53(+)	pain(+)	N/A
10	5.4	0.0398	96(+)	0.0814(+)	0.0243	37.75(+)	pain(-)	N/A
11	15.1	0.0613	152(+)	0.1295(+)	0.0387	59.77(+)	pain(-)	N/A
12	21.3	0.0413	56	0.0477	0.0142	22.02	pain(-)	N/A
13	39.7	0.0384	62(+)	0.0526(+)	0.0157	24.38(+)	pain(+)	86
14	20.6	0.0406	96(+)	0.0817(+)	0.0244	37.75(+)	pain(+)	29
15	80.8	0.0515	61(+)	0.0523(+)	0.0156	23.99(+)	pain(+)	72
16	32.3	0.0258	13	0.0114	0.0034	5.11	pain(-)	N/A
17	32.4	0.0409	27	0.0227	0.0068	10.61	pain(+)	32
18	93.7	0.0370	22	0.0189	0.0056	8.65	pain(+)	34
19	49.4	0.0873	102(+)	0.0865(+)	0.0258	40.11(+)	pain(+)	50
20	48.7	0.0446	32	0.0271	0.0081	12.58	pain(+)	52
21	66.3	0.0572	47	0.0403	0.0120	18.48	pain(-)	N/A
22	54.4	0.0312	20	0.0166	0.0050	7.86	pain(+)	22
23	44.7	0.0582	126(+)	0.1073(+)	0.0321	49.55(+)	pain(+)	89
24	27.4	0.0507	61(+)	0.0518(+)	0.0155	23.99(+)	pain(-)	N/A
25	27.9	0.0690	120(+)	0.1021(+)	0.0305	47.17(+)	pain(+)	N/A
26	19.1	0.0531	84(+)	0.0716(+)	0.0214	33.03(+)	pain(+)	N/A
27	70.2	0.0429	32	0.0273	0.0081	12.58	pain(+)	37
28	71.5	0.0477	35	0.0298	0.0089	13.76	pain(-)	N/A
29	37.8	0.0442	31	0.0262	0.0078	12.19	pain(-)	N/A
30	91.8	0.0483	68(+)	0.0581(+)	0.0173	26.739(+)	pain(+)	76
31	100	0.0408	98(+)	0.0831(+)	0.0248	38.54(+)	pain(+)	67
32	10.1	0.0413	56	0.0477	0.0142	22.02	pain(-)	N/A
33	95.0	0.0470	42	0.0359	0.0107	16.52	pain(-)	N/A
34	28.4	0.0367	83(+)	0.0704(+)	0.0210	32.64(+)	pain(+)	68
35	100	0.0209	19	0.0166	0.0050	7.47	pain(+)	48
36	16.3	0.0440	50	0.0429	0.0128	19.66	pain(-)	N/A
37	26.3	0.0477	38	0.0296	0.0088	13.78	pain(-)	N/A
38	10.7	0.0543	76(+)	0.0648(+)	0.0194	29.88(+)	pain(+)	54

GB	EF (%)	q (min^{-1})	σ^{a}_{max} (mmHg)	f^{max}_{app} (s^{-1})	g^{max}_{app} (s^{-1})	$\Delta[Ca^{2+}]_{max}$ (nM/L)	CCK Test	VAS
39	8.0	0.0549	43	0.0371	0.0111	16.91	pain(+)	44
40	22.2	0.0298	24	0.0191	0.0057	9.44	pain(+)	78
41	36.4	0.0386	35	0.0204	0.0061	13.76	pain(+)	17
42	2.8	0.0888	107(+)	0.0911(+)	0.0272	42.07(+)	pain(-)	N/A
43	14.2	0.0554	201(+)	0.171(+)	0.0512	79.04(+)	pain(+)	N/A
44	15.2	0.0378	70(+)	0.0597(+)	0.0178	27.53(+)	pain(-)	76
45	9.1	0.0377	30	0.0255	0.0076	11.80	pain(-)	N/A
46	6.5	0.0472	40	0.0344	0.0103	15.73	pain(-)	N/A
47	0.34	0.0439	59	0.0502	0.0150	23.20	pain(-)	N/A
48	0.08	0.0442	80(+)	0.0678(+)	0.0203	31.46(+)	pain(+)	N/A
49	1.25	0.0252	40	0.0345	0.0103	15.73	pain(+)	N/A
50	8.75	0.0657	85(+)	0.0524(+)	0.0157	33.42(+)	pain(-)	N/A
51	1.62	0.0403	64(+)	0.0542(+)	0.0162	25.17(+)	pain(+)	N/A
Success rate				0.667	0.667	0.667		

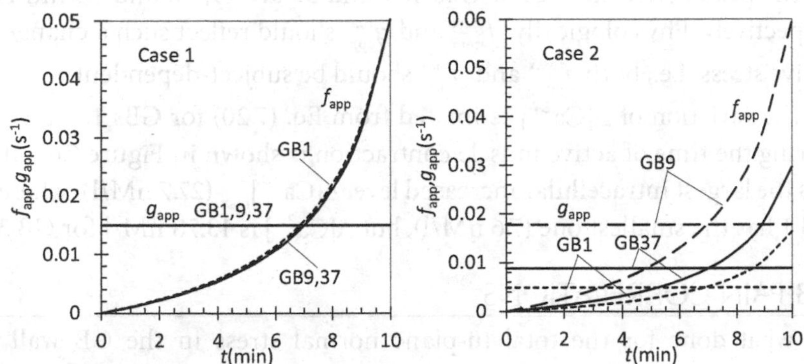

FIGURE 7.5 Attachment and detachment apparent rate constants f_{app} and g_{app} of GBs 1, 9 and 37 versus time for variable g^{max}_{app} (Case 2) and constant g^{max}_{app} (Case 1).

g^{max}_{app} = 0.015 s^{-1} and f^{max}_{app} = 0.05 s^{-1} are held, they are excluded in the table.

In Case 2, g^{max}_{app} varies with Eq. (7.19), accordingly, the peak rate constant f^{max}_{app} changes in the range of (0.011–0.17) s^{-1}, but also the f_{app} and g_{app} profiles with time vary from one subject to another as shown in Figure 7.5 for three typical GBs, such as GBs 1, 9 and 37. In Case 1, however, g^{max}_{app} and

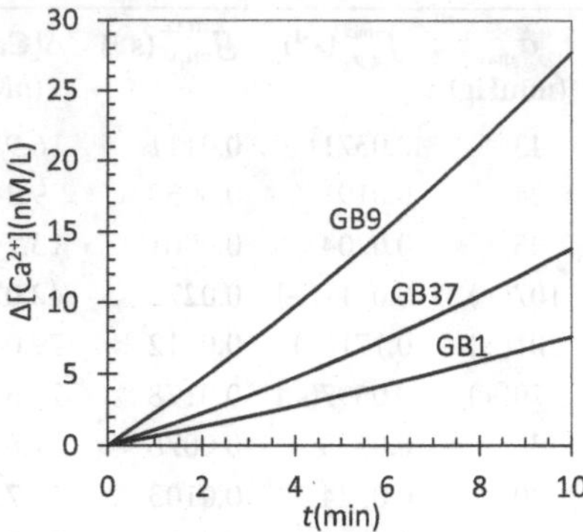

FIGURE 7.6 Increased $\Delta[Ca^{2+}]$ level as function of time in GBs 1, 9 and 37, respectively, for variable g_{app}^{max} (Case 2).

f_{app}^{max} are constant across all the patients. As a result, the f_{app} and g_{app} profiles are identical for all the patients.

The peak active stresses of GBs 1, 9 and 37 are 19, 70 and 38 mmHg, respectively. Physiologically, f_{app}^{max} and g_{app}^{max} should reflect such a change of active stress, i.e., both f_{app}^{max} and g_{app}^{max} should be subject-dependent.

The variation of $\Delta[Ca^{2+}]$ estimated from Eq. (7.20) for GBs 1, 9 and 37 during the time of active muscle contraction is shown in Figure 7.6. GB 9 has the largest intracellular increased level $\Delta[Ca^{2+}]_{max}$ (27.7 nM/l), whereas GB 1 has the smallest one (7.6 nM/l), but $\Delta[Ca^{2+}]$ is 13.78 nM/l for GB 37.

GB PAIN CORRELATIONS

As what done for the total in-plane normal stress in the GB wall in Chapters 5 and 6, the parameters of the smooth muscle contraction are correlated to the GB pain induced by the CCK test. These parameters are f_{app}^{max}, σ_{max}^{a} and $\Delta[Ca^{2+}]_{max}$. However, the difficulty here is that there is no threshold of pain for any of these parameters in the literature. Hence, the threshold from the model that corresponds to the threshold of the total peak normal stress when pain occurs has to be extracted, resulting the thresholds $[f_{app}^{max}] = 0.051\ s^{-1}$, $[\sigma_{max}^{a}] = 60$ mmHg and $[\Delta[Ca^{2+}]_{max}] = 23$ nM/l. The pain prediction results are listed in Table 7.2.

The success rates, defined as ratio of the number of patients with positive correlation to the total number of patients, are 0.667 for all the above three parameters. Generally, the pain correlation of these parameters is lower than using the peak total in-plane normal stress (>75% success rate; see Chapter 6). Thus, these three parameters are not a sensitive GB pain index.

GB motility is applied to describe GB emptying behaviour and can be assessed quantitatively by EF of bile in emptying (Wegstapel et al. 1999; Stads et al. 2006). The factors affecting GB motility include smooth muscle contraction and relaxation (Portincasa et al. 2004). The correlation of predicted peak active stress of 21 patients with biliary pain in CCK provocation test to EF is plotted in Figure 7.7. The figure seems to suggest the correlation between them is not remarked. In similarity, the correlation of f_{app}^{max} and g_{app}^{max} to EF is also the same.

Figure 7.8 illustrates the correlation of pain intensity with respect to the peak active stress and EF. The pain intensity appears positively to correlate

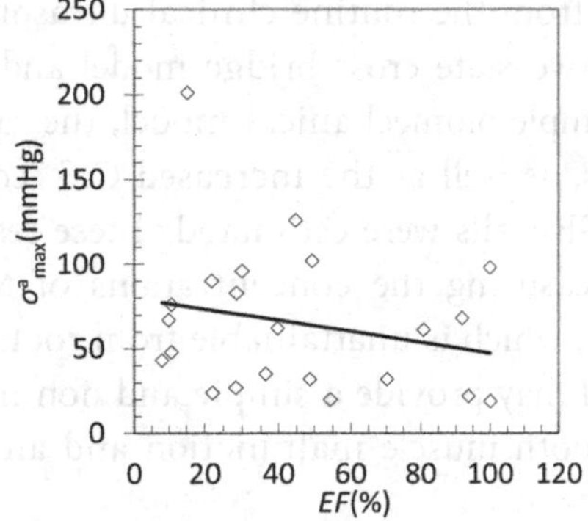

FIGURE 7.7 Peak active stress is correlated to EF of bile for 21 subjects with painful GB.

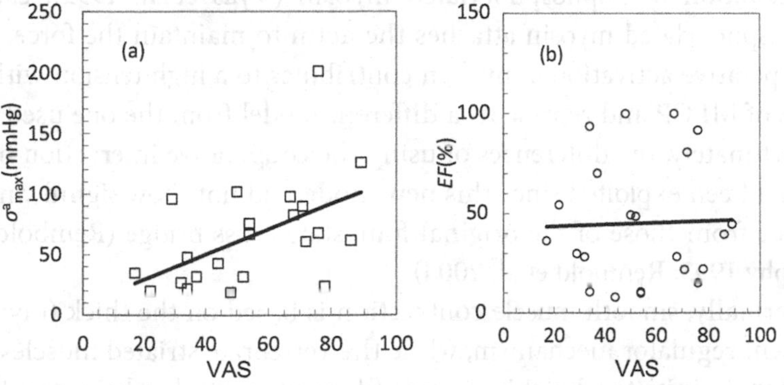

FIGURE 7.8 Correlation of VAS with respect to peak active stress (a) and EF (b) for 21 patients with GB pain identified in CCK provocation tests.

to the active stress rather than EF. This indicates that the active stress plays a more important role in GB pain compared to EF. Similarly, f_{app}^{max} and g_{app}^{max} also demonstrate a positive correlation to the biliary pain intensity.

In the study by Wegstapel et al. (1999), EF was positively correlated to the mean maximum contraction of GB strips *in vitro* but VAS was not. These observations did not agree with the present results. This is because our results are based on *in vivo* GB images taken by ultrasonography and the flow resistance effects in the CD were involved. *In vitro* experiments, such effects were excluded.

SIGNIFICANCE OF THE MODEL

A model was proposed to extract the cross-bridge information of human GB smooth muscle from the routine clinical ultrasonography scans. By using the Huxley's two-state cross-bridge model and the active stresses estimated from a simple biomechanical model, the rate constants of the cross-bridge models, as well as the increased Ca^{2+} concentration of the smooth muscle in GB walls were estimated. These results were achieved without directly measuring the concentrations of Mp + AMp in the smooth muscle cells, which is unattainable from routine clinical diagnoses. Thus, the model may provide a simple and non-invasive way of estimating possible smooth muscle malfunction and an additional clinical assessment tool.

The model used is based on the cross-bridge model proposed by Hai and Murphy (1988a). However, recent work suggests that there may be a cooperative interaction among the contractile proteins in rabbit portal vein smooth muscle. The phosphorylation of myosin light chain results in activation of unphosphorylated myosin (Vyas et al. 1992), causing dephosphorylated myosin attaches the actin to maintain the force. Such a cooperative activation of myosin contributes to a high tension with low levels of MLCP and represents a different model from the one used here. Unfortunately, the differences of using the cooperative interaction model has not been exploited since this new model did not show significant difference from those of the original four-state cross bridge (Rembold and Murphy 1993; Rembold et al. 2004).

Normally, smooth muscle contraction is based on the thick (myosin)-filament regulator mechanism, while the vertebrate striated muscles contraction is initiated by thin (actin)-filament control, where regulatory proteins (troponin and tropomyosin) limit the cross-bridge cycling until

Ca^{2+} binds to troponin (Somlyo et al. 1988). However, in the absence of Ca^{2+}, vertebrate smooth muscle contraction appears to be thin-filament regulated and cooperative (Somlyo et al. 1988; Haeberle 1999), which is achieved by the PKC pathway (Morgan and Gangopadhyay 2001). It has been tested that for GB smooth muscle stimulated by a high CCK dose the pathway is thick-filament regulation. At a low-dose CCK, because the thin-filament PKC pathway is very sensitive to low Ca^{2+} concentration, it adopts thin-filament mechanism (Yu et al. 1994, 1998).

As an extension to their previous four-state cross bridge with an ultra-slow latch-bridge cycle (Hai and Murphy 1988a), Hai and Kim (2005) proposed a new model that can incorporate the thin-filament-based regulation. The new model can achieve a steadily growing isometric active stress with decreasing MLCP concentration against time as shown the *in vitro* experiments for the airway smooth muscle stimulated by 1 μM phorbol dibutyrate (PDBu), which is a specific stimulant for PKC pathway. Thus, this model seems to be able to deal with both thick-filament and thin-filament regulatory mechanisms.

This model was not used in this chapter as the premier interest here was to investigate patients given a high dose of CCK in the CCK provocation tests. Caution is required in defining the high-dose CCK here, as CCK injection of 0.05 μg/kg was performed to patients, which is not directly convertible to the high dose (10^{-2} mM/l) used in laboratories. However, the CCK concentration based on the estimated volume of blood contained per kilogram in men and women could be calculated and the dose of CCK-10 was given. It suggests that the concentrations are likely similar to the high dose by Yu et al. (1994).

The apparent rate constants f_{app}^{max} and g_{app}^{max} in Table 7.1 stand for the averaged behaviour of single cross bridge in smooth muscle (Lecarpentier et al. 2002). They are sensitive to species, i.e., type of smooth muscles, presumably because of variable myosin concentration or degree of phosphorylation of myosin required in the smooth muscle cells. The apparent rate constants estimated here appear to support this observation, although more experiments are needed for acquisition of data.

The estimation of the Ca^{2+} concentration in the current model may be validated when experimental data become available. For example, it was confirmed experimentally that the apparent rate constant g_{app} of skinned rabbit adductor magnus fibres is Ca^{2+} concentration-dependent during isometric contraction (Kerrick et al. 1991). For smooth muscles, however, such dependence is yet to be observed.

REFERENCES

Beuckelmann, D. J., M. Nabauer, and E. Erdmann. 1992. Intracellular calcium handling in isolated ventricular myocytes from patients with terminal heart failure. *Circulation* 85:1046–1055.

Brenner, B. 1988. Effect of Ca^{2+} on cross-bridge turnover kinetics in skinned single rabbit psoas fibers: Implications for regulation of muscle contraction. *Proceedings of the National Academy of Sciences of the United States of America* 85:3265–3269.

Brotschi, E. A., K. C. Crocker, A. N. Gianitsos, et al. 1989. Effect of low extracellular calcium on gallbladder contraction in vitro. *Digestive Diseases and Sciences* 34:360–366.

Bursztyn, L., O. Eytan, A. J. Jaffa, et al. 2007. Mathematical model of excitation-contraction in a uterine smooth muscle cell. *American Journal of Physiology-Cell Physiology* 292:C1816–C1829.

Gwathmey, J. K., and R. J. Hajjar. 1990. Relation between steady-state force and intracellular [Ca2+] in intact human myocardium: Index of myofibrillar responsiveness to Ca^{2+}. *Circulation* 82:1266–1276.

Haeberle, J. 1999. Thin-filament linked regulation of smooth muscle myosin. *Journal of Muscle Research and Cell Motility* 20:363–370.

Hai, C. M. 1991. Length-dependent myosin phosphorylation and contraction of arterial smooth muscle. *Pflügers Archive-European Journal of Physiology* 418:564–571.

Hai, C. M., and H. R. Kim. 2005. An expanded latch-bridge model of protein kinase C-mediated smooth muscle contraction. *Journal of Applied Physiology* 98:1356.

Hai, C. M., and R. A. Murphy. 1988a. Cross-bridge phosphorylation and regulation of latch state in smooth muscle. *American Journal of Physiology-Cell Physiology* 254:99–106.

Hai, C. M., and R. A. Murphy. 1988b. Regulation of shortening velocity by cross-bridge phosphorylation in smooth muscle. *American Journal of Physiology-Cell Physiology* 255:C86–C94.

Hai, C. M., and R. A. Murphy. 1989. Cross-bridge dephosphorylation and relaxation of vascular smooth muscle. *American Journal of Physiology-Cell Physiology* 256:C282–C287.

Harnett, K. M., W. Cao, and P. Biancani. 2005. Signal-Transduction pathways that regulate smooth muscle function I. Signal transduction in phasic (esophageal) and tonic (gastroesophageal sphincter) smooth muscles. *American Journal of Physiology-Gastrointestinal and Liver Physiology* 288:G407–G416.

Huxley, A. 1957. Muscle structure and theories of contraction. *Progress in Biophysics and Biophysical Chemistry* 7:255–318.

Ishizuka, J., M. Murakami, G. A. Nichols, et al. 1993. Age-related changes in gallbladder contractility and cytoplasmic Ca2+ concentration in the guinea pig. *American Journal of Physiology-Gastrointestinal and Liver Physiology* 264:G624–G629.

Kamm, K. E., and J. T. Stull. 1985a. The function of myosin and myosin light chain kinase phosphorylation in smooth muscle. *Annual Review of Pharmacology and Toxicology* 25:593–620.

Kamm, K. E., and J. T. Stull. 1985b. Myosin phosphorylation, force, and maximal shortening velocity in neurally stimulated tracheal smooth muscle. *American Journal of Physiology-Cell Physiology* 249:C238–C247.

Kerrick, W. G., J. D. Potter, and P. E. Hoar. 1991. The apparent rate constant for the dissociation of force generating myosin crossbridges from actin decreases during Ca 2+ activation of skinned muscle fibres. *Journal of Muscle Research and Cell Motility* 12:53–60.

Kroon, M. 2010. A constitutive model for smooth muscle including active tone and passive viscoelastic behaviour. *Mathematical Medicine and Biology* 27:129–155.

Lecarpentier, Y., F. X. Blanc, S. Salmeron, et al. 2002. Myosin cross-bridge kinetics in airway smooth muscle: A comparative study of humans, rats, and rabbits. *American Journal of Physiology-Lung Cellular and Molecular Physiology* 282:L83–L90.

Lee, K. Y., P. Biancani, and J. Behar. 1989. Calcium sources utilized by cholecystokinin and acetylcholine in the cat gallbladder muscle. *American Journal of Physiology-Gastrointestinal and Liver Physiology* 256:785–G788.

Li, W. G., X. Y. Luo, N. A. Hill, et al. 2011. Corss-bridge apparent rate constants of human gallbladder smooth muscle. *Journal of Muscle Research and Cell Motility* 32:209–220.

Mbikou, P., A. Fajmut, M. Brumen, et al. 2011. Contribution of Rho kinase to the early phase of the calcium—contraction coupling in airway smooth muscle. *Experimental Physiology* 96:240–258.

Meiss, R. A. 1975. Graded activation in rabbit mesotubarium smooth muscle. *American Journal of Physiology* 229:455–465.

Morgan, K. G., and S. S. Gangopadhyay. 2001. Invited review: Cross-bridge regulation by thin filament-associated proteins. *Journal of Applied Physiology* 91:953–962.

Murtada, S. I., M. Kroon, and G. A. Holzapfel. 2010. A calcium-driven mechanochemical model for prediction of force generation in smooth muscle. *Biomechanics and Modeling in Mechanobiology* 9:749–962.

Portincasa, P., A. Di Ciaula, N. G. P. van Berge-Hemegouwe. 2004. Smooth muscle function and dysfunction in gallbladder disease. *Current Gastroenterology Reports* 6:151–162.

Ratz, P. H., C. M. Hai, and R. A. Murphy. 1989. Dependence of stress on cross-bridge phosphorylation in vascular smooth muscle. *American Journal of Physiology-Cell Physiology* 256:C96–C100.

Rembold, C. M., and R. C. Murphy. 1993. Models of the mechanism for cross-bridge attachment in smooth muscle. *Journal of Muscle Research and Cell Motility* 14:325–334.

Rembold, C. M., R. L. Wardle, C. J. Wingard, et al. 2004. Cooperative attachment of cross bridges predicts regulation of smooth muscle force by myosin phosphorylation. *American Journal of Physiology-Cell Physiology* 287:C594–C602.

Renzetti, L. M., M. B. Wang, and J. P. Ryan. 1990. Contribution of intracellular calcium to gallbladder smooth muscle contraction. *American Journal of Physiology-Gastrointestinal and Liver Physiology* 259:G1–G5.

Robertson, S. P., J. D. Johnson, and J. D. Potter. 1981. The time-course of Ca2+ exchange with calmodulin, troponin, parvalbumin, and myosin in response to transient increases in Ca^{2+}. *Biophysical Journal* 34:559–569.

Shaffer, E., A. Bomzon, H. Lax, et al. 1992. The source of calcium for CCK-induced contraction of the guinea-pig gall bladder. *Regulatory Peptides* 37:15–26.

Shen, P., B. J. Fang, P. T. Zhu, et al. 2007. Effect of traditional Chinese herbs for nourishing the liver on intracellular free calcium level in gallbladder cells of guinea pigs with gallstones. *Journal of Chinese Integrative Medicine* 5:179–182.

Si, X., L. Huang, H. Luo, et al. 2009. Inhibitory effects of somatostatin on cholecystokinin octapeptide induced bile regurgitation under stress: Ionic and molecular mechanisms. *Regulatory Peptides* 156:34–41.

Singer, H. A., and R. A. Murphy. 1987. Maximal rates of activation in electrically stimulated swine carotid media. *Circulation Research* 60:438–445.

Smythe, A., A. W. Majeed, M. Fitzhenry, et al. 1998. A requiem for the cholecystokinin provocation test? *Gut* 43:571–574.

Solaro, R. J., and H. M. Rarick. 1998. Troponin and tropomyosin: Proteins that switch on and tune in the activity of cardiac myofilaments. *Circulation Research* 83:471–480.

Somlyo, A. V., Y. E. Goldman, T. Fujimori, et al. 1988. Cross-bridge kinetics, cooperativity, and negatively strained cross-bridges in vertebrate smooth muscle: A laser-flash photolysis study. *Journal of General Physiology* 91:165–192.

Stads, S., N. G. Venneman, R. C. Scheffer, et al. 2006. Evaluation of gallbladder motility: Comparison of two-dimensional and three dimensional ultrasonography. *Annals of Hepatology* 6:164–169.

Stålhand, J., A. Klarbring, and G. A. Holzapfel. 2008. Smooth muscle contraction: Mechanochemical formulation for homogeneous finite strains. *Progress in Biophysics and Molecular Biology* 96:465–481.

Thompson, M., A. Kliewer, D. Maass, et al. 2000. Increased cardiomyocyte intracellular calcium during endotoxin-induced cardiac dysfunction in guinea pigs. *Pediatric Research* 47:669–676.

Vyas, T. B., S. U. Mooers, S. R. Narayan, et al. 1992. Cooperative activation of myosin by light chain phosphorylation in permeabilized smooth muscle. *American Journal of Physiology-Cell Physiology* 263:C210–C219.

Washabau, R. J., M. B. Wang, C. Dorst, et al. 1994. Role of myosin light-chain phosphorylation in guinea pig gallbladder smooth muscle contraction. *American Journal of Physiology-Gastrointestinal and Liver Physiology* 266:469–474.

Wegstapel, H., N. C. Bird, R. Chess-Williams, et al. 1999. The relationship between in vivo emptying of the gallbladder, biliary pain, and in vitro contractility of the gallbladder in patients with gallstones: Is biliary colic muscular in origin? *Scandinavian Journal of Gastroenterology* 34:421–425.

Wong, A. 1972. Mechanics of cardiac muscle based on Huxley's model: Simulation of active state and force-velocity relation. *Journal of Biomechanics* 5:107–117.

Wong, A. 1973. Myocardial mechanics: Application of sliding-filament theory to isovolumic concentration of the left ventricle. *Journal of Biomechanics* 6:565–581.

Yang, J., J. W. Clark, R. M. Btyan, et al. 2003a. The myogenic response in isolated rat cerebrovascular arteries: Smooth muscle cell model. *Medical Engineering and Physics* 25:691–709.

Yang, J., J. W. Clark, R. M. Btyan, et al. 2003b. The myogenic response in isolated rat cerebrovascular arteries: Vessel model. *Medical Engineering and Physics* 25:711–717.

Yu, P., Q. Chen, Z. Xiao, et al. 1998. Signal transduction pathways mediating CCK-induced gallbladder muscle contraction. *American Journal of Physiology-Gastrointestinal and Liver Physiology* 275:G203–G211.

Yu, P., G. De Petris, P. Biancani, et al. 1994. Cholecystokinin-coupled intracellular signaling in human gallbladder muscle. *Gastroenterology* 106:763–770.

Yu, P., K. Harnett, K. Harnett, et al. 1993. Interaction between signal transduction pathways contributing to gallbladder tonic contraction. *American Journal of Physiology-Gastrointestinal and Liver Physiology 265*:G1082-G1089.

Yu, S. N., P. E. Crago, and H. J. Chiel. 1997. A nonisometric kinetic model for smooth muscle. *American Journal of Physiology-Cell Physiology* 272:C1025–C1039.

Zulliger, M., A. Rachev, and N. Sterqiopulos. 2004. A constitutive formulation of arterial mechanics including vascular smooth muscle tone. *American Journal of Physiology-Heart and Circulatory Physiology* 287: H1335–H1343.

CHAPTER 8

Quasi-Nonlinear Analysis of the Anisotropic GB Wall

ELASTOGRAPHY OF SOFT BIO-TISSUE

Soft bio-tissue biomechanical property, i.e., elasticity can indicate pathological changes caused from disease, namely carcinoma of the skin (Tilleman et al. 2004), plaque (Schulze-Bauer and Holzapfel 2003; Karimi et al. 2008) and ageing problems (Escoffier et al. 1989; Lee et al. 2010). As a result, the property is potentially applicable in clinical diagnoses (Samani and Plewes 2007). How to evaluate biomechanical properties of soft bio-tissue from medical images has been an increasingly researched topic, resulting in a study, named elastography.

Unusually, one solves an inverse problem for the stress-free configuration or biomechanical property to be determined when the boundary conditions, load and strain or displacement field are known. In a direct problem, however, the biomechanical property, boundary conditions, load and stress-free configuration are known, and one solves the problem for the strain or displacement. Elastography is the inverse problem for determining the biomechanical property of bio-soft tissue.

There is another sort of inverse problem, known as elastostatics, where one seeks the stress-free configuration with known biomechanical properties, strain or displacement field, boundary conditions and loads (Iding et al. 1974; Govindjee and Mihalic 1996, 1998; Lu et al. 2007a,b, 2008;

Gee et al. 2009; Zhou et al. 2010). In this chapter, the elastography approach is concerned.

In elastography, one estimates either elastic moduli (Guo et al. 2010) or a general linear elasticity tensor (Raghupathy and Barocas 2010) directly from solving the equilibrium equations in an iterative manner. Normally, this requires solving direct linear problems repeatedly with successively updated material property constants. The computed strain or displacement field is then used to match the experimental observations (or medical images). This process is continued until an minimum error is achieved in the difference of the estimated quantities from observations (Moulton et al. 1995; Kauer et al. 2002; Seshaiyer and Humphrey 2003; Bosisio et al. 2007; Lei and Szeri 2007; Samani and Plewes 2007; Gokhale et al. 2008; Karimi et al. 2008; Li et al. 2009; Balocco et al. 2010; Kroon 2010).

In stimulations with CCK (CCK-8), a peptide hormone which stimulates the GB muscle to contract and release bile from the GB lumen, the strips from a human GB wall in the circumferential direction demonstrated a slightly stiffer active response than in the longitudinal direction (Bird et al. 1996). Additionally, passive tensile tests were conducted on the strips harvested from the GBs of a human (Su 2005; Karimi et al. 2017) and a pig (Xiao et al. 2013). Different Young's moduli in the strips isolated from the body of GB samples in the circumferential and longitudinal orientations have been demonstrated primarily.

Those measurements suggest that the biomechanical property of the human GB wall is anisotropic and nonlinear. However, this anisotropic and nonlinear behaviour of the human GB wall has thus far been ignored (Liao et al. 2004; Li et al. 2008, 2011).

In this chapter, a quasi-linear elastography inverse approach was used to estimate GB wall elasticity. This means that linear analyses on a series of configurations of human GBs determined by routine ultrasonographic images in the emptying (Smythe et al. 1998). The transmural pressure in GBs, which was estimated from measured volume changes in Chapters 5 and 6, is applied as the loading condition. The computed displacements at several observation points are adjusted against those from clinical measurements to minimise the error. This analysis leads to strain-dependent incremental Young's moduli to characterise the GB wall biomechanical behaviour.

COMPUTATIONAL MODEL

Histological studies delineate a three-layer structure in the wall of the GB of the sea cow, a mammalian species (Caldwell et al. 1969). The mucosal layer is with lamina propria and folds projecting into the GB lumen. The perimuscular layer is covered by serous mesothelium. The muscular layer consists of enormous bundles of smooth muscle cells oriented circumferentially and longitudinally. The layer is responsible for the development of passive and active stresses. Electron microscopy of a transverse longitudinal section of a GB wall presents two large smooth muscle bundles running almost orthogonal to each other in the muscle layer. For the human GB wall, there are three layers such as mucosa, fibromuscular layer, adventitia and serosa, which mainly consist of collagen fibres and fibroblasts as well as blood vessels (Thounaojam et al. 2017). To simplify the model, it is supposed that those layers of the human GB wall form a homogeneous, nonlinear and anisotropic material as a whole.

In the emptying under intravenous infusion of CCK (0.05 μg/kg body weight) (Smythe et al. 1998), GB geometries were monitored by using ultrasonography at different time instants. Based on the images taken, the ellipsoidal membrane models can be generated as described in Chapters 5 and 6. These ellipsoids have three axes with the lengths denoted by D_1, D_2, D_3 and a constant thickness of h_{GB} (see Figure 8.1).

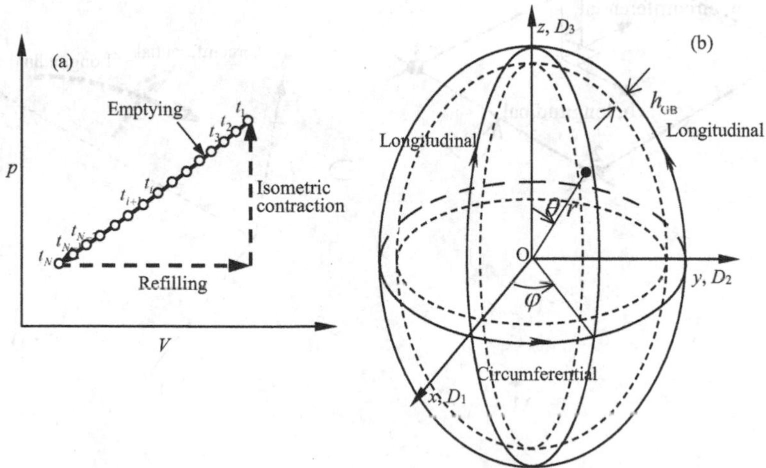

FIGURE 8.1 GB pressure-volume (p-V) diagram (a) and ellipsoid model (b).

The transmural pressure of the GB can be estimated from a 1D model (see Chapter 5):

$$p = p_d + (p_e - p_d)e^{([(t_e - t)/RC]} \tag{8.1}$$

where p_e and p_d are the mean pressures in the GB and duodenum at the end of emptying, t_e is the total time taken for emptying the bile, and C is the GB compliance, the ratio of the incremental volume to the incremental pressure, and R is the flow resistance when the bile streams in CD and CBD.

The direct problem is solved using the FEA package ADINA 8.7.2. Although the material of the human GB wall is nonlinear, it is assumed that the material parameters are constant between each increment of pressure loading. In a typical 4-node membrane element, the definitions of the orthotropic material and local material Cartesian coordinate systems are shown in Figure 8.2.

A spherical coordinate system (θ, φ, r) with the origin O, as shown in Figure 8.1(b), is used for presenting the GB geometry, and the material

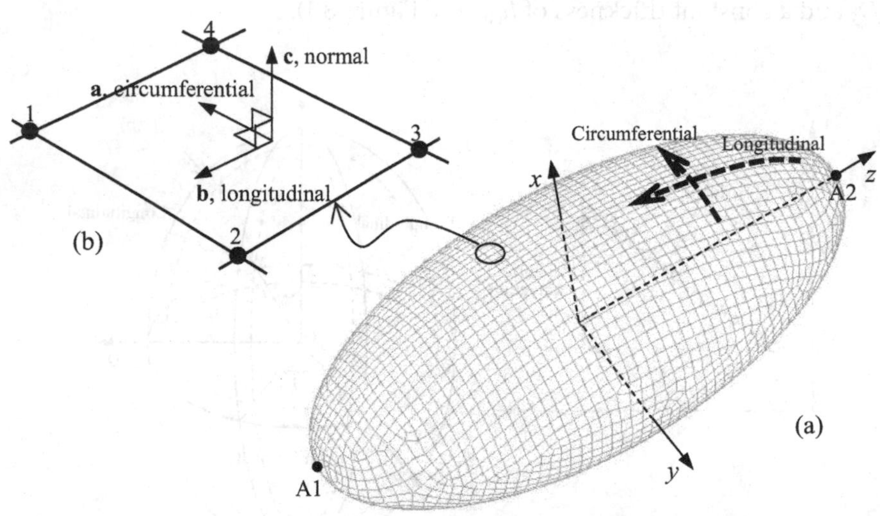

FIGURE 8.2 The mesh on the GB wall (a) and a typical 4-node membrane element used (b) along with the material local Cartesian coordinate system (**a**, **b**, **c**), the boundary conditions are applied at the two apexes A1 and A2.

coordinates (**a, b, c**) located on the mid-surface of the GB are chosen to be in the local θ, φ and their normal n directions.

The GB wall behaves like a membrane, at any time interval t_i to t_{i-1} during a pressure loading, and the constitutive relation is

$$\begin{bmatrix} \varepsilon_{\theta\theta} \\ \varepsilon_{\varphi\varphi} \\ \varepsilon_{rr} \\ 2\varepsilon_{\theta\varphi} \\ 2\varepsilon_{\theta r} \\ 2\varepsilon_{\varphi r} \end{bmatrix} = \begin{bmatrix} 1/E_\theta & -\upsilon_{\varphi\theta}/E_\varphi & -\upsilon_{\varphi r}/E_r & 0 & 0 & 0 \\ -\upsilon_{\varphi\theta}/E_\theta & 1/E_\varphi & -\upsilon_{\varphi\varphi}/E_r & 0 & 0 & 0 \\ -\upsilon_{r\theta}/E_\theta & -\upsilon_{r\varphi}/E_\varphi & 1/E_r & 0 & 0 & 0 \\ 0 & 0 & 0 & 1/G_{\theta\varphi} & 0 & 0 \\ 0 & 0 & 0 & 0 & 1/G_{\theta r} & 0 \\ 0 & 0 & 0 & 0 & 0 & 1/G_{\varphi r} \end{bmatrix} \begin{bmatrix} \sigma_{\theta\theta} \\ \sigma_{\varphi\varphi} \\ 0 \\ \tau_{\theta\varphi} \\ 0 \\ 0 \end{bmatrix} \quad (8.2)$$

where υ_{lm} ($l, m = \theta, \varphi, r$) are the Poisson's ratios, ε_{ll}, σ_{ll} ($l = \theta, \varphi, r$) are the normal strains and stresses, respectively, $2\varepsilon_{lm}$, τ_{lm} ($l \neq m$) are the shear strains and stresses, respectively, and E_l and G_{lm} are the elastic and shear moduli, respectively. Furthermore, the wall material of GB is considered transversely isotropic, namely

$$\upsilon_{\theta\varphi} = \upsilon_{\varphi r} = \upsilon, E_\varphi = E_r, G_{\theta r} = \frac{E_\theta}{2(1+\upsilon)} \text{ and } G_{\theta r} = \frac{E_\varphi}{2(1+\upsilon)} \quad (8.3)$$

A FE mesh is shown in Figure 8.2(a) for GB 1, which was employed to check the mesh independence. When the maximum mesh size (edge length) is decreased from 2 to 1 mm, the maximum first principal stress increases from 8717 to 8752 Pa with only a 0.4% rise. Therefore, 1 mm was used as the maximum edge length for all other GB samples.

The Dirichlet boundary condition is applied at the two apexes of the GB, denoted as A1 and A2 as shown in Figure 8.2(a). Specifically, all the displacements are fixed at A1, and only the displacements in the x and y directions are fixed at A2.

COMPUTATIONAL PROCEDURE

A series of incremental linear analyses was carried out between successive two image configurations, working backwards in the time from the end to the start of GB emptying. The GB wall material property constants are given in each analysis. The incremental pressure loadings are defined as the following [see Figure 8.3(a)],

$$\Delta p_i = p_i - p_{i+1} \tag{8.4}$$

where $i = N - 1, N - 2, \ldots, 3, 2, 1$, and N is the total number of images. $i = 1$ and N denote the start and the end of emptying, respectively.

Such incremental pressure loadings result in a local incremental stress and strain that can figure out a local incremental Young's modulus between two equilibriums i and $i + 1$, see Figure 8.3(b), at a material point in the GB wall mid-surface. Clearly, a local incremental Young's modulus has nothing to do with the stress and strain themselves at very time instance in Figure 8.3(b), but the stress and strain differences between two instants. Consequently, the initial stress and strain can be ignored in each analysis, but the configuration should be updated continuously according to the deformed mesh in the last analysis. In this process, the configurations are updated throughout the emptying, but the initial configuration is at the end of emptying [see Figure 8.3(c)].

To estimate the stress and strain in an equilibrium state i at a material point, the incremental stress and strains can be summarised as follows:

$$\begin{cases} \sigma_i = \sigma_N + \Sigma_{l=N-1}^{i} \Delta\sigma_l \\ \varepsilon_i = \varepsilon_N + \Sigma_{l=N-1}^{i} \Delta\varepsilon_l \end{cases} \tag{8.5}$$

where $i = N - 1, N - 2, \ldots, 3, 2, 1$, σ_N and ε_N are the stress and strain components at the end of emptying. They are the initial stress and strain in the GB wall, and depend upon the stress-free configuration, material true property constants and the pressure loading p_N $(=p_e)$. Unfortunately, they cannot be predicted by the proposed FEA, but by the other method.

To figure out the initial stress, the residual strain is ignored and the configuration at the end of emptying is taken to be the stress-free configuration, then σ_N is estimated by means of the biomechanical model in Chapter 6 based on the GB geometry and pressure at the end of emptying. Both σ_N and ε_N are dead stress and strain because a GB always carries the

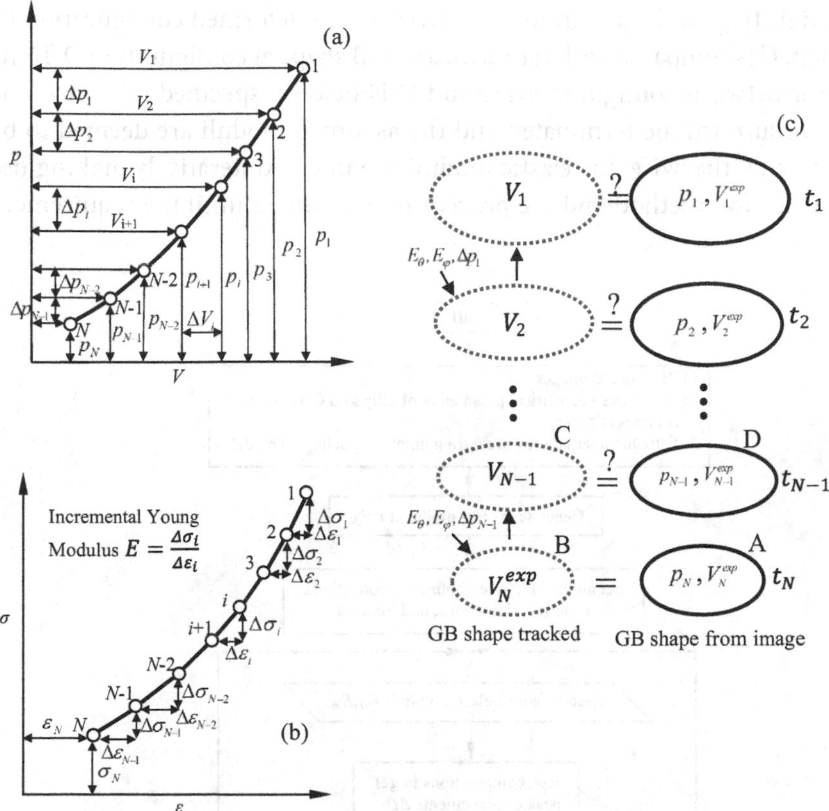

FIGURE 8.3 Incremental method for determining incremental Young's modulus under each incremental pressure loading, the computations are initiated from the initial configuration B, and proceeds to the start of emptying by continuously updating mesh node coordinates according to the results just obtained in each FEA, (a) incremental pressure-volume, (b) incremental stress-strain curve under a series of incremental pressure loadings at the material point of the GB wall surface, (c) computational procedure, V and $V^{\exp}$ are the estimated and measured GB volumes, respectively.

dead pressure loading p_N (=p_e). Both $\Sigma_{l=N-1}^{i}\Delta\sigma_l$ and $\Sigma_{l=N-1}^{i}\Delta\varepsilon_l$ are alive since they disappear after an emptying phase and reappear in the next emptying.

Incremental computations start with configuration B at $t = t_N$ in Figure 8.3(c), where B is assumed to be same as the measured GB shape at configuration A, and a linear FEA is performed with an assumed set of elastic moduli and an incremental pressure loading Δp_{N-1} as well as an initial stress and zero strain. This gives us the deformed configuration C. Then, C is compared with the measured GB shape at configuration D. If the error between configurations C and D is below a specified tolerance, the procedure will be terminated and the assumed moduli are deemed to be correct. Otherwise, the elastic moduli are updated iteratively making use of bisection method and the procedure is repeated until the requirement

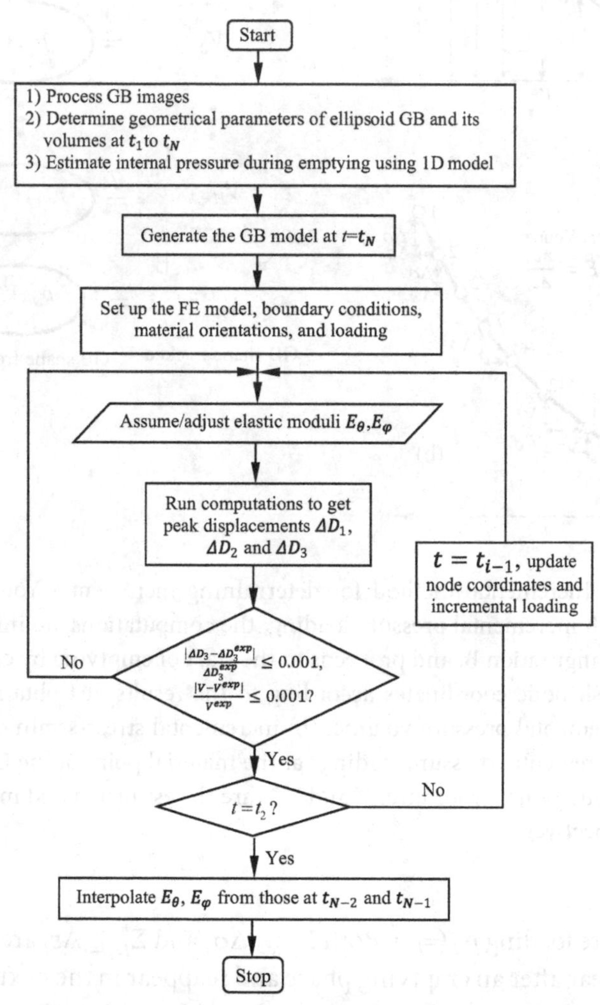

FIGURE 8.4 The flowchart of the inverse estimation for the elastic moduli, where $\Delta D_1^{\exp}$, $\Delta D_2^{\exp}$, $\Delta D_3^{\exp}$ and $V^{\exp}$ are the maximum displacements along the ellipsoid three axes and the GB bile volume from the images, those displacements were calculated from the current configuration and one at t_N.

is satisfied, then the mesh is output. Normally, 5–10 iterations are needed to achieve a convergent tolerance of 10^{-3}. A similar process is carried out for other emptying time intervals with a continuously updated mesh, until $t = t_1$. The material property constants at the final time instance t_N are simply interpolated from those at the time step t_{N-2} and t_{N-1}. The detailed procedure is illustrated in Figure 8.4.

Notice that in the computations, an incremental pressure is applied to the GB configurations updated at $t_{N-1}, t_{N-2}, \ldots, t_3, t_2$, respectively, based on the deformed mesh, and the elastic moduli estimated are different at each time step t_i ($i = N - 1, N - 2, \ldots, 3, 2, 1$). This gives rise to a quasi-linear stress-strain (σ-ε) relation as shown in Figure 8.3(b). The obtained parameters are the incremental elastic moduli as defined by Bergel (1961).

GB SAMPLES

Six GB samples, namely GB 1, 17, 19, 21, 30 and 37, are selected from 51 GBs samples in Chapter 5. These GBs differ from each other in size and emptying behaviour and a typical ultrasonographical image is exhibited in Figure 8.5. The detailed parameters are listed in Table 8.1. In all

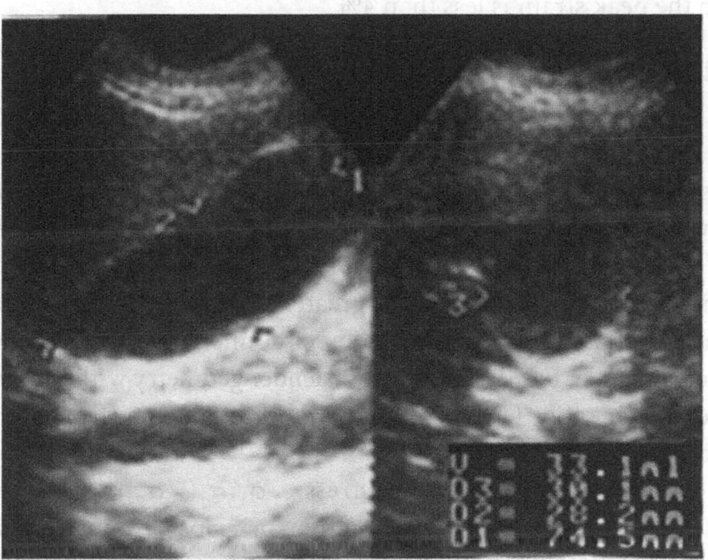

FIGURE 8.5 The ultrasonographical image of GB 21 at the start of emptying, three axes D_1, D_2 and D_3 are indicated, and they and GB volume are used to provide the comparison for the inverse approach, and the computed ellipsoid GB has the same axes within 0.1% error tolerance in D_3 and volume.

TABLE 8.1 Parameters of six human GB samples.

Item	Parameter	GB					
		1	**17**	**19**	**21**	**30**	**37**
Start of emptying	t_1(min)	0.0	0.0	0.0	0.0	0.0	0.0
	p_1 (Pa)	2032.8	2206.5	3512.4	2598.1	2377.7	2361.9
	D_1^1 (mm)	23.4	27.2	34.7	28.2	24.1	30.5
	D_2^1 (mm)	25.0	27.2	35.7	30.1	30.8	30.5
	D_3^1 (mm)	54.1	55.9	92.3	74.5	68.6	53.8
End of emptying	t_{15} (min)	615.3	80.4	55.3	32.9	18.2	105.4
	p_{15} (Pa)	1466.5	1466.5	1466.5	1466.5	1466.5	1466.5
	D_1^{15} (mm)	14.2	18.0	22.6	17.5	17.1	19.1
	D_2^{15} (mm)	15.4	17.5	24.7	20.4	18.5	22.7
	D_3^{15} (mm)	43.6	39.4	61.4	53.1	48.3	34.7
EF at 30 min		poorest	poor	good	good	best	poor

computations, the following parameters are kept used, i.e., $h_{GB} = 2.5$ mm, $p_e = 11$ mmHg, $p_d = 6$ mmHg, $C = 2.73$ ml/mmHg, see Chapters 5 or 6, and $v = 0.49$. Particularly, $N = 15$ was held, which ensures that the estimated error in the peak strain is less than 4%.

ESTIMATED ELASTIC MODULI

Figure 8.6 shows the longitudinal and circumferential elastic moduli, E_θ, E_φ and their ratio, E_φ/E_θ, for six GB samples. The estimated elastic moduli decrease almost linearly with the GB volume. In other words, GBs are stiffer when they are more stretched. This effect agrees with the behaviour of other soft bio-tissues (Fung 1967; Wu and Yao 1976). Furthermore, the ratio of elastic moduli, E_φ/E_θ, is greater than unity, implying that the human GB wall is anisotropic and stiffer circumferentially than longitudinally. In particular, GB 19 has the most slender geometry with the smallest initial diameter ratios (D_1^1/D_3^1, D_2^1/D_3^1) and presents the strongest anisotropic behaviour, i.e., the largest E_φ/E_θ.

The computed peak principal stresses $\sigma_1, \sigma_2 (= \sigma_\theta, \sigma_\varphi)$ are plotted in Figure 8.7 against the corresponding principal cumulative strains $\varepsilon_1, \varepsilon_2 (= \varepsilon_\theta, \varepsilon_\varphi)$. Both peak principal stress and strain are estimated based on the peak principal incremental stress $\Delta\sigma_i$ and strain $\Delta\varepsilon_i$ as well as the initial stress σ_N and strain ε_N by using Eq. (8.5). The initial stress σ_N is predicted approximately with the analytical biomechanical model in

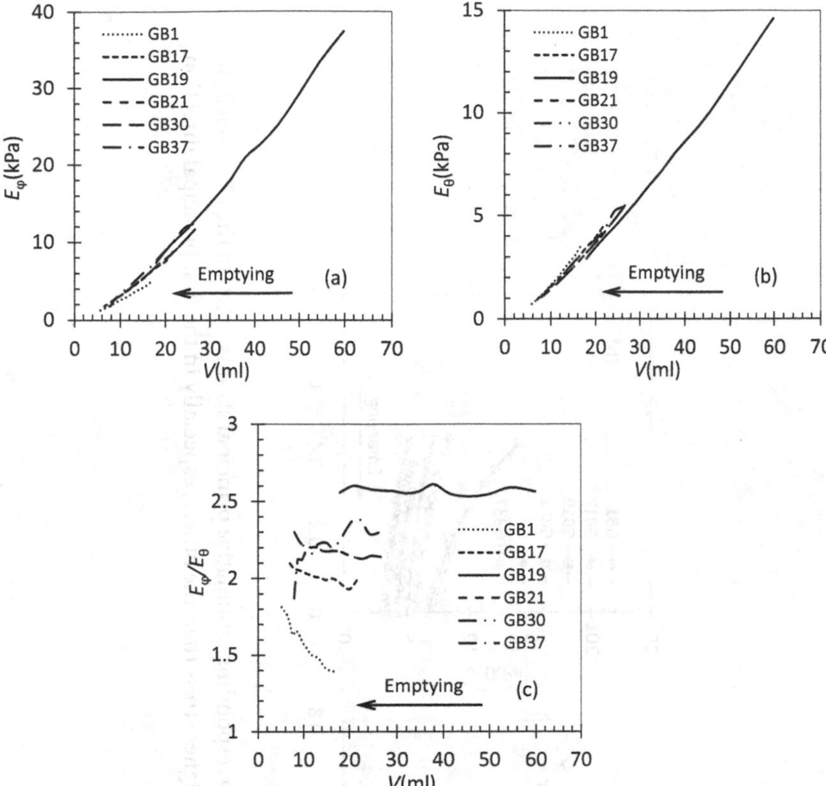

FIGURE 8.6 Elastic moduli in the circumferential and longitudinal directions and their ratios for six GB samples, (a) E_φ, (b) E_θ and (c) E_φ/E_θ.

Chapters 5 or 6 applied at the end of emptying without residual stress. The initial strain ε_N cannot be estimated, since the true biomechanical properties and residual strain in GB walls have remained unknown so far. The principal cumulative strains here are just the summaries of all the peak principal incremental strains.

To facilitate comparison with other published work, the stress-strain curves can be well fitted with exponential functions

$$\sigma_i = c_1 e^{c_2 \varepsilon_i}, \quad i = 1,2 \tag{8.6}$$

where c_1 and c_2 are the fitting parameters listed in Table 8.2.

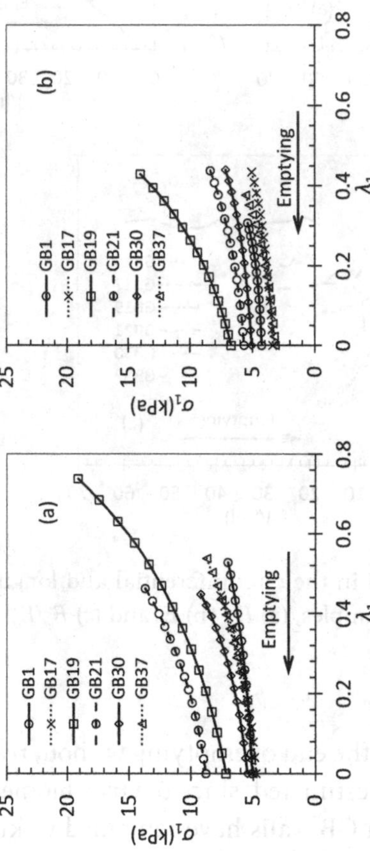

FIGURE 8.7 The peak principal stresses σ_1, σ_2 versus corresponding cumulative principal strains $\varepsilon_1, \varepsilon_2$ of the GB models, (a) σ_1-ε_1 curve, (b) σ_2-ε_2 curve, GB 19, 21 and 30 are subject to a higher stress than the others, especially in the first principal direction.

TABLE 8.2 Fitting parameters in Eq. (8.6).

Curve	Parameter	GB					
		1	17	19	21	30	37
σ_1-ε_1	c_1 (kPa)	4.9107	4.5685	6.8065	8.4885	5.2348	4.8119
	c_2 (-)	0.6233	1.0573	1.3267	0.9636	1.1649	1.0314
σ_2-ε_2	c_1 (kPa)	4.1447	3.2403	6.4665	5.4929	4.9155	3.0737
	c_2 (-)	0.7497	0.9342	1.7121	0.8898	0.7772	1.4573

PEAK DISPLACEMENTS, VOLUME AND STRESS PATTERNS

A comparison of the peak displacements between the images and computed GBs is shown in Figure 8.8. Obviously, good agreement between them is achieved, in particular for GBs 17, 21 and 37. For GBs 1, 19 and 30, however, there exist obvious discrepancies in the tracked changes in the short axes (ΔD_1 and ΔD_2), where the estimated ΔD_1 fits better with the measured ΔD_2 and vice versa.

GBs 17, 21 and 37 deform in a way, so that their cross section is more or less circular, i.e., they are more like an ellipsoid of revolution ($D_1 \approx D_2 < D_3$). This behaviour can be better described by the homogeneous constitutive model here, while the other samples deform with slightly ellipse cross section, and hence are not homogeneous. A heterogeneous constitutive model will be presented in Chapter 10.

A comparison of GB bile volume calculated from the GB images and that based on the quasi-nonlinear analysis is made Figure 8.9. Clearly, the tracked GB bile volume is in very good agreement with that from images at the same pressure. It is unsurprised because a constant compliance ($C = \mathrm{d}V/\mathrm{d}p$, the reciprocal of the slope of a curve in Figure 8.9) and a linear material have been applied in the model.

A detailed stress pattern on GB 37 at the start of emptying is shown in Figure 8.10 and compared with the results predicted with the analytical method for a linear membrane model in Chapters 5 or 6. The present results have similar stress patterns as those from the linear membrane models, but with slight difference in the maximum peak stress.

A summary of the peak principal stresses is listed in Table 8.3. The difference in the first principal stress, σ_1, is less than 40% for all the samples except GB 30. The poorest case is GB 30 where the GB shape cannot be tracked properly by the model; see GB 30 in Figure 8.8. The ratio between

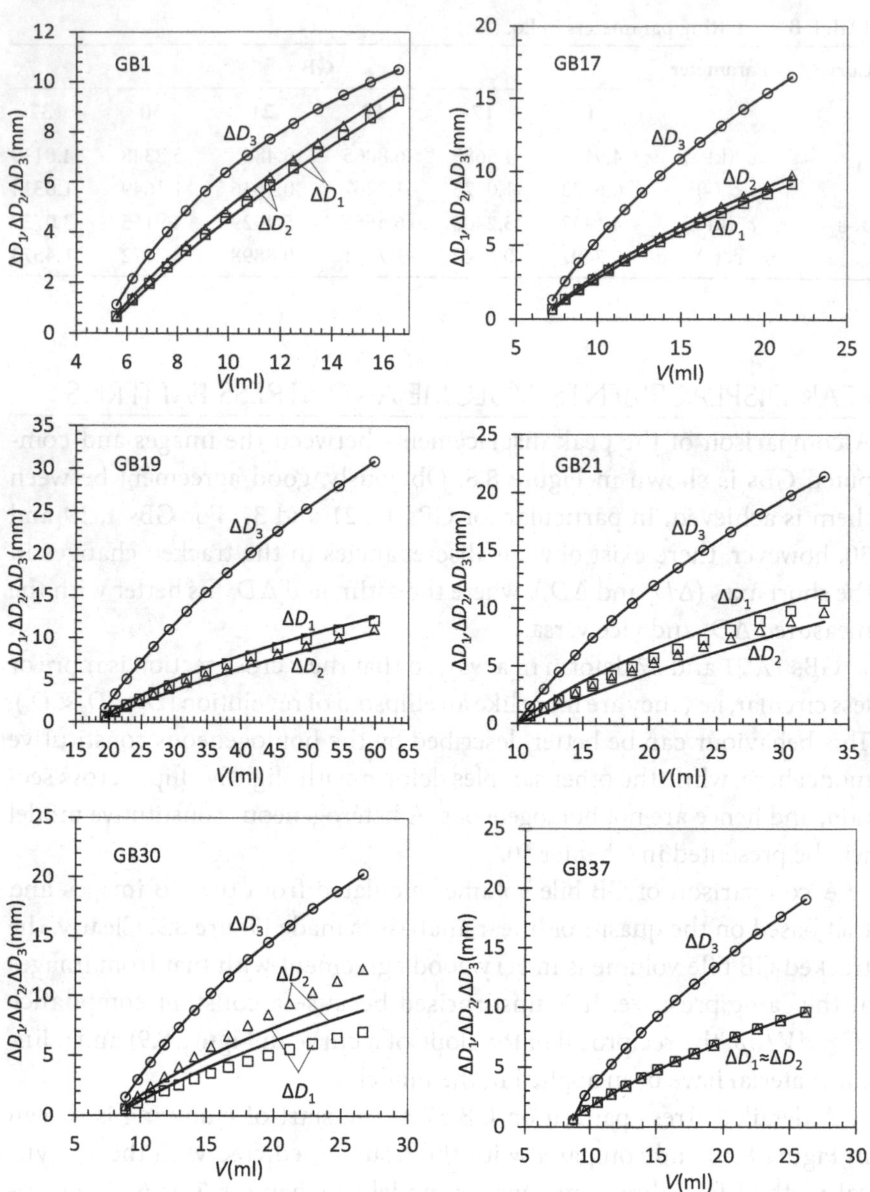

FIGURE 8.8 Comparison of the peak displacements along the three axes between images and computed results for six GB samples. The solid lines are displacements computed, the symbols are from images, square – ΔD_2, delta – ΔD_2, circle – ΔD_3.

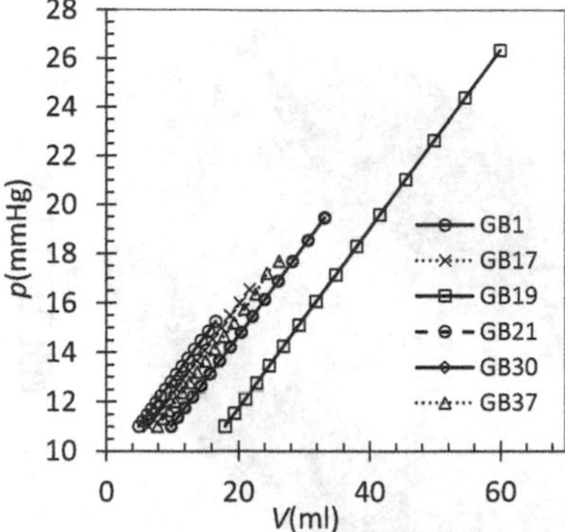

FIGURE 8.9 *p*-*V* diagram of GBs 1, 17, 19, 21, 30 and 37 based on GB bile volume calculated from images (line) and tracked one in a series of quasi-nonlinear analyses (symbols).

the present approach to the linear model in the second principal stress, σ_2, however, varies a bit significantly. The present model is homogeneous and cannot cope with the incremental elastic modulus to be different in the circumferential direction.

Additionally, in FEA, GB wall thickness in the initial configuration was assigned to be 2.5 mm. With computation preceding, the GB volume size increases and the thickness gets thinner. In the analytical solutions, the GB wall thickness always is 2.5 mm whatever the GB volume size is. Thus, the peak stresses in FEA should be higher in the analytical solutions. This issue will be made clear in Chapter 9.

GB COMPLIANCE

The load applied inside the GB wall relies on Eq. (8.1) that is dependent of GB compliance in a quasi-nonlinear analysis. To validate the equation, two kinds of experiments were conducted. One *in vitro* experiment was performed on four GBs removed from the patients. The GBs was stimulated with a certain amount of CCK initially, then water was infused into the GB, and the pressure and water volume are recorded simultaneously, eventually, a series of pressure-volume curves could be obtained. Figure 8.11(a) shows the compliances of these GBs estimated from the curves. The compliances seem volume-dependent but quiet noisy, and an averaged compliance $C = 2.843$ ml/mmHg is worked out, which is much

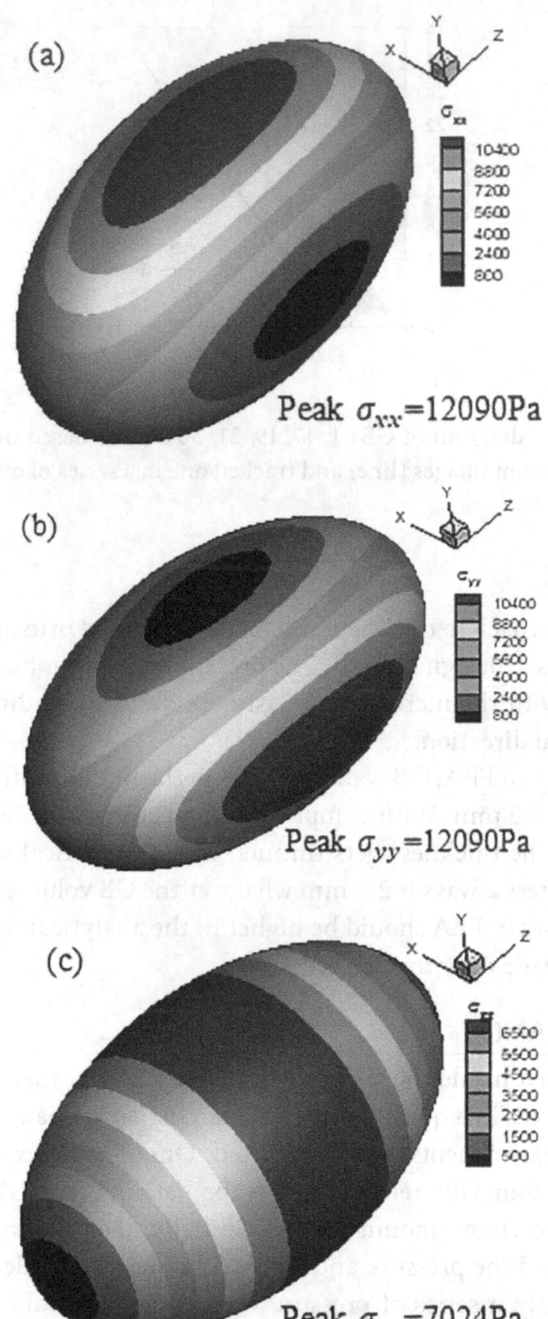

FIGURE 8.10 Comparison of the stress component distributions between the analytical solution of the linear model and the present approach for GB 37 at the start of emptying, (a)–(c) for analytical solutions, (d)–(f) for numerical results.

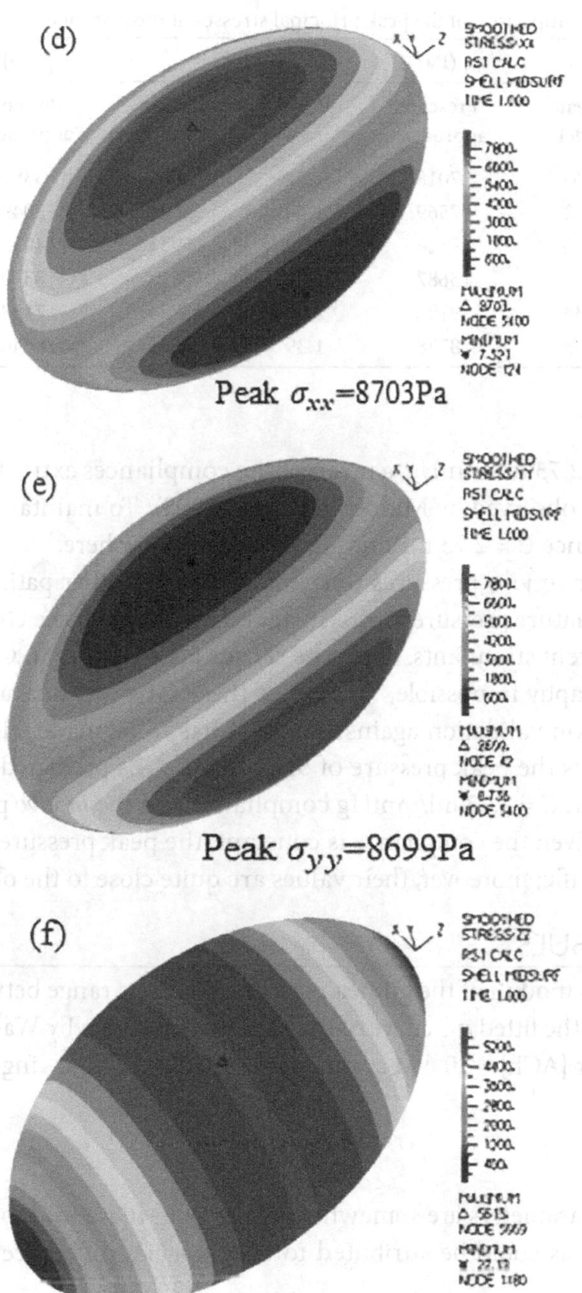

FIGURE 8.10 Comparison of the stress component distributions between the analytical solution of the linear model and the present approach for GB 37 at the start of emptying, (a)–(c) for analytical solutions, (d)–(f) for numerical results.

TABLE 8.3 Comparison of the peak principal stresses at the start of emptying.

GB	σ_1 (Pa)			σ_2 (Pa)		
	Linear model	Present approach	Ratio	Linear model	Present approach	Ratio
1	9170	7018	1.31	7907	5381	1.47
17	10582	7569	1.40	6001	5049	1.19
19	23265	19126	1.22	16951	14053	1.21
21	14478	13687	1.06	13578	8370	1.62
30	23733	9196	2.58	13934	7158	1.95
37	12093	8728	1.39	7204	5546	1.20

close to $C = 2.73$ ml/mmHg, a mean of the compliances extracted from the *in vivo* data observed by Middelfart et al. (1998). To maintain consistent, the compliance $C = 2.73$ ml/mmHg was kept in use here.

The other *in vivo* pressures were monitored on other patients' GBs by using a miniature pressure sensor installed inside when the GBs were subject to different stimulants. Since the sensor and catheter in GBs made an ultrasonography impossible, the size of these GBs was unavailable; thus, an indirection validation against Eq. (8.1) has to be made. Figure 8.11(b) demonstrates the peak pressure of 51 patients' GBs predicted in terms of Eq. (8.1) with $C = 2.73$ ml/mmHg compliance and the *in vivo* pressure just measured. Even the compliance is constant, the peak pressures are clearly patient-specific; moreover, their values are quite close to the observations.

OTHER ISSUES

The Young's moduli of the guinea pig GB wall strips range between 11 and 88 kPa, and the fitted σ-ε curve of the experimental data by Washabau et al. (1991) under [ACh] = 10^{-4} M stimulation, results the following relation

$$\sigma = 13.825e^{6.6375\varepsilon} \tag{8.7}$$

where the parameters are somewhat higher than these from our model in Table 8.2. This could be attributed to inter-species difference and guinea pig GB strips.

The exponential stress-strain relation similar to Eq. (8.6) has been established in many soft tissues in their passive state, e.g., porcine CBD (Duch et al. 2002; Duch et al. 2004), rat small intestine (Dou et al. 2002, 2006),

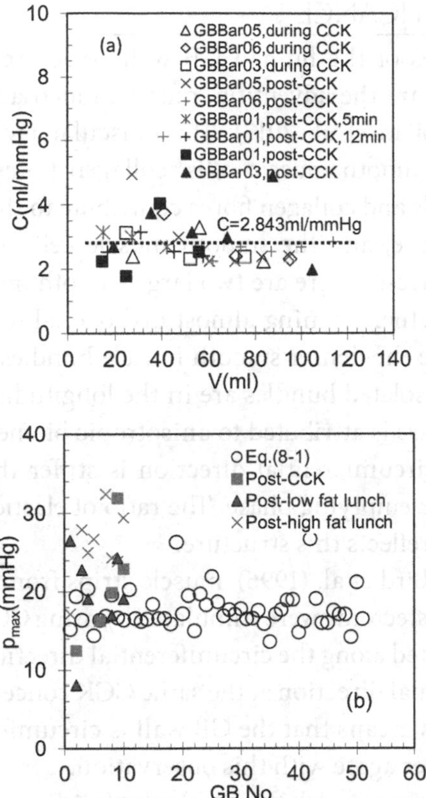

FIGURE 8.11 *In vitro* measured GB compliance of four GBs under CCK applied condition (a), the peak pressures *in vivo* observed for six patients and the peak pressures predicted by using Eq. (8.1) with C = 2.73 ml/mmHg for 51 GBs (b).

rat and rabbit stomach (Zhao et al. 2005), bladder (Colding-Jørgensen and Steven 1993) and cat/canine heart muscle (Yeatman et al. 1969; Mirsky and Parmley 1973).

At the start of GB emptying, the ratio of the active stress over the passive stress is found to be between 0.2 and 0.6 for human GBs, and it gradually drops off to zero as the emptying progresses. A similar ratio (~0.4) was found in the descending thoracic aortic wall of dog by Barra et al. (1993). Importantly, Barra et al. (1993) also reported that the total, active and passive stresses all were fitted well with an exponential function in the form of Eq. (8.6).

CLINICAL SIGNIFICANCE

Histological images of the human GB wall structure are illustrated in Figure 8.12. There are the adventitia, muscle, mucosa and epithelium in the wall (Thounaojam et al. 2017). The muscular layer is composed of many bundles of smooth muscle cells, collagen fibres and elastin. The smooth muscle cells and collagen fibres contribute to the passive response in the refilling phase, and the smooth muscle cells are responsible for generating active stress. There are two large smooth muscle bundles with collagen fibres together running almost orthogonal to each other in the muscle layer, where the denser smooth muscle bundles run circumferentially, and several isolated bundles are in the longitudinal direction. This histological structure is attributed to anisotropic biomechanical property and suggests the circumferential direction is stiffer than the longitudinal direction in the emptying phase. The ratio of elastic moduli $E_\varphi/E_\theta > 1$ determined above reflects this structure.

In the study by Bird et al. (1996), muscle strips from a human GB wall removed at cholecystectomy were stimulated by using CCK-8. The response of the strips harvested along the circumferential direction is stronger than along the longitudinal direction at the same CCK concentration, as shown in Figure 8.13. This means that the GB wall is circumferentially stiff. The results in the chapter agree with this observation.

The quasi-nonlinear model, the elastic moduli E_φ and E_θ hold constant in both the circumferential direction and in the longitudinal direction, respectively. Hence, the GB wall is homogeneous material along these directions in the model. Strictly speaking, the mechanical model of the GB wall proposed here is applicable to GB samples in an ellipsoid of revolution such as GBs 17, 21 and 37 only.

However, it was shown that the response of the strips cut from a GB neck was significantly smaller than that from the GB body (Bird et al. 1996). Young's modulus might vary along the longitudinal direction as well. Therefore, the GB wall should be an inhomogeneous or functionally graded material. An interesting study concerning the vascular soft tissue with different local Young's moduli has been conducted (Khalil et al. 2006). How to model such a complicated material is rather challenging.

More sophisticated inhomogeneous constitutive modelling is required to extend the present work. Furthermore, GB wall thickness was decreased from 4.24 mm at the fundus to 1.91 mm at the neck across the GB body in the study by Su (2005). These two issues will be addressed in Chapter 10.

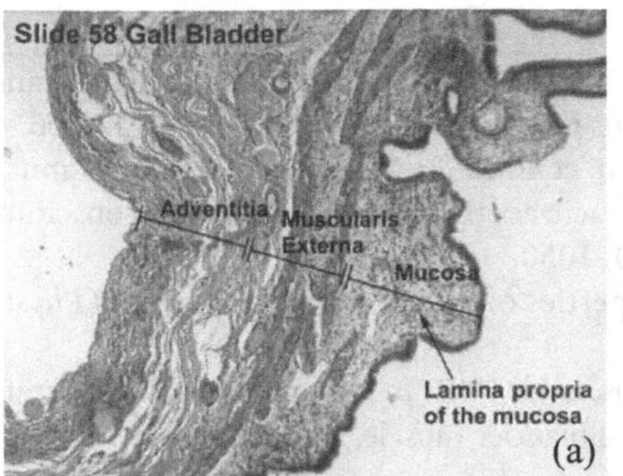

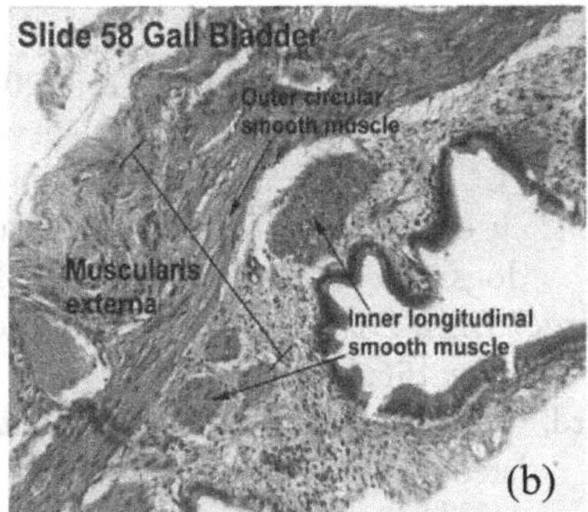

FIGURE 8.12 The histological structure of a human GB wall, where two families of orthogonal fibres are shown in the circumferential and longitudinal directions, the two pictures are credited to Wiechmann (2013) and with his permission for use here.

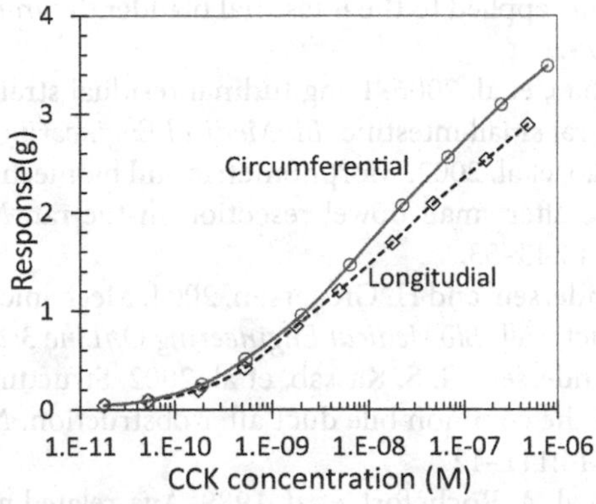

FIGURE 8.13 The response of strips harvested from a GB wall longitudinally and circumferentially to CCK-8 concentrations applied, the curves represent the means of 37 (longitudinal) and 38 (circumferential) strip samples, the data values are based on Bird et al. (1996).

Like all soft tissues, the GB wall presents viscoelastic behaviour because the Young's modulus depends on the frequency of loading applied (Sun 2005). This fact means that the refilling and emptying phases may suffer from viscous resistance in the GB wall. As a result, the viscosity of the GB wall may alter GB physiological cycle paths, i.e., GB motor function. The GB wall viscoelastic behaviour will be modelled in Chapter 11.

REFERENCES

Balocco, S., O. Camara, E. Vivas, et al. 2010. Feasibility of estimating regional mechanical properties of cerebral aneurysms in vivo. *Medical Physics* 37:1689–1706.

Barra, J. G., R. L. Armentano, J. Levenson, et al. 1993. Assessment of smooth muscle contribution to descending thoracic aortic elastic mechanics in conscious dogs. *Circulation Research* 73:1040–1050.

Bergel, D. H. 1961. The static elastic properties of the arterial wall. *Journal of Physiology* 156:445–457.

Bird, N., H. Wegstapel, R. Chess-Williams, et al. 1996. In vitro contractility of simulated human and non-stimulated gallbladder muscle. *Neurogastroenterology & Motility* 8:63–68.

Bosisio, M. R., M. Talmant, W. Skalli, et al. 2007. Apparent Young's modulus of human radius using inverse finite-element method. *Journal of Biomechanics* 40:2022–2028.

Caldwell, F. T., E. B. Sherman, and K. Levitsky. 1969. The composition of bladder bile and the histological pattern of the gallbladder and liver of the sea cow. *Comparative Biochemistry and Physiology* 28:437–440.

Colding-Jørgensen, M., and K. Steven. 1993. A model of the mechanics of smooth muscle reservoirs applied to the intestinal bladder. *Neurourology and Urodynamics* 12:59–79.

Dou, Y., Y. Fan, J. Zhao, et al. 2006. Longitudinal residual strain and stress-strain relationship in rat small intestine. *BioMedical Engineering OnLine* 5:37.

Dou, Y., X. Lu, J. Zhao, et al. 2002. Morphometric and biomechanical remodelling in the intestine after small bowel resection in the rat. *Neurogastroenterology & Motility* 14:43–53.

Duch, B. U., H. L. Andersen, and H. Gregersen. 2004. Mechanical properties of the porcine bile duct wall. *BioMedical Engineering OnLine* 3:23.

Duch, B. U., H. L. Andersen, G. S. Kassab, et al. 2002. Structural and mechanical remodelling of the common bile duct after obstruction. *Neurogastroenterology & Motility* 14:111–122.

Escoffier, C., J. de Rigal, A. Rochefort, et al. 1989. Age-related mechanical properties of human skin: An in vivo study. *Journal of Investigative Dermatology* 93:353–357.

Fung, Y. C. 1967. Elasticity of soft tissues in simple elongation. *American Journal of Physiology* 213:1532–1544.

Gee, M. W., C. Reeps, H. H. Eckstein, et al. 2009. Prestressing in finite deformation abdominal aortic aneurysm simulation. *Journal of Biomechanics* 42:1732–1739.

Gokhale, N. H., P. E. Barbone, and A. A. Obersi. 2008. Solution of the nonlinear elasticity imaging inverse problem: The compressible case. *Inverse Problems* 24:045010.

Govindjee, S., and P. A. Mihalic. 1996. Computational methods for inverse finite elastostatics. *Computer Methods in Applied Mechanics and Engineering* 136:47–57.

Govindjee, S., and P. A. Mihalic. 1998. Computational methods for inverse deformations in quasi incompressible finite elasticity. *International Journal for Numerical Methods in Engineering* 43:821–838.

Guo, Z., S. You, X. Wan, et al. 2010. A FEM-based direct method for material reconstruction inverse problem in soft tissue elastography. *Computers & Structures* 88:1459–1468.

Iding, R. H., K. S. Pister, and R. L. Taylor. 1974. Identification of nonlinear elastic solids by a finite element method. *Computer Methods in Applied Mechanics and Engineering* 4:121–142.

Karimi, A., A. Shojaei, and P. Tehrani. 2017. Measurement of the mechanical properties of the human gallbladder. *Journal of Medical Engineering & Technology* 41:191–202.

Karimi, R., T. Zhu, B. E. Bouma, et al. 2008. Estimation of nonlinear mechanical properties of vascular tissues via elastography. *Cardiovascular Engineering* 8:191–202.

Kauer, M., V. Vuskovic, J. Dual, et al. 2002. Inverse finite element characterization of soft tissues. *Medical Image Analysis* 6:275–287.

Khalil, A. S., B. E. Bouma, and M. R. Mofrad. 2006. A combined FEM/genetic algorithm for vascular soft tissue elasticity estimation. *Cardiovascular Engineering* 6:93–102.

Kroon, M. 2010. A numerical framework for material characterisation of inhomogeneous hyperelastic membranes by inverse analysis. *Journal of Computational and Applied Mathematics* 234:563–578.

Lee, T., R. R. Garlapati, K. Lam, et al. 2010. Fast tool for evaluation of iliac crest tissue elastic properties using the reduced-basis methods. *ASME Journal of Biomechanical Engineering* 132:121009.

Lei, F., and A. Z. Szeri. 2007. Inverse analysis of constitutive models: Biological soft tissues. *Journal of Biomechanics* 40:936–940.

Li, J., Y. Cui, R. E. English, et al. 2009. Ultrasound estimation of breast tissue biomechanical properties using a similarity-based non-linear optimization approach. *Journal of Strain Analysis for Engineering Design* 44:363–374.

Li, W. G., X. Y. Luo, N. A. Hill, et al. 2008. Correlation of mechanical factors and gallbladder pain. *Computational and Mathematical Methods in Medicine* 9:27–45.

Li, W. G., X. Y. Luo, R. W. Odgden, et al. 2011. A mechanical model for CCK-induced acalculous gallbladder pain. *Annals of Biomedical Engineering* 39:786–800.

Liao, D., Duch, B. U., Stodkilde-Jorfensen, H., et al. 2004. Tension and stress calculations in a 3-D Fourier model of gall bladder geometry obtained from MR images. *Annals of Biomedical Engineering* 32:744–755.

Lu, J., X. Zhou, and M. L. Raghavan. 2007a. Computational method of inverse elastostatics for anisotropic hyperelastic solids. *International Journal for Numerical Methods in Engineering* 69:1239–1261.

Lu, J., X. Zhou, and M. L. Raghavan. 2007b. Inverse elastostatic stress analysis in pre-deformed biological structures: Demonstration using abdominal aortic aneurysms. *Journal of Biomechanics* 40:693–696.

Lu, J., X. Zhou, and M. L. Raghavan. 2008. Inverse method of stress analysis for cerebral aneurysms. *Biomechanics and Modeling in Mechanobiology* 7:477–486.

Middelfart, H. V., P. Jensen, L. Hojgaard, et al. 1998. Pain patterns after distension of the gallbladder in patients with acute cholecystitis. *Scandinavian Journal of Gastroenterology* 33:982–987.

Mirsky, I., and W. W. Parmley. 1973. Assessment of passive elastic stiffness for isolated heart muscle and the intact heart. *Circulation Research* 33:233–243.

Moulton, M. J., L. L. Creswell, R. L. Actis, et al. 1995. An inverse approach to determining myocardial material properties. *Journal of Biomechanics* 28:935–948.

Raghupathy, R., and V. H. Barocas. 2010. Generalized anisotropic inverse mechanics for soft tissues. *ASME Journal of Biomechanical Engineering* 132:081006.

Samani, A., and D. Plewes. 2007. An inverse problem solution for measuring the elastic modulus of intact ex vivo breast tissue tumours. *Physics in Medicine and Biology* 52:1247–1260.

Schulze-Bauer, C. A., and G. A. Holzapfel. 2003. Determination of constitutive equations for human arteries from clinical data. *Journal of Biomechanics* 36:165–169.

Seshaiyer, P., and J. D. Humphrey. 2003. A sub-domain inverse finite element characterization of hyperelastic membranes including soft tissues. *Journal of Biomechanical Engineering* 125:363–371.

Smythe, A., A. W. Majeed, M. Fitzhenry, et al. 1998. A requiem for the cholecystokinin provocation test? *Gut* 43:571–574.

Su, Y. 2005. The mechanical properties of human gallbladder. BEng diss., University of Sheffield, Sheffield, UK.

Thounaojam, K., A. K. Pfoze, N. S. Singh, et al. 2017. A histological study of human foetal gallbladder. *International Journal of Anatomy and Research* 5:4648–4653.

Tilleman, T. R., M. M. Tilleman, and M. H. Neumann. 2004. The elastic properties of cancerous skin: Poisson's ratio and Young's modulus. *Israel Medical Association Journal* 6:753–755.

Washabau, R. J., M. B. Wang, C. L. Dorst, et al. 1991. Effect of muscle length on isometric stress and myosin light chain phosphorylation in gallbladder smooth muscle. *American Journal of Physiology-Gastrointestinal and Liver Physiology* 260:G920–G924.

Wiechmann, A. F. 2013. Liver, gall bladder, and pancreas. www.ouhsc.edu/histology/Glass%20slides/58_08.jpg, 58_08.jpg (accessed August 16, 2020).

Wu, H. C., and R. F. Yao. 1976. Mechanical behavior of the human annulus fibrosus. *Journal of Biomechanics* 9:1–7.

Xiao, L., C. K. Chui, and C. L. Teo. 2013. Reality based modeling and simulation of gallbladder shape deformation using variational methods. *International Journal of Computer Assisted Radiology and Surgery* 8:857–865.

Yeatman, L. A., W. W. Parmley, and E. H. Sonnenblick. 1969. Effects of temperature on series elasticity and contractile element motion in heart muscle. *American Journal of Physiology* 217:1030–1034.

Zhao, J., D. Liao, and H. Gregersen. 2005. Tension and stress in the rat and rabbit stomach are location-and direction-dependent. *Neurogastroenterology & Motility* 17:388–398.

Zhou, X., M. L. Raghavan, and R. E. Harbaugh, et al. 2010. Patient-specific wall stress analysis in cerebral aneurysms using inverse shell model. *Annals of Biomedical Engineering* 38:478–489.

CHAPTER 9

Fully Nonlinear Analysis of the GB Wall

FINITE STRAIN ANALYSIS

In Chapter 8, a quasi-linear inverse approach was developed to estimate GB wall elasticity. In the approach, incremental linear analyses on a series of configurations of human GB images scanned by routine ultrasonography in the emptying were performed. The transmural pressure in the GB, which was estimated from the measured GB volume changes, was applied as loading condition. The biomechanical property constants such as incremental Young's moduli were determined in such a way that the computed displacements at several observation points were adjusted against those from the images to reach the minimum error.

The determined incremental Young's moduli have included the active contraction effect of the smooth muscle in GB walls in the emptying. In this context, the anisotropic property of the GB wall should involve passive and active contributions. In Chapter 8, both the contributions were clarified by making use of a quasi-nonlinear method since the patient-specific active contribution is hardly modelled explicitly now. Because establishing a constitutive law for passive the GB wall is slightly easy, the anisotropic property of the GB wall in the refilling will be determined in a fully nonlinear manner to reflect the passive contribution of the tissue in the wall to the property herein.

A finite-strain fully nonlinear fibre-reinforced biomechanical constitutive model for soft bio-tissues in the passive state was proposed for the human GB wall based on the work of Holzapfel et al. (2000). To avoid modelling the active stress caused by smooth muscle cross-bridge phosphorylation in the emptying phase (Hai and Murphy 1988a, 1988b; Li et al. 2011a), the inverse computations are carried out in the refilling phase where only the passive stress is involved. The GB ellipsoid shape at a GB volume in the refilling phase was assumed to be equal to the shape at the same GB volume in the emptying phase, which has been determined from a series of ultrasonographic images (Smythe et al. 1998). The patient-specific constants in the constitutive law are decided inversely by using the FEM package-Abaqus, two custom codes of MATLAB and Python. The details of the model and method can be found in the study by Li et al. (2013).

P-V DIAGRAM AND PRESSURE PROFILE

CCK provocation tests usually are conducted on patients who have experienced repeated attacks of biliary pain in the absence of gallstones (Cozzolino et al. 1963). After an overnight fast, the patients were injected with an intravenous infusion of saline (control) followed by an intravenous infusion of CCK (0.05 μg/kg body weight). Ultrasonography of their GBs was taken to monitor initial volume, changes in shape and volume as well as wall thickness at 5- or 15-min intervals for 60 min. The test was only considered positive when the patients usual GB pain was reproduced following CCK infusion (Smythe et al. 1998).

A pressure-volume diagram in CCK provocation tests is illustrated in Figure 9.1. At point D, the sphincter of Oddi is closed, the patient is fasting and the GB cavity/bile volume and pressure are at their minimum levels. Between D and A, a low but positive pressure difference between the liver and the GB exists (Herring and Simpson 1907), and hepatic bile is secreted slowly into the GB. During this refilling, the GB volume and pressure increase and the smooth muscle in the GB wall is passively stretched. At point A, CCK starts to be infused, causing the GB to contract. The pressure in the GB rises up to point B rapidly in 3 to 10 min and exceeds the pressure in the CBD. During this period of time, the sphincter of Oddi relaxes and the pressure in the CBD is lowered. The pressure in the GB is now higher than that in the CBD, and the emptying starts until the volume drops to $V_C = (30–50)\%\ V_A$ (point C), and $p_c = p_A$. From point C to D, the GB wall is in isometric relaxation, with the pressure reduced from p_C to p_D owing to the smooth muscle relaxation and the cycle is then repeated.

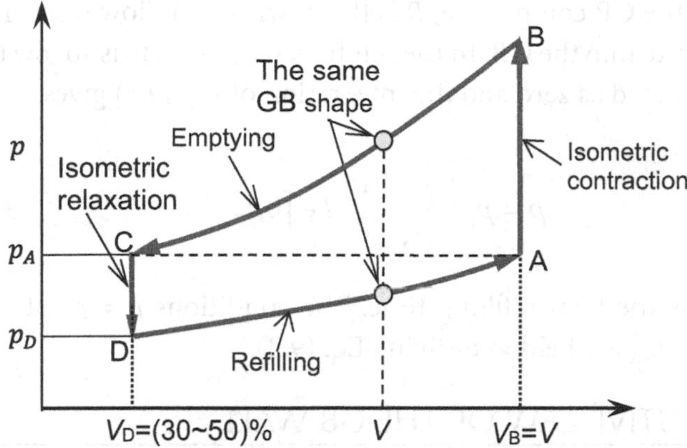

FIGURE 9.1 A pressure-volume diagram of the GB in the CCK provocation test, only the passive refilling is modelled here.

In the refilling phase, the GB wall is subject to a passive stretch. For simplicity, it is assumed that the GB has the same shape at a given volume, regardless of whether it is in the refilling or emptying phase, as indicated in Figure 9.1. This allows us to use the ultrasonographic images taken during the emptying. In Chapters 5 to 8, the simplified refilling-emptying cycle of triangle ABC is focused by neglecting the smooth muscle relaxation between C and D.

Thus far, there have been limited intraluminal pressure measurements *in vivo* in human GBs in passive refilling. However, when the guinea pigs suffered from dietary cholesterol and indomethacin caused from cholelithiasis, GB pressure has been monitored *in vivo* in the animal by means of an infusion/withdrawal pump, catheter and pressure transducer (Brotschi et al. 1984). The intraluminal pressure of human GB with acute cholecystitis was also measured *in vivo* (Borly et al. 1996). Based on these measurements, in this chapter, the basal intraluminal GB pressure is chosen to be p_D = 3.5 mmHg (466.6 Pa) (Borly et al. 1996) and p_A = 11 mmHg (1466.5 Pa) (Li et al. 2008, 2011b).

The pressure inside the GB can be related to the volume change using the Windkessel model (Li et al. 2008; Westerhof et al. 2009)

$$\frac{dp}{dt}+\frac{p}{RC}=\frac{Q}{C} \tag{9.1}$$

where C is the GB compliance, R is the downstream flow resistance and Q is the flow rate into the GB. In the refilling, the flow rate is so low that it can be approximated as zero and the integration of Eq. (9.1) gives

$$p = p_{\mathrm{D}}\left(\frac{p_{\mathrm{A}}}{p_{\mathrm{D}}}\right)^{\frac{t}{t_{\mathrm{DA}}}},\ t \in [0, t_{\mathrm{DA}}] \tag{9.2}$$

where t_{DA} is the total refilling time. The conditions $p = p_{\mathrm{D}}$ at $t = 0$, and $p = p_{\mathrm{A}}$ at $t = t_{\mathrm{DA}}$ are held in deriving Eq. (9.2).

CONSTITUTIVE LAW OF THE GB WALL

A general constitutive law was initially proposed by Humphrey and Yin (1987) for human soft tissues, based on a combination of existing phenomenological and micro-structural features. In their work, the human soft tissue is simplified to a mixture of a homogeneous matrix and a number of families of distributed thin extensible fibres, where the fibres and the matrix share the same gross deformation field. The total strain density energy is the sum of the different strain density energy functions of non-muscle matrix, nonlinear straight elastin and wavy collagen fibres.

For the walls of arteries, a simpler version was put forward by Holzapfel et al. (2000), where the elastin was part of the homogeneous, incompressible matrix material. The collagen is embedded in as fibre bundles with preferred orientations. The matrix material and collagen fibres have different strain energy functions, as such the total strain energy function is written as (Holzapfel et al. 2000)

$$\psi = c(I_1 - 3) + \frac{k_1}{2k_2}\Sigma_{i=4,6}[\mathrm{e}^{(k_2(I_i-1)2} - 1] \tag{9.3}$$

where the first term represents the contribution of the matrix, whereas the last two terms denote the collagen fibre effect, c is the matrix material property constant, while k_1 and k_2 are the material constants for collagen fibres. Note that in this model, the two families of fibres share the same mechanical properties. For incompressible arterial walls, the invariants I_1, I_4 and I_6 have the following forms (Holzapfel et al. 2000)

$$\begin{cases} | I_1 = tr(\mathbf{C}), I_4 = C_{ij}n_{4i}n_{4j}, I_6 = C_{ij}n_{6i}n_{6j} \\ | \mathbf{C} = (2\mathbf{E} + \mathbf{I}), \quad i = 1,2,3, j = 1,2,3 \end{cases} \tag{9.4}$$

where **C** is the Cauchy-Green deformation tensor, $\mathbf{C} = \mathbf{F}^T\mathbf{F}$ and **F** is the deformation tensor, n_{4i} (i = 1, 2, 3) are the direction cosines for the first family of fibres, and n_{6i} (i = 1, 2, 3) are the direction cosines for the second family of fibres. I_4 and I_6 are the squared stretch ratios in the directions of the two families of fibres. **E** is the Green-Lagrange strain tensor, and **I** is a 3 × 3 identity tensor.

The material parameters c, k_1 and k_2 are usually determined *in vitro* by biaxial tube inflation experiment (Holzapfel and Weizsäcker 1998; Holzapfel et al. 2000) or uniaxial tensile test (Holzapfel et al. 2005) for the aorta or artery walls. A more recent investigation indicates the neo-Hookean strain energy function represented by the first term of Eq. (9.3) can capture the biomechanical response of elastin (Watton et al. 2009).

In Chapter 8, the four-layer histological structure in the human GB wall is illustrated in Figure 8.12. These layers include adventitia, muscle, mucosa and epithelium (Thounaojam et al. 2017). The muscular layer, which consist of many bundles of smooth muscle cells, collagen fibres and elastin, contributes to the passive response to the refilling. There are two large smooth muscle bundles with collagen fibres running almost orthogonal to each other in the layer. Also, the denser smooth muscle bundles run circumferentially but just several isolated bundles are in the longitudinal direction.

Considering the GB wall's histological structure above, it is simplified that the human GB wall is composed of homogeneous elastic matrix material and two families of collagen fibres and smooth muscle cells in the circumferential and the longitudinal directions. Each family of fibres has different biomechanical property constants, see Figure 9.2(b) for the ellipsoid GB model, and the strain energy function reads

$$\psi = c(I_1 - 3) + \frac{k_1}{2k_2}[\mathrm{e}^{k_2(I_4-1)^2} - 1] + \frac{k_3}{2k_4}[\mathrm{e}^{k_4(I_6-1)^2} - 1] \tag{9.5}$$

And the directional cosines n_{4i} and n_{6i} of the two orthogonal families of fibres in Eq. (9.4) are determined from

$$\begin{cases} n_{41} = -\sin\varphi_4, n_{42} = \cos\varphi_4, n_{43} = 0 \\ n_{61} = \cos\theta_6, \cos\varphi_6, n_{62} = \cos\theta_6 \sin\varphi_6, n_{63} = -\sin\theta_6 \end{cases} \tag{9.6}$$

in which φ_4 is the azimuth angle of the circumferential family of fibres, but θ_6 and φ_6 are the zenith and azimuth angles of the longitudinal family of fibres, respectively, as shown in Figure 9.2(a).

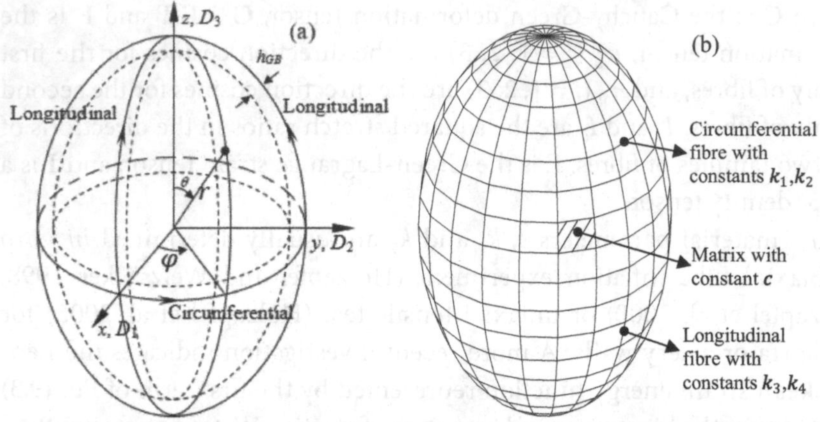

FIGURE 9.2 The ellipsoid shell model of human GB (a) and two orthogonal families of fibres are aligned in the longitudinal and circumferential directions (b).

Since experimental information on smooth muscle cells response is lack for the human GB, the passive contribution from the smooth muscle cells is not separated from that of the collagen fibres. The last two terms in Eq. (9.5) present the combined contribution from smooth muscles and fibres. The values of material parameters c, k_1, k_2, k_3 and k_4 will be determined from the inverse approach.

ZERO-STRESS CONFIGURATION

To examine the human GB stress-free configuration, two digital photos were taken against a GB, which was excised from a patient with gallstones, when it was full of bile and then after the bile was emptied completely (zero-pressure), as shown in Figure 9.3.

By noting the length between the mark and the apex of the GB fundus as well as the width of the fundus based on the scale in the photos, and assuming the GB is ellipsoidal with a thickness of 2.5 mm, it was found that the GB size shrunk by 50.6% after it was emptied, implying that the volume of zero-pressure configuration is around 50% of the filled GB, i.e., $V_C = V_D \approx 50\% \ V_A$ in Figure 9.1.

This zero-pressure configuration is also found to be approximately stress-free since no opening angles were detected after longitudinal or circumferential cuts were made on several GBs excised during cholestectomy.

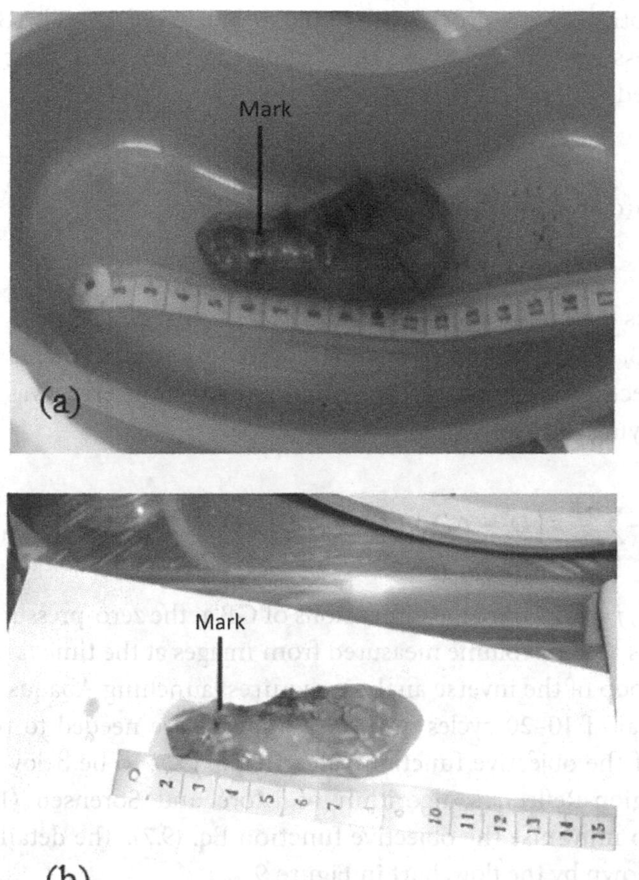

FIGURE 9.3 Excised human GB with bile (a) and the GB without bile (the zero-loading configuration) (b), a 50% decrease in volume is observed when bile is expelled from the GB.

COMPUTATIONAL METHOD

The FE model of each GB was generated using Abaqus 6.11-2. A minimum of 6100 quadrilateral elements were used, and for each GB, the edge length was kept to be between 1 and 1.5 mm. The 4-node shell element with large-strain formulation, which has excellent stability in the numerical solution procedure, was adopted. In solving the inverse problem, the constitutive model and the material parameters were specified in a user-subroutine in the forward problem simulation.

The optimisation process was controlled using MATLAB. During each pressure-loading step, the following objective function was minimised

$$\mathrm{fun}(c,k_1,k_2,k_3,k_4)=\sum_{i=1}^{N}\begin{pmatrix}\left|\Delta D_1-\Delta D_1^{\exp}\right|_i+\left|\Delta D_2-\Delta D_2^{\exp}\right|_i\\+\left|\Delta D_3-\Delta D_3^{\exp}\right|_i\end{pmatrix}^2 \tag{9.7}$$

where N is the number of the pressure increments, $N = 15$ here. ΔD_j, $D_j^{\exp}$, $j = 1, 2, 3$, are the computed and measured peak displacements in the x, y and z directions, respectively. In addition, the volume error was evaluated by employing the expression

$$\chi=\frac{1}{N}\sum_{i=1}^{N}\left|\frac{1}{6}\pi\left(D_1^0+\Delta D_1\right)_i\left(D_2^0+\Delta D_2\right)_i\left(D_3^0+\Delta D_3\right)_i-V_i^{\exp}\right|/V_1^{\exp} \tag{9.8}$$

where $D_j^{\exp}$, $j = 1, 2, 3$, are the dimensions of GB at the zero-pressure loading, and $V_i^{\exp}$ is the GB volume measured from images at the time t_i.

Each loop of the inverse analysis requires launching Abaqus six times, and a total of 10–20 cycles of inverse analysis are needed to reduce the change of the objective function value in Eq. (9.7) to be below 10^{-3}. The Trust-Region-Reflective algorithm of Moré and Sorensen (1983) was applied to minimise the objective function Eq. (9.7). The detailed procedure is shown by the flowchart in Figure 9.4.

VALIDATION OF THE INVERSE APPROACH

At first, the inverse approach was validated by using optical measurements for human arteries (Avril et al. 2010). As done by Avril et al. (2010), the experimental artery was treated as a thin-walled membrane without residual stress and with two spiral families of collagen fibres, as shown in Figure 9.5(a). The experimental internal pressure loading profile is plotted in terms of artificial time in Figure 9.5(b). The dimensions of the zero-pressure configuration of the artery are shown in Figure 9.5(a) as well. Since the inverse problem is static, the length of the artificial time does not affect the computational results. All the displacements of the bottom of the artery were fixed to zero, and a 10% longitudinal strain and a zero radial displacement were applied at the top of the artery. The constitutive model specified by Eq. (9.3) was applied as the constitutive model. Finally,

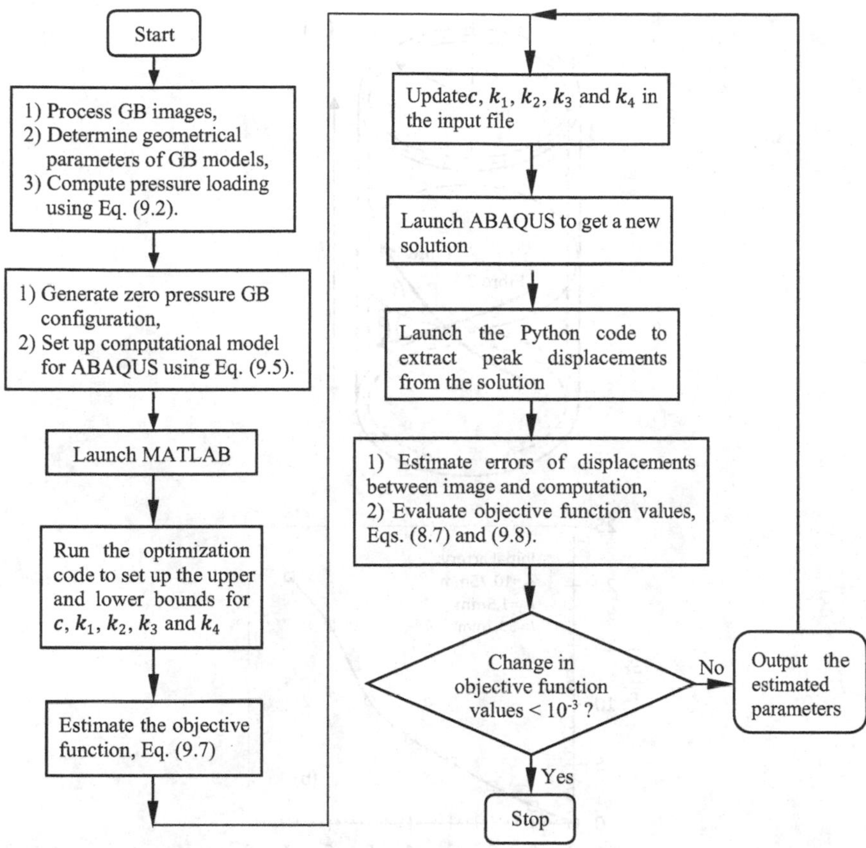

FIGURE 9.4 Flowchart for the inverse method.

four material parameters β, c, k_1 and k_2 were inversely estimated for the artery.

The estimated material parameters β, c, k_1, k_2 and the peak radial displacement Δr against internal pressure are illustrated in Figure 9.5(c). The agreement between the model prediction and the experimental data is very good with the maximum error of 2.03% in Δr.

For comparison, the peak radial displacement computed using the Fung strain energy density function by Fung et al. (1979) is also plotted with the dashed line in Figure 9.5c. Figure 9.5(c) shows that the model Eq. (9.3) seems to give better agreement with the experimental measurements compared with the Fung model.

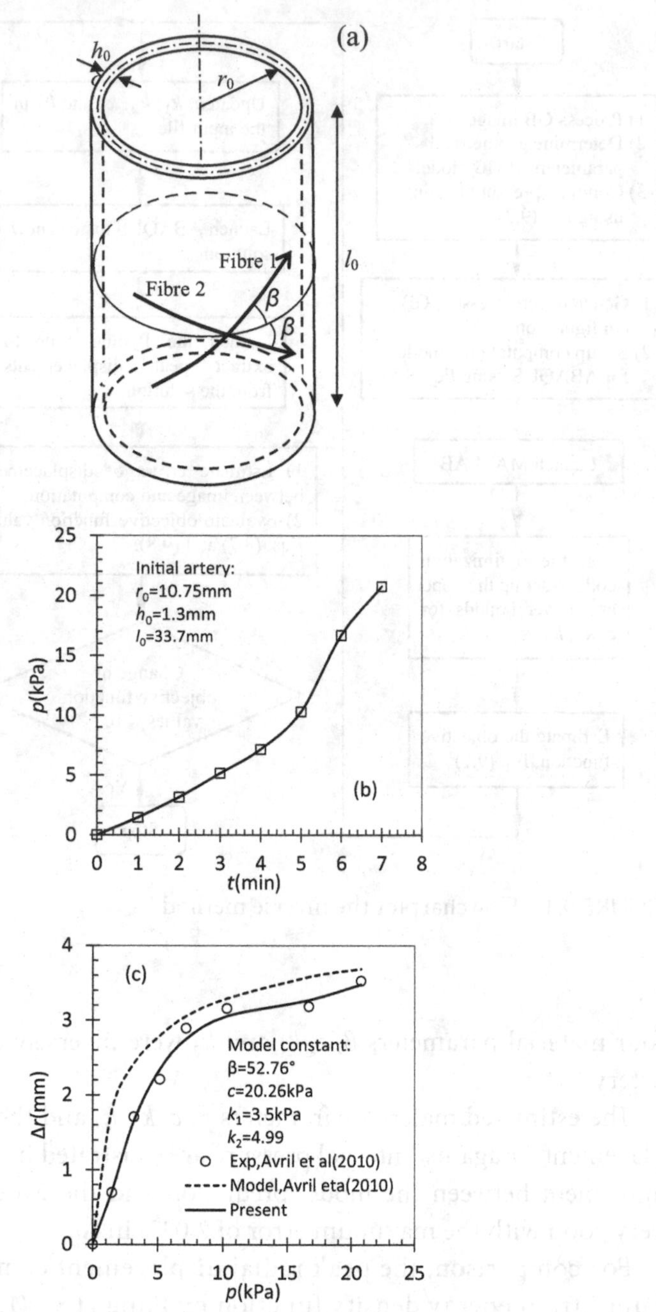

FIGURE 9.5 The arterial geometry from the study by Avril et al. (2010) (a), internal pressure versus time curve (b), and a comparison of the peak radial displacement versus pressure (c), the symbols for the experiment.

TEN GB SAMPLES

Ten human GB models were chosen to investigate the nonlinear effect of the human GB wall. These GBs differ in size and emptying behaviour; see the detailed parameters listed in Table 9.1. For simplicity, a uniform wall thickness [h_{GB} = 2.5 mm, see Figure 9.3(a)] is assumed for all the models. The EF recorded 30 min in the emptying phase is also shown in Table 9.1.

TABLE 9.1 Parameters of the human GB samples.

Item	Parameter	GB									
		1	3	4	17	19	21	29	37	39	43
End of refilling	p_A (Pa)	1466.5									
	D_1^A (mm)	23.4	26.8	32.9	27.2	34.7	28.2	28.1	30.2	33.2	37.6
	D_2^A (mm)	25.0	27.9	35.2	27.2	35.7	30.1	28.9	30.8	33.5	38.0
	D_3^A (mm)	54.1	70.7	57.5	55.9	92.3	74.5	56.1	53.8	53.9	82.1
Zero-pressure configuration	p_D (Pa)	466.6									
	D_1^D (mm)	16.8	21.0	24.8	21.4	26.8	20.8	20.1	24.7	24.2	28.1
	D_2^D (mm)	18.2	21.2	25.8	20.7	29.2	24.2	22.7	24.7	26.1	29.7
	D_3^D (mm)	51.7	59.3	54.9	46.7	72.9	62.9	49.9	41.2	47.5	70.3
Data from images	ΔD_1^{exp} (mm)	6.6	5.8	7.5	5.8	7.9	4.0	8.0	5.8	9.0	9.5
	ΔD_2^{exp} (mm)	6.8	6.6	7.7	6.5	6.5	9.3	6.2	5.8	7.4	8.3
	ΔD_3^{exp} (mm)	2.4	11.4	10.0	9.2	19.4	11.6	6.2	12.6	6.4	11.8
EF in 30 min (%)		4.5	11.4	13.3	32.4	49.4	66.3	37.8	77.0	60.1	2.7

TABLE 9.2 Material parameters inversely estimated for the GB samples.

GB	c (kPa)	k_1 (kPa)	k_2 (–)	k_3 (kPa)	k_4 (–)	χ (%)
1	2.3349	0.5977	0.8512	1.3952	1.0430	2.2
3	1.8375	4.7385	1.3538	0.8694	0.9568	3.9
4	2.1817	2.9539	0.7230	0.6578	1.0458	3.0
17	1.6810	2.9213	0.1161	0.4784	1.5530	2.5
19	2.2772	6.2427	0.1106	0.2182	0.8042	3.0
21	2.2309	3.0375	0.0176	0.2213	0.7755	3.4
29	2.0624	1.6658	0.7148	0.8237	1.1547	2.6
37	1.9243	4.3563	0.4350	0.1451	1.3890	2.5
39	2.4066	1.7295	0.5803	0.8437	1.1167	2.5
43	2.9435	3.3794	0.0741	0.3327	1.0654	2.7

GB WALL MATERIAL PARAMETERS

The estimated material parameters of the ten GB samples are tabulated in Table 9.2. These parameters are in the ranges of $c \in [1.68, 2.94]$ kPa, $k_1 \in [0.60, 6.24]$ kPa, $k_2 \in [0.018, 1.36]$, $k_3 \in [0.15, 1.40]$ kPa and $k_4 \in [0.78, 1.55]$.

In general, the matrix material parameter c shows little variation from one subject to another; however, the other parameters seem to be more patient-dependent. Moreover, most of the GB samples have stiffer circumferential fibres (indicated by a larger value of k_1) compared with the longitudinal ones (indicated by a smaller value of k_3), except for GB 1 (Table 9.1).

For comparison, the material parameters of the rabbit carotid artery in Table 9.3 obtained by Holzapfel et al. (2000) based on the experimental data of Fung et al. (1979) are listed. The material parameters of the human GB seem to be in similar ranges to those of the media of the rabbit carotid artery.

PEAK DISPLACEMENTS

The estimated peak displacements and the p-V curves are shown in Figure 9.6 for five of the GB samples. All the other samples exhibit similar patterns and are not shown. In particular, the p-V curves in Figure 9.6 show that the GB volume estimated from the modelling agrees very well with that of the images. Also, the peak displacement ΔD_3 is consistent with the observations. However, there are some discrepancies in ΔD_1 and ΔD_2, especially for GB 1, 19 and 21. These are presumably due to the simplification of using an ellipsoid model. While such a simplification is adopted clinically, it has been pointed out that the error between the actual image and an ellipsoid assumption is around 12.5% (Pauletzki et al. 1996).

In Figure 9.6, it is observed that there is a kink in each curve, which occurs at the second loading step. The reason for this is that the reference configuration chosen is subject to a pressure of 3.5 mmHg (the start of the refilling) and ignored the initial stress. If the parameter properties are known, one

TABLE 9.3 Parameters of rabbit carotid artery estimated by Holzapfel et al. (2000).

Artery	β **(°)**	c **(kPa)**	k_1 **(kPa)**	k_2
Media	29.0	3.0000	2.3632	0.8393
Adventitia	62.0	0.3000	0.5620	0.7112

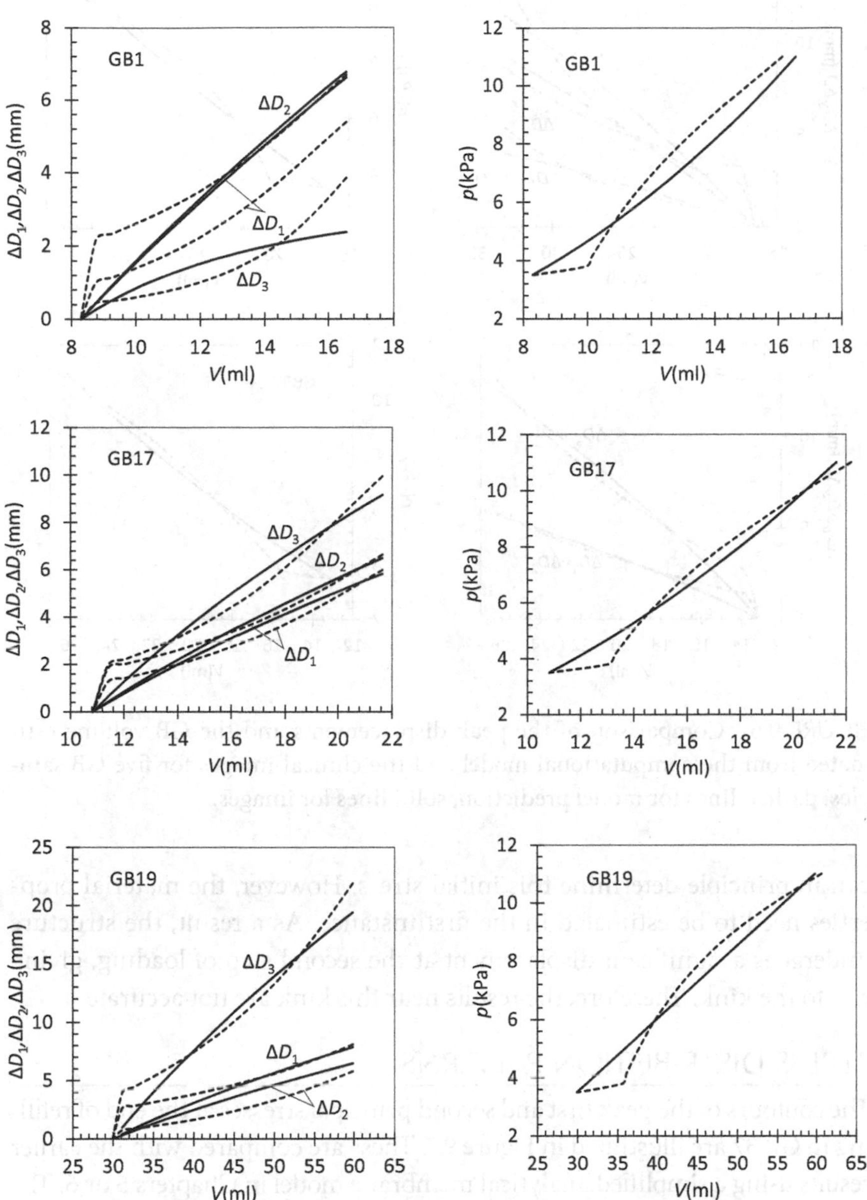

FIGURE 9.6 Comparison of the peak displacements and the GB volume estimated from the computational model and the clinical images for five GB samples, dashed lines for model prediction, solid lines for images.

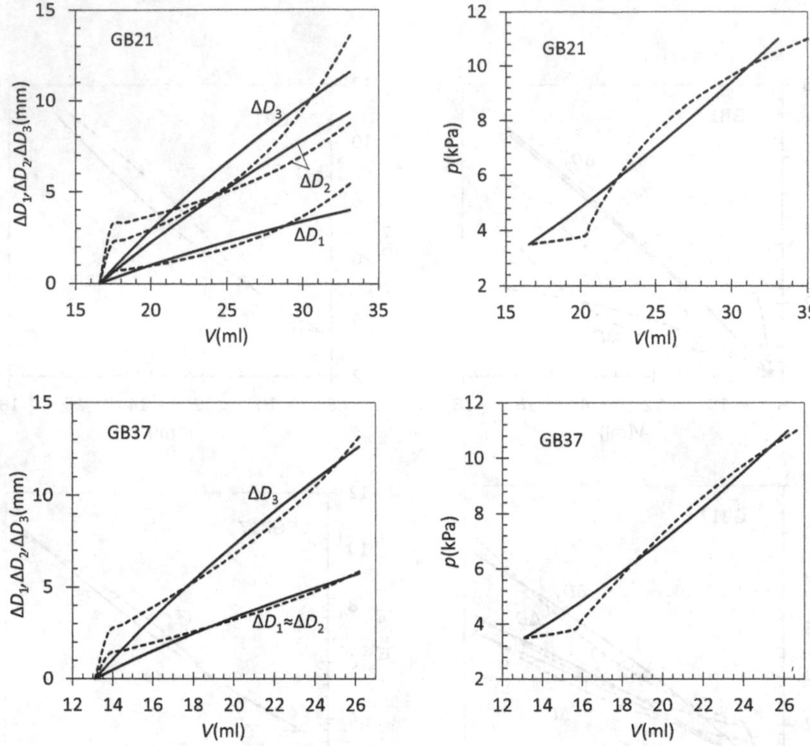

FIGURE 9.6 Comparison of the peak displacements and the GB volume estimated from the computational model and the clinical images for five GB samples, dashed lines for model prediction, solid lines for images.

can in principle determine this initial stress. However, the material properties need to be estimated in the first instance. As a result, the structure undergoes a significant displacement at the second step of loading, giving rise to the kink. Therefore, the results near this kink are not accurate.

STRESS DISTRIBUTION PATTERNS

The contours of the peak first and second principal stresses at the end of refilling in GB 37 are illustrated in Figure 9.7. These are compared with the earlier results using a simplified analytical membrane model in Chapters 5 or 6. The stress patterns of the present nonlinear model are similar to those of the analytical model, but the maximum and minimum principal stresses of the nonlinear material are increased to about 1.67 times those of the analytical model. The comparison for all the GB samples is summarised in Table 9.5.

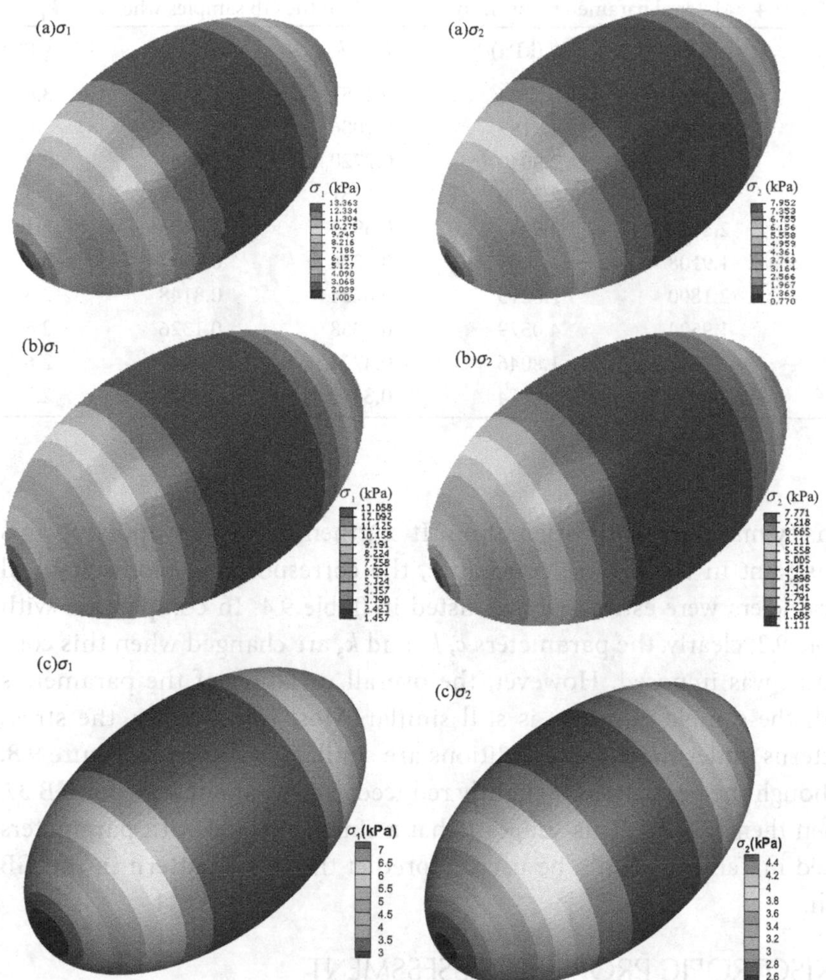

FIGURE 9.7 Contours of the first and second principal stresses for GB 37, (a) the present model, when $k_2 \neq k_4$, (b) the present model, when $k_2 = k_4$ and (c) results from the analytical model in Chapters 5 and 6.

Two Families of Fibres with Identical Properties

In the proposed model, it was supposed that two families of fibres were with different stiffness or biomechanical property in the *GB Wall Material Parameters* section. An alternative, perhaps more natural choice, would be to assume that all collagen fibres are of the same quality but their

TABLE 9.4 Material parameters inversely estimated for the GB samples when $k_2 = k_4$.

GB	c (kPa)	k_1 (kPa)	$k_2 = k_4$	k_3 (kPa)	χ (%)
1	2.2923	0.7274	0.4251	1.8183	3.6
3	1.9750	4.8187	0.9086	0.6619	3.6
4	2.1930	2.9040	0.7720	0.6623	2.9
17	1.6901	2.5200	0.6251	0.5537	2.5
19	2.2187	5.9592	0.3933	0.3483	3.1
21	1.9108	3.1405	0.7274	0.6937	3.7
29	2.1800	2.0213	0.0234	0.8148	3.3
37	1.9524	4.0579	0.9358	0.1326	2.6
39	2.4542	1.8046	0.4772	0.8539	2.6
43	2.9539	3.0274	0.3583	0.4368	2.7

density may vary with orientation. If so, then $k_2 = k_4$. By applying this constraint in the inverse procedure, the corresponding biomechanical parameters were estimated and listed in Table 9.4. In comparison with Table 9.2, clearly, the parameters c, k_1 and k_3 are changed when this constraint was imposed. However, the overall variation of the parameters with the sample numbers is still similar. Most importantly, the stress patterns under these two conditions are similar, as shown in Figure 9.8, although the peak stress is slightly reduced, e.g., by about 2.2% for GB 37, when then $k_2 = k_4$. This suggests that a simpler model with parameters listed in Table 9.4 could be used to predict the stress pattern in the GB wall.

ANISOTROPIC PROPERTY ASSESSMENT

In Chapter 8, the anisotropic property of the GB wall in the emptying was assessed by using quasi-nonlinear incremental method. The active contribution of GB smooth muscle has been included in the anisotropic property. In this chapter, the GB wall is only in the passive state, the results represent the passive anisotropic property of the wall.

As performed in Chapter 8, the ratio of the circumferential Young's modulus E_φ and the longitudinal Young's modulus E_θ determined by the strain energy function Eq. (9.5) along with the model parameters presented in Table 9.2 is worked out to characterise the anisotropic property. The circumferential and longitudinal Cauchy stresses in a GB wall are defined as

$$\begin{cases} \sigma_\varphi = \lambda_\varphi \dfrac{\mathrm{d}\psi}{\mathrm{d}\lambda_\varphi} = \lambda_\varphi \left(\dfrac{\partial \psi}{\partial I_1} \dfrac{\partial I_1}{\partial \lambda_\varphi} + \dfrac{\partial \psi}{\partial I_4} \dfrac{\partial I_4}{\partial \lambda_\varphi} \right) \\ \sigma_\theta = \lambda_\varphi \dfrac{\mathrm{d}\psi}{\mathrm{d}\lambda_\theta} = \lambda_\theta \left(\dfrac{\partial \psi}{\partial I_1} \dfrac{\partial I_1}{\partial \lambda_\theta} + \dfrac{\partial \psi}{\partial I_6} \dfrac{\partial I_6}{\partial \lambda_\theta} \right) \end{cases} \tag{9.9}$$

where σ_φ and σ_θ are the circumferential and longitudinal Cauchcy stresses, λ_φ and λ_θ are the circumferential and longitudinal stretch ratios, I_1 is the first invariant of stretches, $I_1 = \lambda_\varphi^2 + \lambda_\theta^2 + \lambda_h^2, \lambda_h$ is the wall thick stretch ratio, $\lambda_h = (\lambda_\varphi \lambda_\theta)^{-1}$ based on the material incompressible condition: $\lambda_\varphi\ \lambda_\theta\ \lambda_h = 1$, I_4 and I_6 are the second invariants, $I_4 = \lambda_\varphi^2$, $I_6 = \lambda_\theta^2$. The expressions of two stresses are detailed as follows:

$$\begin{cases} \sigma_\varphi = 2c \left(\lambda_\varphi^2 - \dfrac{1}{\lambda_\varphi^2 \lambda_\varphi^2} \right) + 2k_1 \lambda_\varphi^2 (I_4 - 1) \exp[k_2 (I_4 - 1)^2] \\ \sigma_\theta = 2c \left(\lambda_\theta^2 - \dfrac{1}{\lambda_\varphi^2 \lambda_\varphi^2} \right) + 2k_3 \lambda_\varphi^2 (I_6 - 1) \exp[k_4 (I_6 - 1)^2] \end{cases} \tag{9.10}$$

The circumferential and longitudinal Young's moduli are defined as

$$\begin{cases} E_\varphi = \mathrm{d}\sigma_\varphi / \mathrm{d}\lambda_\varphi \\ E_\theta = \mathrm{d}\sigma_\theta / \mathrm{d}\lambda_\theta \end{cases} \tag{9.11}$$

and their corresponding slightly complicated expressions read as

$$\begin{cases} E_\varphi = 2c \left(2\lambda_\varphi + \dfrac{1}{\lambda_\varphi^3 \lambda_\varphi^2} \right) + 4k_1 [\lambda_\varphi^3 + \lambda_\varphi (I_4 - 1) + 2k_2 \lambda_\varphi^3 (I_4 - 1)^2] \exp[k_2 (I_4 - 1)^2] \\ E_\theta = 2c \left(2\lambda_\theta + \dfrac{1}{\lambda_\theta^3 \lambda_\theta^2} \right) + 4k_3 [\lambda_\theta^3 + \lambda_\theta (I_6 - 1) + 2k_4 \lambda_\theta^3 (I_6 - 1)^2] \exp[k_4 (I_6 - 1)^2] \end{cases} \tag{9.12}$$

Based on the model parameters of ten GB samples in Table 9.2, the two Young's moduli and their ratio were calculated with Eq. (9.12) when the two stretch ratios vary in the range [1, 1.4], the Young's modulus ratios of ten GBs are plotted in Figure 9.8.

GBs 3, 19 and 37 suffer from the largest longitudinal change in geometry $\Delta D_3^{\exp}$ in comparison with the circumferential changes, $\Delta D_1^{\exp}$ and $\Delta D_2^{\exp}$ as shown in Table 9.1 during the refilling; thus, they exhibit the

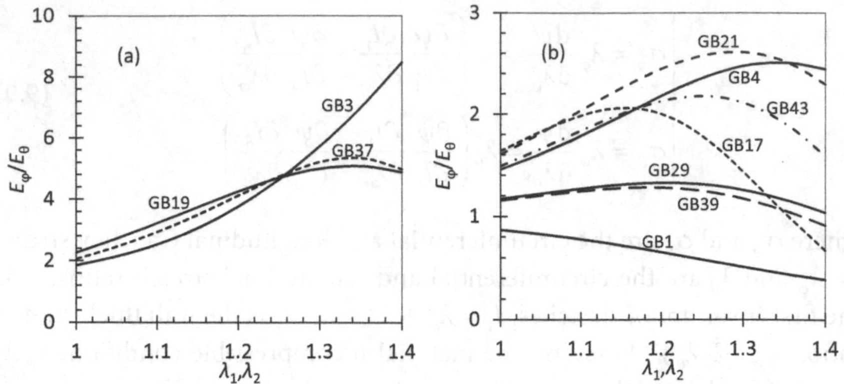

FIGURE 9.8 The ratios of the circumferential Young's modulus to the longitudinal Young's modulus for 10 GB samples based on the strain energy function Eq. (9.5) and the model constants in Table 9.2.

TABLE 9.5 The maximum first principal stresses estimated.

GB	σ_1 (kPa)	$\lambda_1 \times \lambda_2$	h^{def} (kPa)	σ_1^{lin} (kPa)	σ_1/σ_1^{li}	σ_1^{def} (kPa)
1	9.75	1.54	1.62	6.63	1.47	10.21
3	12.38	1.54	1.62	7.58	1.63	11.69
4	13.66	1.64	1.52	8.54	1.60	14.01
17	11.89	1.69	1.48	7.05	1.69	11.91
19	17.11	1.79	1.40	9.74	1.76	17.43
21	14.41	1.78	1.40	8.19	1.76	14.58
29	12.17	1.65	1.51	7.39	1.65	12.19
37	12.78	1.72	1.46	7.53	1.67	12.95
39	13.26	1.67	1.49	7.95	1.67	13.31
43	17.09	1.73	1.44	9.97	1.72	17.20

σ_1-calculated with nonlinear model, the deformed wall thickness $h^{def} = \dfrac{h}{\lambda_1 \times \lambda_2}$, σ_1^{lin}-calculated stress with linear model and $h_{GB} = 2.5$ mm, σ_1^{def}-calculated stress with linear model and h^{def}.

strongest anisotropic behaviour in the passive biomechanical property of the GB wall, i.e., the highest Young's modulus ratio and the stiffest in the circumferential direction.

For GB 1, since its longitudinal change in geometry is smaller than the changes in the circumferential direction, so it still exhibits an anisotropic property with a Young's modulus ratio $E_\varphi/E_\theta < 1$, i.e., the longitudinal direction is stiffer than the circumferential direction.

The Young's modulus ratio of GBs 29 and 39 is more than one but close to one, they show a negligible anisotropic property in passive state because their ΔD_1^{exp}, ΔD_2^{exp} and ΔD_3^{exp} are very similar in values as illustrated in Table 9.1.

The rest GBs experience an intermediate change in geometry in the longitudinal direction compared with those in the circumferential direction, so their Young's modulus ratio is ranged roughly from 0.7 to 2.6, but demonstrates significant variations with the stretch ratio, especially for GB 17. This fact suggests a GB geometrical change in the refilling should play an important role in the determination of anisotropic property of the GB wall or vice versa.

Based on ten GB samples, in most cases, the GB wall exhibits a circumferentially stiff anisotropic property in the passive state. The anisotropic biomechanical property in passive state has been confirmed by *in vitro* experiments (Su 2005; Xiong et al. 2013; Karimi et al. 2017). In those experiments, the Young's modulus of the human GB wall in the passive state was dependent on orientation.

EFFECT OF GB WALL THICKNESS ON STRESS LEVEL

As the peak stress is strongly correlated to GB pain in as shown in Chapters 5 and 6, it is important to check a reliable stress estimate based on routine clinical ultrasonography. In this chapter, the stress level evaluated is 1.67 times the corresponding level by the analytical linear membrane model in Chapters 5 or 6, see Figure 9.7(c) and Table 9.5. In Chapter 8, the same phenomenon also occurs.

The increase in the stress level is due to the change of GB wall thickness during the nonlinear analysis. At the end of refilling, the thickness of the GB wall is reduced from its initial value of 2.5 mm to around (1.4–1.6) mm. This significant change of wall deformation is not captured by the analytical model in Chapters 5 and 6 because the constitutive law is not needed. If, however, the reduced wall thickness is applied in this analytical model, then the final stress from the model agrees with the nonlinear model, as illustrated in the last right column of Table 9.5.

In Table 9.5, λ_1 and λ_2 are the first- and second-principal in-plane stretch ratios, $h_{\text{GB}} = 2.5$ mm is the initial GB wall thickness, $1/(\lambda_1\lambda_2)(=\lambda_3)$ is the stretch ratio of the thickness by making use of the incompressibility condition of soft tissue, $\lambda_1\lambda_2\lambda_3 = 1$ and $h^{\text{def}} = h_{\text{GB}}/(\lambda_1\lambda_2)$ is the thickness of the GB wall at the end of refilling. Since the GB size and shape are nearly the same in both the models, the narrowed thickness in the GB wall must be responsible for the increase in the stress level.

EFFECT OF WALL THICKNESS ON GB PAIN THRESHOLD

If the linear modelling underestimates the stress level in the GB wall, then how to explain the seemly good correlation between the 'pain' predicted by the linear model and the clinical observation; here pain is defined as when the peak stress is over a stress threshold, see Chapters 5 and 6.

In Chapters 5 and 6, the stress threshold used for GB pain is evaluated from the pressure threshold measured on CBD inflation in men by saline (Gaensler 1951), and a linear model (Case et al. 1999) is used to estimate the stress from the recorded pressure of the CBD. To illustrate this important point, The hoop stress was calculated in the pig CBD wall based on the experimental data recorded under two conditions: one is the changes in configuration are unconsidered and the other one is the changes are taken into account (Duch et al. 2004). The thin-walled linear model stress for the hoop stress gives (Case et al. 1999)

$$\sigma_\varphi = \frac{pd_{\text{CBD0}}}{2h_{\text{CBD0}}} \tag{9.13}$$

where p is the pressure in a CBD in kPa, and d_{CBD0} = 7.6 mm and h_{CBD0} = 1 mm are the diameter and thickness of the CDB at the zero-pressure loading (Duch et al. 2004).

In linear elasticity, the stress-pressure relationship is completely described by Eq. (9.13). However, in the nonlinear finite strain modelling, both the diameter and wall thickness change with pressure (Duch et al. 2004) and the stress should be estimated using the current (changed) configuration, namely

$$\sigma_\varphi = \frac{pd_{\text{CBD}}}{2h_{\text{CBD}}} \tag{9.14}$$

This stress-pressure relation is plotted in Figure 9.9 and compared with the results for the linear model Eq. (9.13). Evidently, the stress estimated by Eq. (9.9) is less than that by Eq. (9.14). Indeed, at the peak pressure (5 kPa or 37.4 mmHg), the ratio of the nonlinear to linear stress is 1.57. This is consistent with the ratios we obtained in Table 9.5.

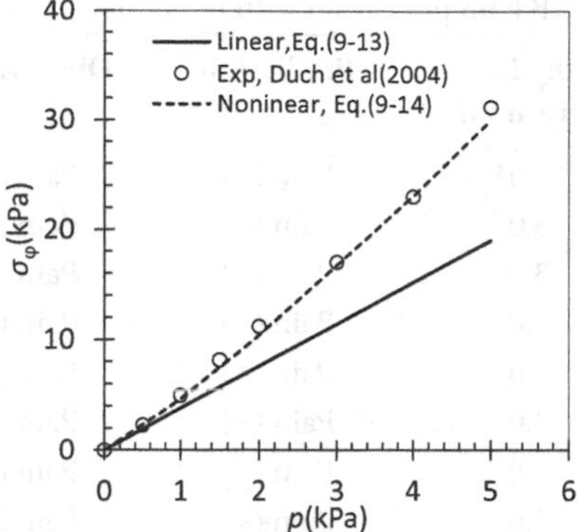

FIGURE 9.9 The hoop stress versus the internal pressure (estimated from the experimental data in Duch et al. 2004).

Importantly, Table 9.5 indicates that all 10 GB samples show the same level of stress elevation using the nonlinear model, thus tentatively suggesting that the overall trend of the peak stress is captured using the linear model. This is presumably the reason for the good correlations between the linear stress and linearly estimated stress threshold for pain observed in Chapters 5 and 6.

CLINICAL SIGNIFICANCE

The inverse nonlinear analysis approach has provided us with more insight into the stress-strain relations and the material parameters (Bertuzzi et al. 1992). However, the nonlinear analysis method may provide useful information for acalculous GB pain diagnosis.

It is postulated that a too large Young's modulus ratio E_φ/E_θ can trigger GB pain due to different stress responses in the circumferential and longitudinal directions. The threshold is simply specified to be a constant $[E_\varphi/E_\theta] = 3.0$ for all the patients, and if a ratio E_φ/E_θ is higher than 3.0, then GB pain occurs, otherwise no pain emerges at all. The predicted GB pain profile is listed and compared with that observed in Table 9.6.

It is surprising that there are eight patients exhibiting agreement between prediction and observation in the GB pain profile. As a result, the corresponding success rate is as good as 80%. If the pain threshold $[E_\varphi/E_\theta] = 2.0$ is held, then the success rate is reduced by 60%. If the pain threshold $[E_\varphi/E_\theta] = 4.0$ yields, then the success rate remains at 80%. This fact demonstrates that the

TABLE 9.6 Acalculous GB pain prediction with the peak E_φ/E_θ for 10 patients.

GB	Peak E_φ/E_θ	E_φ/E_θ threshold	Prediction	Observation	Agreement
1	0.87	3.0	Pain (−)	Pain (−)	Yes
3	8.47	3.0	Pain (+)	Pain (−)	No
4	2.51	3.0	Pain (−)	Pain (−)	Yes
17	2.05	3.0	Pain (−)	Pain (+)	No
19	5.06	3.0	Pain (+)	Pain (+)	Yes
21	2.60	3.0	Pain (−)	Pain (−)	Yes
29	1.33	3.0	Pain (−)	Pain (−)	Yes
37	5.32	3.0	Pain (+)	Pain (+)	Yes
39	1.28	3.0	Pain (−)	Pain (−)	Yes
43	2.17	3.0	Pain (−)	Pain (−)	Yes
Success rate					8/10 = 0.8

Pain (+) indicates that GB is painful, Pain (−) means that GB is not painful. The observation data were extracted from Table 6.1 in Chapter 6.

Young's modulus ratio E_φ/E_θ has the capability of being a potential GB pain index. The Young's modulus ratio reflects the constitutive law characteristics of an individual GB actually and needs to be studied further.

In the chapter, the selected GB samples are with nearly the same vales of D_1 and D_2 to improve the accuracy of the method. The reason for this is that the GB wall is assumed homogeneous and the GB tends to be a deformed into revolutionary ellipsoid under an internal pressure loading for a membrane ellipsoidal GB.

Frankly, this nonlinear approach is complicated, time-consuming and impractical to apply to large cohort of GB samples. In Chapter 10, this homogeneous assumption will be removed, and a simple and quick nonlinear analytical method is put forward.

REFERENCES

Avril, S., P. Badel, and A. Duprey. 2010. Anisotropic and hyperelastic identification of in vitro human arteries from full-field optical measurements. *Journal of Biomechanics* 43:2978–2985.

Bertuzzi, A., A. Gandolfi, A. V. Greco, et al. 1992. Material identification of guinea-pig gallbladder wall. In *14th Annual International Conference of the IEEE Engineering in Medicine and Biology Society*, 29 October–1 November, Paris, France.

Borly, L., L. Hojgaard, S. Gronvall, et al. 1996. Human gallbladder pressure and volume: Validation of a new direct method for measurements of gallbladder pressure in patients with acute cholecystitis. *Clinical Physiology and Functional Imaging* 16:145–156.

Brotschi, E. A., W. W. Lamorte, and L. F. Williams. 1984. Effect of dietary cholesterol and indomethacin on cholelithiasis and gallbladder motility in guinea pig. *Digestive Diseases and Sciences* 29:1050–1056.

Case, J., L. Chilver, and C. T. Ross. 1999. *Strength of Material and Structures*. London: Arnold.

Cozzolino, H. J., F. Goldstein, R. R. Greening, et al. 1963. The cystic duct syndrome. *JAMA* 185:920–924.

Duch, B. U., H. Andersen, and H. Gregersen. 2004. Mechanical properties of the porcine bile duct wall. *BioMedical Engineering OnLine* 3(1):23.

Fung, Y. C., K. Fronek, and P. Patitucci. 1979. Pseudoelasticity of arteries and the choice of its mathematical expression. *American Journal of Physiology-Heart and Circulatory Physiology* 237:H620–H631.

Gaensler, E. A. 1951. Quantitative determination of the visceral pain threshold in man. *Journal of Clinical Investigation* 30:406–420.

Hai, C. M., and R. A. Murphy. 1988a. Cross-bridge phosphorylation and regulation of latch state in smooth muscle. *American Journal of Physiology-Cell Physiology* 254:99–106.

Hai, C. M., and R. A. Murphy. 1988b. Regulation of shortening velocity by cross-bridge phosphorylation in smooth muscle. *American Journal of Physiology-Cell Physiology* 255:C86–C94.

Herring, P., and S. Simpson. 1907. The pressure of bile secretion and the mechanism of bile absorption in obstruction of the bile duct. *Proceedings of the Royal Society of London-Series B* 79:517–532.

Holzapfel, G. A., T. C. Gasser, and R. W. Ogden. 2000, A new constitutive framework for arterial wall mechanics and a comparative study of material models. *Journal of Elasticity* 61(1):1–48.

Holzapfel, G. A., G. Sommer, C. T. Gasser, et al. 2005. Determination of layer-specific mechanical properties of human coronary arteries with nonatherosclerotic intimal thickening and related constitutive modeling. *American Journal of Physiology-Heart and Circulatory Physiology* 289:H2048–H2058.

Holzapfel, G. A., and H. W. Weizsäcker. 1998. Biomechanical behavior of the arterial wall and its numerical characterization. *Computers in Biology and Medicine* 28:377–392.

Humphrey, J. D., and F. C. Yin. 1987. A new constitutive formulation for characterizing the mechanical behavior of soft tissues. *Biophysical Journal* 52:563–570.

Karimi, A., A. Shojaei, and P. Tehrani. 2017. Measurement of the mechanical properties of the human gallbladder. *Journal of Medical Engineering & Technology* 41:191–202.

Li, W. G., N. A. Hill, R. W. Ogden, et al. 2013. Anisotropic behaviour of human gallbladder walls. *Journal of the Mechanical Behavior of Biomedical Materials* 20:363–375.

Li, W. G., X. Y. Luo, N. A. Hill, et al. 2008. Correlation of mechanical factors and gallbladder pain. *Computational and Mathematical Methods in Medicine* 9:27–45.

Li, W. G., X. Y. Luo, N. A. Hill, et al. 2011a. Corss-bridge appearent rate constants of human gallbladder smooth muscle. *Journal of Muscle Research and Cell Motility* 32:209–220.

Li, W. G., X. Y. Luo, R. W. Odgden, et al. 2011b. A mechanical model for CCK-induced acalculous gallbladder pain. *Annals of Biomedical Engineering* 39:786–800.

Moré, J. J., and D. C. Sorensen. 1983. Computing a trust region step. *SIAM Journal on Scientific and Statistical Computing* 4:553–572.

Pauletzki, J., M. Sackmann, J. Holl, et al. 1996. Evaluation of gallbladder volume and emptying with a novel three-dimensional ultrasound system: Comparison with the sum-of-cylinders and the ellipsoid methods. *Journal of Clinical Ultrasound* 24:277–285.

Smythe, A., A. W. Majeed, M. Fitzhenry, et al. 1998. A requiem for the cholecystokinin provocation test? *Gut* 43:571–574.

Su, Y. 2005. The mechanical properties of human gallbladder. BEng diss., University of Sheffield, Sheffield, UK.

Thounaojam, K., A. K. Pfoze, N. S. Singh, et al. 2017. A histological study of human foetal gallbladder. *International Journal of Anatomy and Research* 5:4648–4653.

Watton, P. N., Y. Ventikos, and G. A. Holzapfel. 2009. Modelling the mechanical response of elastin for arterial tissue. *Journal of Biomechanics* 42:1320–1325.

Westerhof, N., J. W. Lankhaar, and B. E. Westerjof. 2009. The arterial Windkessel. *Medical & Biological Engineering & Computing* 47:131–141.

Xiong, L., C. K. Chui, and C. L. Teo. 2013. Reality based modeling and simulation of gallbladder shape deformation using variational methods. *International Journal of Computer Assisted Radiology and Surgery* 8:857–865.

CHAPTER 10

Heterogeneous Biomechanical Behaviour of the GB Wall

GB HETEROGENEOUS PROPERTY

The human GB wall is commonly regarded as a homogeneous anisotropic nonlinear material in the passive state of bile refilling phase (Li et al. 2012, 2013). This assumption was held in the previous chapters as well. However, recent work based on *in vitro* test on a healthy lamb GB suggested that this might not be proper, showing a heterogeneous property (Genovese et al. 2014). The heterogeneous property means the biomechanical property of a soft tissue depends on the location observed.

In addition, acalculous biliary disease can lead to increased material heterogeneity in the GB wall (Bird et al. 1996). Thus, in this chapter, the GB wall's heterogeneous property is addressed by extending the homogeneous anisotropic nonlinear biomechanical model for a human GB wall in Chapter 9. As a result, an inverse pointwise method is proposed to identify the heterogeneous anisotropic property at three different points on the human GB wall based on the ellipsoid membrane model shown in Chapters 5–9 and an in-house developed program in MATLAB. Their details are shown by Li et al. (2017).

IMAGE-BASED STRETCH AND STRESS

A series of ultrasonic images of acalculous human GB had been scanned in 10-min interval for 60 min during the emptying phase in the hospital. A typical example of this kind of images is illustrated in Figure 10.1, marked with three axes D_1, D_2 and D_3 ($D_1 \le D_2 \le D_3$). From these images, the corresponding ellipsoid models, as shown in Figure 10.2a, which are used to estimate the GB volume, can be generated.

The passive biomechanical property of the GB wall that exhibits in the refilling phase is addressed in the chapter. The refilling phase is the reverse process of the emptying phase that shares the same ellipsoid shape with the emptying phases at the same GB volume, as in Chapter 9. The heterogeneous and anisotropic biomechanical property of the human GB wall in the refilling phase will be determined inversely at points 1, 2 and 3 on the GB ellipsoid model.

Point 1 is an intersected point of two ellipses, one is along the equator and the other is in the longitudinal direction in a meridian plane, point 2 is also on the equator but 90° apart from point 1, and point 3 is at the apex as shown in Figure 10.2a. In the spherical coordinate system (r, φ, θ), the coordinates of points 1, 2 and 3 are ($D_1/2$, 0, $\pi/2$), ($D_1/2$, $\pi/2$, 0) and ($D_1/2$, 0, π), respectively.

Observing that the volume of the GB model was reduced by 50% by the end of emptying in Chapter 9, the configuration at the end of emptying was chosen as the reference configuration.

Fifteen GB geometrical models were generated throughout the refilling phase based on GB images using the geometrical similarity (Li et al. 2011).

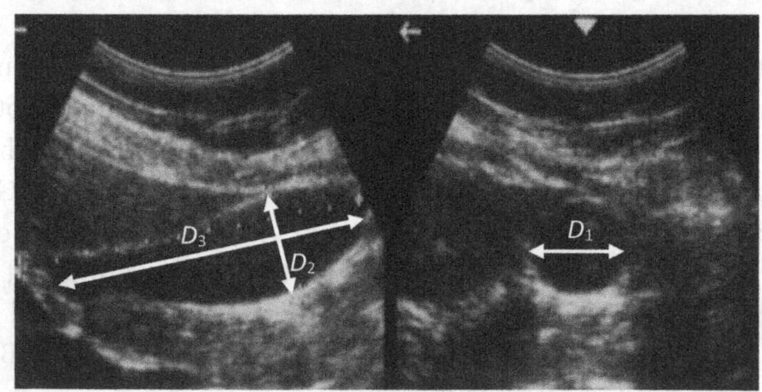

FIGURE 10.1 A typical ultrasonic image of human GB in the emptying phase.

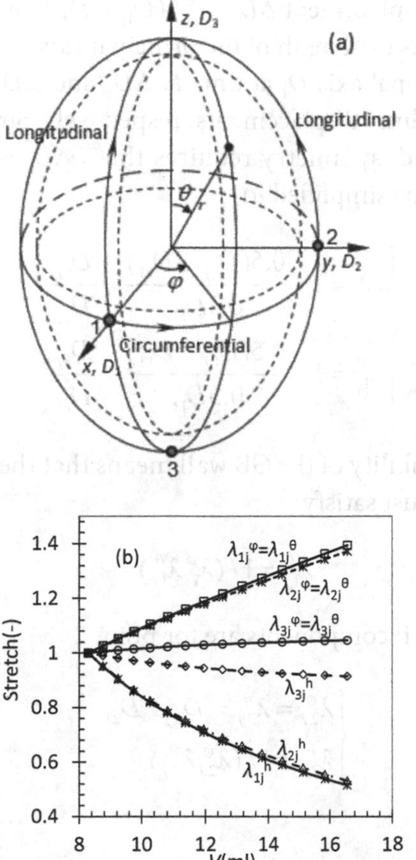

FIGURE 10.2 The imaged-based ellipsoid model for GB in refilling, (a) ellipsoid model with three control points on the surface, (b) the stretch components estimated from the ellipsoid model for GB 1 listed in Table 9.2.

The GB wall circumferential and longitudinal in-plane stretches at point 1 at a time instant t_j is calculated with

$$\begin{cases} \lambda_{1j}^{\varphi} = 1 + \dfrac{\Delta D_r}{\dfrac{D_{11}}{2}} + \dfrac{D_{11}}{2} \dfrac{\partial \Delta D_{\varphi}}{\partial \varphi} \\ \lambda_{1j}^{\theta} = 1 + \dfrac{\Delta D_r}{\dfrac{D_{11}}{2}} + \dfrac{D_{11}}{2} \dfrac{\partial \Delta D_{\theta}}{\partial \theta} \end{cases} \tag{10.1}$$

where the radial displacement $\Delta D_r = 0.5(D_{1j} - D_{11})$, and $j = 1, 2, 3, \ldots, N$, where $N = 15$, D_{11} is the length of the principal axis D_1 at time t_1, D_{1j} is the length of the principal axis D_1 at time t_j, ΔD_φ and ΔD_θ are the circumferential and longitudinal displacements, respectively. Since point 1 is in the axis of the ellipsoid, symmetry requires that $\partial \Delta D_\varphi / \partial \varphi = \partial \Delta D_\theta / \partial \theta = 0$. So, Eq. (10.1) can be simplified to

$$\begin{cases} \lambda_{1j}^{\varphi} = 1 + \dfrac{0.5(D_{1j} - D_{11})}{0.5D_{11}} = \dfrac{D_{1j}}{D_{11}} \\ \lambda_{1j}^{\theta} = 1 + \dfrac{0.5(D_{1j} - D_{11})}{0.5D_{11}} = \dfrac{D_{1j}}{D_{11}} \end{cases} \tag{10.2}$$

The incompressibility of the GB wall means that the stretch component of GB thickness must satisfy

$$\lambda_{1j}^{h} = 1 / (\lambda_{1j}^{\varphi} \lambda_{1j}^{\theta}) \tag{10.3}$$

Similarly, the stretch components are for point 2

$$\begin{cases} \lambda_{2j}^{\varphi} = \lambda_{2j}^{\theta} = D_{2j} / D_{21} \\ \lambda_{2j}^{h} = 1 / (\lambda_{2j}^{\varphi} \lambda_{2j}^{\theta}) \end{cases} \tag{10.4}$$

and for point 3

$$\begin{cases} \lambda_{3j}^{\varphi} = \lambda_{3j}^{\theta} = D_{3j} / D_{31} \\ \lambda_{3j}^{h} = 1 / (\lambda_{3j}^{\varphi} \lambda_{3j}^{\theta}) \end{cases} \tag{10.5}$$

where D_{21} and D_{2j} are the lengths of minor axis D_2 at time t_1 and t_j, respectively, while D_{31} and D_{3j} are the lengths of major axis D_3 at time t_1 and t_j, respectively.

These stretch components at t_j ($j = 1, \ldots, N$) and point i ($i = 1, 2, 3$) can be presented simply as

$$\begin{cases} \lambda_{ij}^{\varphi} = \lambda_{ij}^{\theta} = D_{ij} / D_{i1} \\ \lambda_{ij}^{h} = 1 / (\lambda_{ij}^{\varphi} \lambda_{ij}^{\theta}) \end{cases} \tag{10.6}$$

The stretches during the refilling phase are plotted against the GB volume in Figure 10.2b at points 1, 2 and 3 for GB 1. The GB volume changed with time exponentially based on an earlier model expressed by Eq. (5.6) in Chapter 5, where two parameters are determined analytically using the measured GB volume and pressure at the moments t_1 and t_N.

The expressions of in-plane stress components in the GB wall during the refilling phase were the same as these in the emptying phase since the GB material is assumed an elastic membrane, see Chapter 6

$$\begin{cases} \sigma_{ij}^{\theta\exp} = p_j F_\theta F_n \\ \sigma_{ij}^{\varphi\exp} = p_j F_\varphi / F_n \end{cases} \tag{10.7}$$

where p_j is the refilling pressure at time t_j, and F_θ, F_φ and F_n are the functions describing the geometry of the GB

$$\begin{cases} F_\theta = \dfrac{D_{3j}\kappa_{1j}\kappa_{2j}}{4h_{\mathrm{GB}ij}}\left(1 - \dfrac{\kappa_{1j}^2 - \kappa_{2j}^2}{\kappa_{1j}^2\kappa_{2j}^2}\cos 2\varphi_i\right) \\ F_\varphi = \dfrac{D_{3j}}{4\kappa_{1j}\kappa_{2j}h_{ij}}\left[\begin{array}{l} \kappa_{1j}^2\kappa_{2j}^2 + \left(\kappa_{1j}^2 + \kappa_{2j}^2 - 2\kappa_{1j}^2\kappa_{2j}^2\right)\sin^2\theta_i \\ +\left(\kappa_{1j}^2 - \kappa_{2j}^2\right)\cos^2\theta_i\cos 2\varphi_i \end{array}\right] \\ F_n = \dfrac{\sqrt{\kappa_{1j}^2\cos^2\theta_i\cos^2\varphi_i + \kappa_{2j}^2\cos^2\theta_i\sin^2\varphi_i + \sin^2\theta_i}}{\sqrt{\kappa_{1j}^2\sin^2\varphi_i + \kappa_{2j}^2\cos^2\varphi_i}} \end{cases} \tag{10.8}$$

where $\kappa_{1j} = D_{1j}/D_{3j}$, $\kappa_{2j} = D_{2j}/D_{3j}$, $h_{\mathrm{GB}ij}$ is the GB wall thickness at point i and time t_j, and $D_{ij} = \lambda_{ij}^{\theta} D_{i1}$, $h_{ij} = \lambda_{ij}^{h} h_{i1}$. The internal pressure p_j is given by, see Chapter 9

$$p_j = p_1\left(\frac{p_N}{p_1}\right)^{\frac{t_j}{t_N}}, t_j \in [0, t_N] \tag{10.9}$$

where p_N is the mean final bile pressure in a GB after the refilling phase chosen to be 1466.5 Pa (11 mmHg), p_1 is the bile pressure when the refilling starts, p_1 = 466.6 Pa (3 mmHg) and t_N is the total time of the refilling phase. These values are estimated from *in vivo* measurements; see Chapter 9.

THE CONSTITUTIVE LAW

To determine the heterogeneous material parameters of a human GB wall, the structure-based anisotropic constitutive law used in Chapter 9 was extended, so that the material parameters are location-dependent. At each point, the GB wall is assumed to be composed of homogeneous matrix and two families of fibres along the circumferential and longitudinal directions, respectively, as shown in Figure 10.3. The strain energy functions are

$$\psi_i = c^i(I_1 - 3) + \frac{k_1^i}{2k_2^i}[e^{k_2^i}((\lambda_i^{\varphi})^2 - 1)^2 - 1] + \frac{k_3^i}{2k_4^i}[e^{k_4^i}((\lambda_i^{\theta})^2 - 1)^2 - 1] \quad (10.10)$$

where the parameters c^i, k_m^i (i = 1, 2, 3; m = 1–4) are location-dependent. The total number of material property constants in Eq. (10.10) for points 1–3 should be 15 in general. However, at point 3, there are no circumferential fibres, so the second term in Eq. (10.10) disappears, i.e., k_1^3 and k_2^3 vanish. Hence, there are 13 parameters to be determined totally.

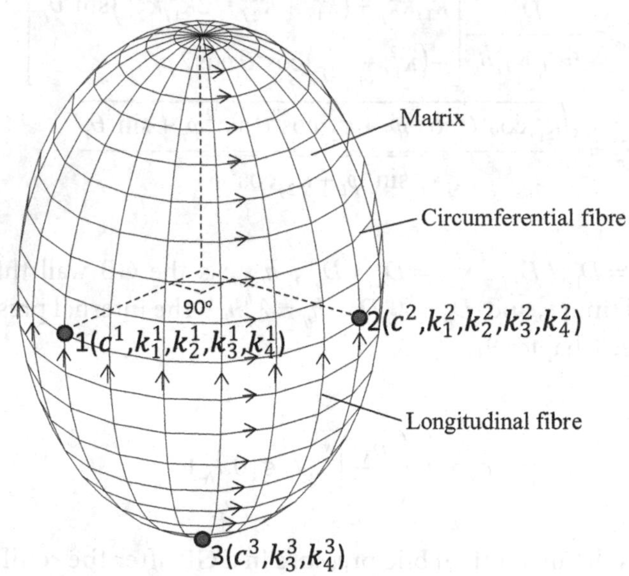

FIGURE 10.3 The GB wall is composed of matrix and two families of fibres, the 13 material parameters are location-dependent, varying from points 1 to 3.

The in-plane Cauchy stress components at t_j are

$$\begin{cases} \sigma_{ij}^{\varphi\text{mod}} = 2c^i\left(\lambda_{ij}^{\varphi 2} - \lambda_{ij}^{h2}\right) + 2\lambda_{ij}^{\varphi 2}k_1^i\left(\lambda_{ij}^{\varphi 2} - 1\right)\exp\left(k_2^i\left(\lambda_{ij}^{\varphi 2} - 1\right)^2\right) + \sigma_{i1}^{\varphi\text{exp}} \\ \sigma_{ij}^{\theta\text{mod}} = 2c^i\left(\lambda_{ij}^{\theta 2} - \lambda_{ij}^{h2}\right) + 2\lambda_{ij}^{\theta 2}k_3^i\left(\lambda_{ij}^{\theta 2} - 1\right)\exp\left(k_4^i\left(\lambda_{ij}^{\theta 2} - 1\right)^2\right) + \sigma_{i1}^{\theta\text{exp}} \end{cases} \tag{10.11}$$

where $\sigma_{i1}^{\varphi\text{exp}}$ and $\sigma_{i1}^{\theta\varphi\text{exp}}$ ($i = 1, 2, 3$) are interpreted as the initial stresses imbedded in the GB wall, which are estimated using Eqs. (10.7)–(10.9) with the internal pressure p_1.

ESTIMATE OF MATERIAL PARAMETERS

The material parameters in Eq. (10.10) are determined by minimising the following objective function

$$\text{fun} = \Sigma_{i=1}^{3}\Sigma_{j=1}^{N}\left[(\sigma_{ij}^{\varphi\text{mod}} - \sigma_{ij}^{\varphi\text{exp}})^2 + (\sigma_{ij}^{\theta\text{mod}} - \sigma_{ij}^{\theta\text{exp}})^2\right] \tag{10.12}$$

The minimisation was performed using the Trust-Region-Reflective algorithm in MATLAB (More and Sorensen 1983), which terminated when the objective function value is less than 10^{-6}. In addition, the following root-mean-square error (RMSE) is calculated to assess the curve-fitting quality

$$\chi = \sqrt{\frac{\frac{1}{6N}\Sigma_{i=1}^{3}\Sigma_{j=1}^{N}\left[(\sigma_{ij}^{\varphi\text{mod}} - \sigma_{ij}^{\varphi\text{exp}})^2 + (\sigma_{ij}^{\theta\text{mod}} - \sigma_{ij}^{\theta\text{exp}})^2\right]}{\frac{1}{6N}\Sigma_{i=1}^{3}\Sigma_{j=1}^{N}\left[\sigma_{ij}^{\varphi\text{exp}} - \sigma_{ij}^{\theta\text{exp}}\right]}} \times 100 \tag{10.13}$$

The optimisation process was conducted at points 1, 2 and 3 simultaneously rather than separately at each point. To secure a global minimum, the initial guesses of the parameters were chosen randomly within a suitable range, such as [0.01, 10] for c^1, $k_1^1, k_2^1, k_4^1, c^2, k_1^2, k_2^2, k_4^2, c^3$ and k_4^3, but [0.01, 3] for k_3^1, k_3^2, and k_3^3. In those ranges, the optimised material parameters did not occur at the boundaries, and the curve-fitting error was in the minimum. Eighty initial guesses were generated randomly, followed by 80 optimisation processes. The mean property constants and curve-fitting errors were chosen to be the results.

VARIABLE WALL THICKNESS

The GB wall thickness is related to the stress magnitude determined from the experimental images, as shown in Eqs. (10.7) and (10.8). This means that even though the pressure is the same, stresses can be different due to a varied thickness. This leads to different material parameters in the strain energy function in Eq. (10.10).

A 3D *in vivo* measurement of wall thickness of the GB was not yet available (Engel et al. 1980; Sanders 1980; Ugwu and Agwu 2010; Mohammed et al. 2010; Prasad et al. 2008; Oluseyi 2018); however, varying thickness of the GB wall was measured *in vitro* with a digital slide calliper (Su 2005; Khan et al. 2012). A contour of GB wall thickness is illustrated in Figure 10.4 (Li et al. 2017). It is observed that the thickness of the GB apex in the fundus increases to around 5 mm maximum and the wall of the neck is as thin as 2 mm. The ratio of the maximum thickness over the thickness of the body is 1.2.

In the study by Khan et al. (2012), 62 GB samples were divided into three age groups: (10–20), (21–40) and (41–70) years. The thicknesses of these GBs were measured manually at the fundus, body and neck. The maximum thickness was found on the neck, and the thinnest wall is located at the fundus. For the (41–70)-year group, which coincides with

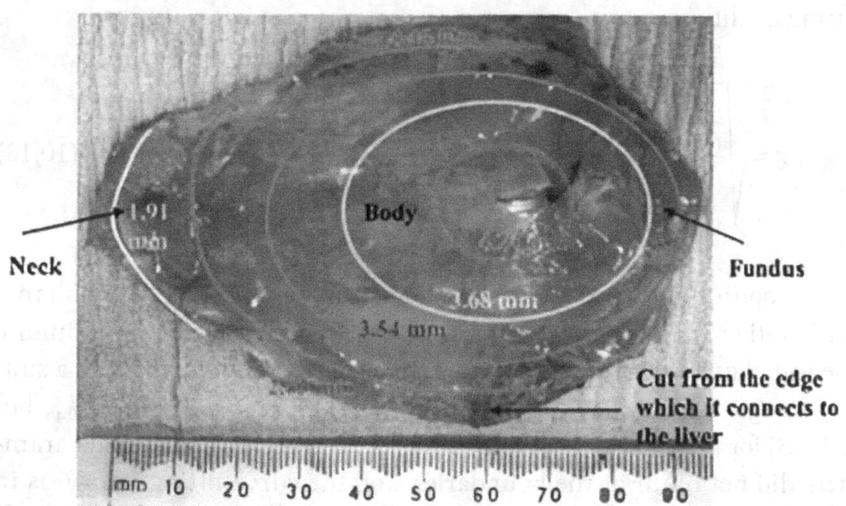

FIGURE 10.4 A GB wall thickness profile, showing the thickest wall at the GB apex and thinnest wall near the neck (the image is from Li et al. 2017).

the patient's age group for GB surgery in the paper, the ratio of the thickness at the fundus over the thickness of the body is 0.9. This is contrary to the finding by Su (2005). These ratios are used to examine the effect of a varying thickness.

GB SAMPLES SELECTED

The input data for the models were the geometrical parameters based on ultrasonic images and internal pressures at the start and end of emptying/refilling phase of ten GB samples in Chapter 9. These geometrical parameters and internal pressures are shown in Table 9.2. Additionally, the geometrical parameters and pressure profile at 15 or more time instants between the start and the end of refilling phase were interpolated according to the method in Chapter 8 and Eq. (10.9). A uniform wall thickness was assumed at the start of refilling phase, i.e., $h_{11} = h_{21} = h_{31} = 2.5$ mm, see Chapter 9. The stretch components of GB 1 over time are shown in Figure 10.2(b). Note that the stretch-volume profiles are patient-specific.

THE SOLUTION OF INVERS PROBLEM

From Table 9.1, the ellipsoid model geometrical parameters at the start and the end of the refilling phase, which were the same as the end (30 min after CCK) and start of the emptying phase from the routine ultrasonographical images taken in the hospital, were generated, as shown in Figure 10.1. As the least-square method required more scattered points than the number of parameters to be estimated, the ellipsoid model was interpolated over 15 or more time points for the emptying phase. The internal bile pressure is then given by Eq. (10.9). These data were used to obtain the initial guess for the optimisation process. The stretch and stress components were then computed, and the objective function and the RMSE were evaluated and compared with a given criterion of 10^{-6}. If the criterion was not satisfied, a new guess based on the Trust-Region-Reflective algorithm would be generated, and the procedure repeated until the convergent result is reached.

EFFECTS OF MATERIAL HETEROGENEITY

The material parameters of heterogeneity inversely determined are listed in Table 10.1 and compared with those from the corresponding homogeneous model in Chapter 9. The error χ in the homogeneous model reflects the error in GB volume between the image observation and homogeneous model prediction.

TABLE 10.1 **Heterogeneous material parameters of 10 GB samples determined and compared with the homogeneous model (Chapter 9) in uniform thickness.**

GB	Model	Point i	c^i (kPa)	k_1^i (kPa)	k_2^i (–)	k_3^i (kPa)	k_4^i (–)	χ (%)
1	Heterogeneous	1	0.2180 ±0.0131	2.1992 ± 0.0206	0.3105 ± 0.0064	0.6391 ± 0.0203	0.5603 ± 0.0267	7.1827 ± 0.0310
		2	0.0160 ± 0.0016	2.4538 ± 0.0025	0.2861 ± 0.0008	0.1041 ± 0.0030	1.5253 ± 0.0385	
		3	1.4801 ± 0.0040	N/A	N/A	2.4807 ± 0.0176	4.8695 ± 0.4743	
	Homogeneous	1, 2, 3	2.3349	0.5977	0.8512	1.3952	1.0430	2.2
3	Heterogeneous	1	0.1742 ± 0.0076	3.6549 ± 0.0142	0.7242 ± 0.0053	0.8486 ± 0.0141	1.4082 ± 0.0281	1.9123 ± 0.0152
		2	0.0885 ± 0.0084	3.3645 ± 0.0150	0.4526 ± 0.0047	0.1987 ± 0.0144	0.9714 ± 0.1228	
		3	0.3487 ± 0.0001	N/A	N/A	0.9275 ± 0.0012	4.2149 ± 0.0085	
	Homogeneous	1, 2, 3	1.8375	4.7385	1.3538	0.8694	0.9568	3.9
4	Heterogeneous	1	0.2229 ± 0.0086	4.0107 ± 0.0159	0.4920 ± 0.0045	1.1588 ± 0.0160	0.8603 ± 0.0185	3.2065 ± 0.0092
		2	0.1669 ± 0.0091	3.9481 ± 0.0168	0.4622 ± 0.0047	0.5402 ± 0.0162	1.2017 ± 0.0454	
		3	0.9313 ± 0.0004	N/A	N/A	0.6766 ± 0.0043	6.2768 ± 0.0462	
	Homogeneous	1, 2, 3	2.1817	2.9539	0.7230	0.6578	1.0458	3.0
17	Heterogeneous	1	0.1484 ± 0.0080	3.3757 ± 0.0151	0.8418 ± 0.0065	0.6389 ± 0.0144	1.8043 ± 0.0455	5.1699 ± 0.0131
		2	0.2152 ± 0.0116	2.9225 ± 0.0205	0.5591 ± 0.0074	0.5387 ± 0.0204	0.9941 ± 0.0516	
		3	0.8548 ± 0.0004	N/A	N/A	0.1702 ± 0.0017	5.1634 ± 0.1110	
	Homogeneous	1, 2, 3	1.6810	2.9213	0.1161	0.4784	1.5530	2.5
19	Heterogeneous	1	0.5583 ± 0.0195	4.4503 ± 0.0362	0.4628 ± 0.0094	0.6047 ± 0.0349	1.0724 ± 0.1010	3.6488 ± 0.0141
		2	0.0978 ± 0.0078	6.7567 ± 0.0169	0.6573 ± 0.0049	0.0512 ± 0.0025	3.2080 ± 0.0823	
		3	0.5965 ± 0.0003	N/A	N/A	0.6651 ± 0.0042	1.1060 ± 0.0216	
	Homogeneous	1, 2, 3	2.2772	6.2427	0.1106	0.2182	0.8042	3.0
21	Heterogeneous	1	0.2321 ± 0.0088	6.1282 ± 0.0227	3.2180 ± 0.0189	1.7915 ± 0.0157	7.2733 ± 0.0529	7.8681 ± 0.0058
		2	0.0576 ± 0.0002	2.7506 ± 0.0005	0.0628 ± 0.0002	0.0101 ± 0.0001	0.1280 ± 0.0228	
		3	0.0188 ± 0.0005	N/A	N/A	1.8847 ± 0.0046	5.3105 ± 0.0167	
	Homogeneous	1, 2, 3	2.2309	3.0375	0.0176	0.2213	0.7755	3.4

GB	Model	Point i	c^i (kPa)	k_1^i (kPa)	k_2^i (–)	k_3^i (kPa)	k_4^i (–)	χ (%)
29	Heterogeneous	1	0.3411 ± 0.0158	2.3990 ± 0.0247	0.2708 ± 0.0066	0.5445 ± 0.0249	0.4547 ± 0.0363	5.1186 ± 0.0165
		2	0.1083 ± 0.0081	3.5589 ± 0.0156	0.8243 ± 0.0065	0.4838 ± 0.0135	2.6977 ± 0.0584	
		3	1.2416 ± 0.0013	N/A	N/A	0.5071 ± 0.0059	3.1976 ± 0.2470	
	Homogeneous	1, 2, 3	2.0624	1.6658	0.7148	0.8237	1.1547	2.6
37	Heterogeneous	1	0.2382 ±0.0126	4.1246 ±0.0250	1.2410 ±0.0120	1.0673 ±0.0257	1.7882 ±0.0579	6.3186 ±0.0054
		2	0.2411 ±0.0117	4.1189 ±0.0233	1.2435 ±0.0113	1.0612 ±0.0236	1.7999 ±0.0555	
		3	0.9575 ±0.0002	N/A	N/A	0.0149 ±0.0005	6.7694 ±0.0517	
	Homogeneous	1, 2, 3	1.9243	4.3563	0.4350	0.1451	1.3890	2.5
39	Heterogeneous	1	0.3117 ±0.0121	2.9031 ±0.0199	0.2749 ±0.0051	0.8384 ±0.0197	0.5063 ±0.0218	5.5799 ±0.0092
		2	0.2700 ±0.0127	3.6645 ±0.0238	0.6873 ±0.0084	0.8535 ±0.0221	1.8349 ±0.0444	
		3	1.8285 ±0.0019	N/A	N/A	0.1972 ±0.0062	3.0987 ±0.2607	
	Homogeneous	1, 2, 3	2.4066	1.7295	0.5803	0.8437	1.1167	2.5
43	Heterogeneous	1	0.3644 ±0.0161	4.3110 ±0.0283	0.2549 ±0.0056	0.7731 ±0.0279	0.5924 ±0.0431	3.7013 ±0.0150
		2	0.2382 ±0.0095	5.3561 ±0.0183	0.4154 ±0.0043	0.7740 ±0.0170	1.6044 ±0.0410	
		3	1.3775 ±0.0005	N/A	N/A	0.2512 ±0.0036	3.6817 ±0.2138	
	Homogeneous	1, 2, 3	2.9435	3.3794	0.0741	0.3327	1.0654	2.7

For all the GBs, the material parameter associated with the matrix in the heterogeneous model is around 10 times that of the homogeneous model. For GBs 3, 4, 17, 29, 37, 39 and 43, the mean values of the fibre-related material parameters at points 1 and 2, k_1^2, and k_3^2, basically agree with k_1 and k_3 in the homogeneous model, i.e., $k_2^1 \approx$ (1–2) k_1 and $k_3^2 \approx$ (1–2) k_3. For GBs 1, 19 and 21, k_1^2 and k_3^2 are different from k_1 and k_3 in the homogeneous model.

On the one hand, the material parameters in the heterogeneous model at points 1 and 2 are similar, implying the heterogeneity of the GB wall along the circumference is small. This is especially true for GB 37 which

TABLE 10.2 The first principal stresses in 10 GB wall samples estimated by using the homogeneous and heterogeneous models.

Stress (kPa)	GB sample									
	1	3	4	17	19	21	29	37	39	43
Homogeneous	9.75	12.38	13.66	11.89	17.11	14.41	12.17	12.78	13.26	17.09
Heterogeneous	10.86	12.40	14.49	11.40	16.25	14.09	14.41	11.41	14.96	17.85

has $D_1 \approx D_2$, and points 1 and 2 share the same parameters. On the other hand, the parameters at point 3 differ substantially from the other two, suggesting a strong heterogeneity from GB body to fundus.

The first principal stresses of all the GB samples are compared in Table 10.2 with those predicted by the homogeneous model in Chapter 9. These stresses are extracted at point 1 since the length of an ellipsoid major axis is the shortest through that point, resulting in the highest stress level there based on Eqs. (10.7) and (10.8) in the φ direction. The homogeneous model underestimates the stresses in the wall of GBs 1, 3, 4, 29, 39 and 43, and overestimates them for the remaining GBs. As a result, the relative error in the first principal stresses vary in a range of −11.4% to +10.8% in comparison with the stresses in the homogeneous model.

VARIATION OF GB WALL THICKNESS

In Table 10.1, there are some small errors ranged between 5.2% and 7.8% for the parameters estimated for five GBs 1, 17, 21, 37 and 39. To identify the cause for the errors, the stress-volume curves of GBs 3 and 39 at points 1, 2 and 3 are shown in Figure 10.5. The predicted stress agrees well with the observations at point 1 and 2, but not so well at point 3.

In Figure 10.5, the GB wall heterogeneity mainly occurs in the apex region, resulting in poor agreement in the stress. Therefore, the GB wall thickness at the apex is altered to examine the effect of varying thickness. First, the apex thickness was changed to 3 mm, based on the ratio of 1.2 found by Su (2005), and kept at 2.5 mm at points 1 and 2. The extracted pointwise mechanical properties are listed in Table 10.3.

The relative changes in these 13 parameters are tabulated in Table 10.4. The increased thickness at the apex by 20% has a considerable effect on the material parameters with a change up to 30%, in particular, on k_4^1, c^2, k_4^2, c^3, k_3^3 and k_4^3, which are associated with the properties of the matrix and the longitudinal fibres. This is very different to the membrane theory, in

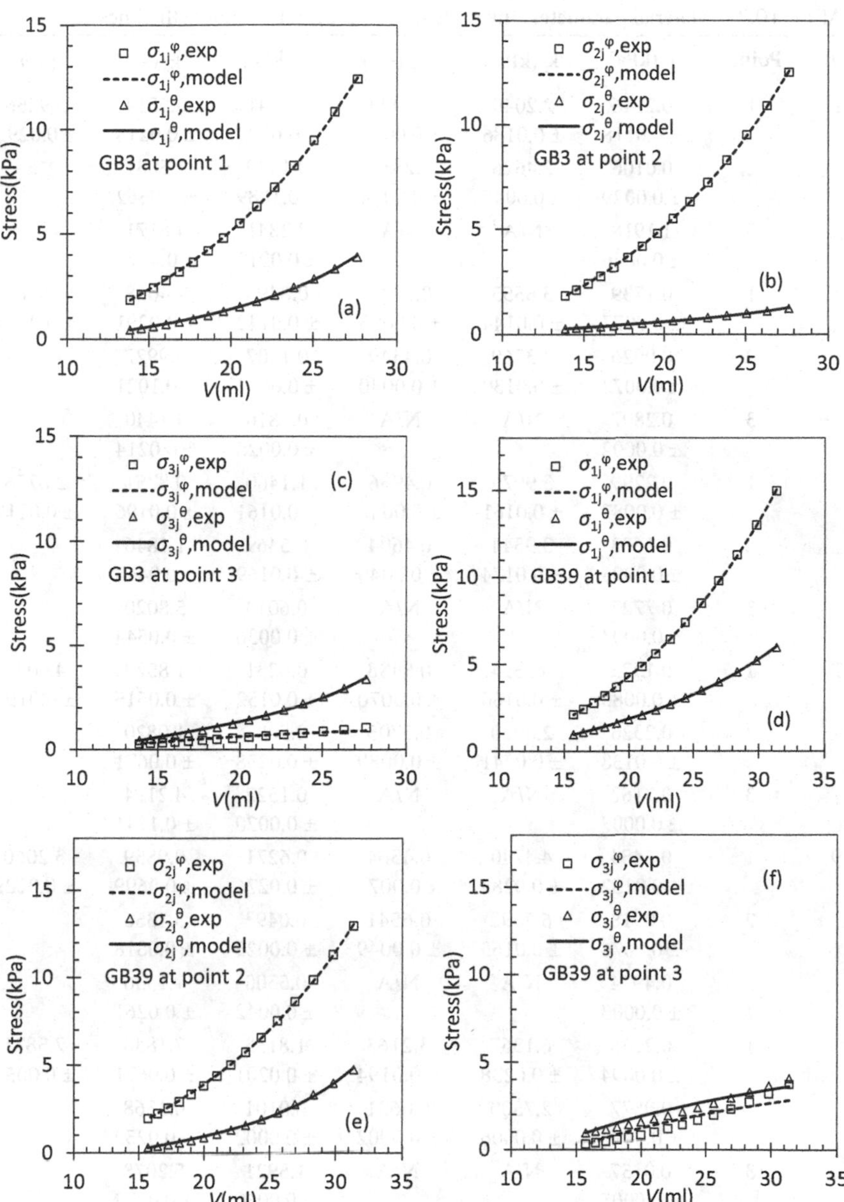

FIGURE 10.5 Comparison of the modelled (lines) and estimated (symbols) circumferential and longitudinal stresses with the image-based ellipsoid membrane mechanic model at points 1, 2 and 3, for GB 3(a)–(c) and GB 39(d)–(f).

TABLE 10.3 Material parameters inversely determined with variable thicknesses.

GB	Point i	c^i (kPa)	k_1^i (kPa)	k_2^i (–)	k_3^i (kPa)	k_4^i (–)	χ (%)
1	1	0.2146 ± 0.0118	2.2040 ± 0.0186	0.3090 ± 0.0056	0.6441 ± 0.0185	0.5514 ± 0.0218	5.9256 ± 0.0297
	2	0.0106 ± 0.0029	2.4626 ± 0.0047	0.2836 ± 0.0014	0.1089 ± 0.1089	1.5081 ± 0.0662	
	3	1.1918 ± 0.0046	N/A	N/A	2.2841 ± 0.0212	4.8171 ± 0.4371	
3	1	0.1739 ± 0.0077	3.6555 ± 0.0144	0.7239 ± 0.0053	0.8493 ± 0.0142	1.4068 ± 0.0291	1.8261 ± 0.0173
	2	0.0926 ± 0.0072	3.3569 ± 0.0130	0.4549 ± 0.0040	0.1907 ± 0.0123	0.9927 ± 0.1021	
	3	0.2897 ± 0.0002	N/A	N/A	0.7816 ± 0.0026	4.1440 ± 0.0214	
4	1	0.2299 ± 0.0087	3.9979 ± 0.0161	0.4956 ± 0.0046	1.1460 ± 0.0161	0.8751 ± 0.0196	2.8756 ± 0.0119
	2	0.1633 ± 0.0095	3.9544 ± 0.0174	0.4604 ± 0.0049	0.5469 ± 0.0169	1.1846 ± 0.0468	
	3	0.7727 ± 0.0003	N/A	N/A	0.6013 ± 0.0036	5.8020 ± 0.0544	
17	1	0.1572 ± 0.0086	3.3594 ± 0.0160	0.8488 ± 0.0070	0.6231 ± 0.0152	1.8528 ± 0.0515	4.6031 ± 0.0195
	2	0.2326 ± 0.0138	2.8920 ± 0.0241	0.5705 ± 0.0089	0.5087 ± 0.0238	1.0820 ± 0.0671	
	3	0.7162 ± 0.0007	N/A	N/A	0.1522 ± 0.0020	4.2194 ± 0.1741	
19	1	0.5454 ± 0.0152	4.4740 ± 0.0284	0.4564 ± 0.0073	0.6271 ± 0.0276	0.9859 ± 0.0699	3.2060 ± 0.0129
	2	0.0918 ± 0.0076	6.7692 ± 0.0165	0.6541 ± 0.0049	0.0493 ± 0.0025	3.1356 ± 0.0818	
	3	0.4974 ± 0.0003	N/A	N/A	0.5500 ± 0.0042	1.1336 ± 0.0261	
21	1	0.2266 ± 0.0094	6.1367 ± 0.0238	3.2163 ± 0.0194	1.8135 ± 0.0201	7.1830 ± 0.0671	7.5871 ± 0.0058
	2	0.0577 ± 0.0002	2.7500 ± 0.0006	0.0631 ± 0.0002	0.0101 ± 0.0002	0.1368 ± 0.0252	
	3	0.0157 ± 0.0005	N/A	N/A	1.5921 ± 0.0060	5.2078 ± 0.0272	
29	1	0.3272 ± 0.0156	2.4211 ± 0.0244	0.2648 ± 0.0065	0.5663 ± 0.0245	0.4261 ± 0.0358	4.4857 ± 0.0208
	2	0.1032 ± 0.0063	3.5691 ± 0.0120	0.8197 ± 0.0049	0.4906 ± 0.0105	2.6659 ± 0.0461	
	3	1.0348 ± 0.0013	N/A	N/A	0.4320 ± 0.0059	2.6961 ± 0.2625	

GB	Point i	c^i (kPa)	k_1^i (kPa)	k_2^i (−)	k_3^i (kPa)	k_4^i (−)	χ (%)
37	1	0.2469 ± 0.0126	4.1076 ± 0.0252	1.2490 ± 0.0122	1.0492 ± 0.0253	1.8288 ± 0.0601	5.5412 ± 0.0075
	2	0.2414 ± 0.0140	4.1183 ± 0.0280	1.2440 ± 0.0136	1.0606 ± 0.0283	1.8085 ± 0.0664	
	3	0.7965 ± 0.0002	N/A	N/A	0.0153 ± 0.0005	6.3465 ± 0.0628	
39	1	0.2999 ± 0.0131	2.9229 ± 0.0217	0.2699 ± 0.0054	0.8575 ± 0.0216	0.4888 ± 0.0222	4.8833 ± 0.0117
	2	0.2650 ± 0.0123	3.6739 ± 0.0230	0.6839 ± 0.0081	0.8623 ± 0.0215	1.8161 ± 0.0420	
	3	1.5221 ± 0.0023	N/A	N/A	0.1718 ± 0.0083	2.8898 ± 0.2615	
43	1	0.3508 ± 0.0156	4.3353 ± 0.0275	0.2502 ± 0.0054	0.7968 ± 0.0275	0.5590 ± 0.0371	3.2537 ± 0.0129
	2	0.2305 ± 0.0099	5.3712 ± 0.0192	0.4119 ± 0.0044	0.7883 ± 0.0181	1.5736 ± 0.0399	
	3	1.1491 ± 0.0005	N/A	N/A	0.2174 ± 0.0031	3.0536 ± 0.2349	

TABLE 10.4 Relative changes in the parameters due to varied wall thickness.

GB	Point i	$\Delta c^i/c^i$(%)	$\Delta k_1^i / k_1^i$ (%)	$\Delta k_2^i / k_2^i$ (%)	$\Delta k_3^i / k_3^i$ (%)	$\Delta k_4^i / k_4^i$ (%)	$\Delta\chi$ (%)
1	1	−1.5229	0.2183	−0.4831	0.7824	−1.5884	−1.2571
	2	−33.7500	0.3586	−0.8738	4.61095	−1.12765	
	3	−19.4784	N/A	N/A	−7.9252	−1.0761	
3	1	−0.1722	0.01642	−0.0414	0.0825	−0.0994	−0.0862
	2	4.6328	−0.2259	0.5082	−4.0262	2.1927	
	3	−16.9200	N/A	N/A	−15.7305	−1.6821	
4	1	3.14042	−0.3192	0.7317	−1.1046	1.7203	−0.3309
	2	−2.1570	0.1596	−0.38944	1.2403	−1.4230	
	3	−17.0300	N/A	N/A	−11.1292	−7.5644	
17	1	5.9299	−0.4829	0.8316	−2.4730	2.6880	−0.5668
	2	8.0855	−1.0436	2.0390	−5.5690	8.8422	
	3	−16.2143	N/A	N/A	−10.5758	−18.2825	
19	1	−2.3106	0.5325	−1.3829	3.7043	−8.0660	−0.4428
	2	−6.1350	0.1850	−0.4868	−3.7109	−2.2569	
	3	−16.6136	N/A	N/A	−17.3057	2.4955	

(*Continued*)

TABLE 10.4 Relative changes in the parameters due to varied wall thickness.

GB	Point i	$\Delta c^i/c^i$(%)	$\Delta k_1^i/k_1^i$ (%)	$\Delta k_2^i/k_2^i$ (%)	$\Delta k_3^i/k_3^i$ (%)	$\Delta k_4^i/k_4^i$ (%)	$\Delta\chi$ (%)
21	1	−2.3697	0.1387	−0.0528	1.2280	−1.2415	-0.2810
	2	0.1736	−0.0218	0.47771	0	6.8750	
	3	−16.4894	N/A	N/A	−15.5250	−1.9339	
29	1	−4.0751	0.9212	−2.2157	4.0037	−6.2898	-0.6329
	2	−4.7091	0.2866	−0.5581	1.4055	−1.1788	
	3	−16.6559	N/A	N/A	−14.8097	−15.6836	
37	1	3.6524	−0.4122	0.6446	−1.6959	2.2704	-0.7774
	2	0.1244	−0.0146	0.0402	−0.0565	0.4778	
	3	−16.8146	N/A	N/A	2.6846	−6.2472	
39	1	−3.7857	0.6820	−1.8188	2.2781	−3.4565	-0.6966
	2	−1.8519	0.2565	−0.4947	1.0310	−1.0246	
	3	−16.7569	N/A	N/A	−12.8803	−6.7415	
43	1	−3.7322	0.56367	−1.8439	3.0656	−5.6381	-0.4476
	2	−3.2326	0.28192	−0.8426	1.8476	−1.9197	
	3	−16.5808	N/A	N/A	−13.4554	−17.0601	

Δc^i, Δk_1^i, Δk_2^i, Δk_3^i, Δk_4^i and $\Delta\chi$ are the differences of these parameters and error between the case of h_{31} = 3.0-mm-thick apex wall and the case of 2.5 mm uniform GB wall, c^i, k_1^i, k_2^i, k_3^i, and k_4^i are the parameters for the 2.5 mm uniform GB wall.

which the Young's modulus is independent of the membrane thickness (Timoshenko and Woinowsky-Krieger 1959).

An increased h_{31} could lower the error in the stress between the model production and the observation. If h_{31} is increased to 5.0 mm, the error reduces by 2.5%. Further increase in thickness does not decrease the error much, see Figure 10.6; 5-mm apex thickness seems to agree with measurement by Su (2005). Finally, if h_{31} is reduced to be 10% thinner than h_{11} and h_{21}, 2.25 mm, according to Khan et al. (2012), then the errors in the stresses are greater, as shown in Figure 10.6. Thus, the observation that apex thickness was thinner than the GB body in the study by Khan et al. (2012) did not agree with the results from the cohort of GB samples used here.

Note that in the study by Khan et al. (2012), post-mortem samples from unclaimed bodies were used, so the samples were not fresh. When left *in situ* the bile will start to break down the GB wall – a process known as autolysis. Therefore, the results for wall thickness might not be reliable. Fresh samples by Su (2005), however, were obtained from the operating

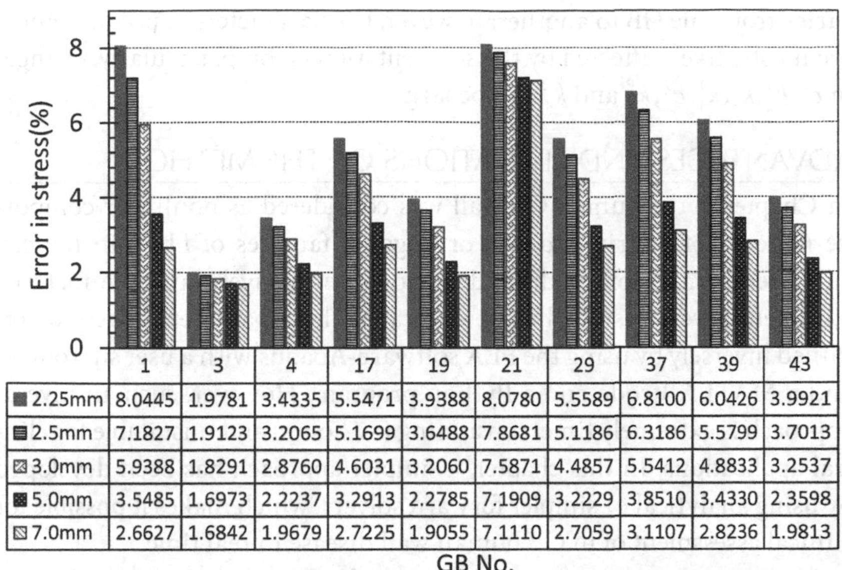

	1	3	4	17	19	21	29	37	39	43
2.25mm	8.0445	1.9781	3.4335	5.5471	3.9388	8.0780	5.5589	6.8100	6.0426	3.9921
2.5mm	7.1827	1.9123	3.2065	5.1699	3.6488	7.8681	5.1186	6.3186	5.5799	3.7013
3.0mm	5.9388	1.8291	2.8760	4.6031	3.2060	7.5871	4.4857	5.5412	4.8833	3.2537
5.0mm	3.5485	1.6973	2.2237	3.2913	2.2785	7.1909	3.2229	3.8510	3.4330	2.3598
7.0mm	2.6627	1.6842	1.9679	2.7225	1.9055	7.1110	2.7059	3.1107	2.8236	1.9813

FIGURE 10.6 Effect of GB wall thickness at the apex on the error in stress, the thickness at the apex is varied to be 2.25, 2.5, 3.0, 5.0 and 7.0 mm, respectively, while the thickness at the other two points 1 and 2 is kept in 2.5 mm.

theatre and washed immediately; hence, the results by Su (2005) should be more reliable.

EFFECTS OF SEGMENTATION ERROR

The segmentation error of estimating the GB diameter from a GB image is usually around 4.31–7.21% (Bocchi and Rogai 2011). To address the effect of this error on the GB wall material parameters, a random error (or noise) of (4.31–7.21)% in three principal axis lengths is introduced, i.e.,

$$\chi_{seg} = 0.0431 + \text{rand} \times (0.0721 - 0.0431) \tag{10.14}$$

where rand is the inner random function in MATLAB to generate a random number in value 0–1. Then the inverse heterogeneous problem code was run with these noisy data for a number of GB samples, say GBs 1, 3 and 17.

When noise is considered, the error in the curve fitting increased mostly by (3.1–6.7)%, some can be as high as 12.3%, in comparison with the case without the noise. The influence of segmentation error on the parameters varies from one GB to another; however, the parameters at points 1 and 2 are mostly likely affected by the segmentation error, particularly, changes in $c^1, k_2^1, k_3^1, k_4^1, c^2, k_3^2$ and k_4^2 can be large.

ADVANTAGES AND LIMITATIONS OF THE METHOD

In Chapter 9, the human GB wall was considered as nonlinear composite material of matrix and two orthogonal families of fibres in the circumferential and longitudinal directions, respectively, and the material parameters were assumed to be constant. These parameters were determined inversely by using the FEA software-Abaqus with a user subroutine and a MATLAB code and a Python program. However, such an inverse approach is extremely time-consuming (~7 hours) and unsuitable for clinical applications. In this chapter, a simpler approach has been developed by using analytical or simpler forward solvers, which makes it possible for clinical assessment of in GB human wall disease in real time.

Moreover, the previous model from the homogeneous analytical membrane model in Chapters 5 or 6 was extended to a heterogeneous model. The inverse estimation of the heterogeneous property constants had an error less than 7% for the ten human GB samples, and the computational time was reduced by 20 times (~30 min). Furthermore, variable GB wall thickness can reduce the error to be less than 4%.

In the chapter, the stretches at points 1–3 were determined analytically during GB emptying phase. This approach for obtaining stretch/strain at points in GB wall tissue remains not validated. Usually, the stretch/strain at points in the human left ventricle can be measured by speckle-tracking echocardiography (Edvardsen et al. 2002; Marwick 2006; Marwick et al. 2009; Crosby et al. 2009; Maffessanti et al. 2009; Tanaka et al. 2010; Hoit 2011; Kleijn et al. 2011). Unfortunately, this *in vivo* technique has not been applied in GB strain measurements so far. Hence, the approach proposed in the chapter for stretch/strain at a point in GB walls, fails to be validated.

The reference configuration of human GBs was only determined by one *in vitro* observation. The size of a reference configuration may not be exactly 50% of the size of totally filled GB. It is possible that the GB reference configuration can be estimated using GB *EF* in CCK-cholescintigraphy (CCK-CS) (Ozden and DiBaise 2003) or fatty meal cholescintigraphy (FM-CS) (Al-Muqbel et al. 2010) examination for GB patients' *in vivo* clinical diagnosis.

When human GBs suffer from diseases such as acute cholecystitis, acalculous cholecystitis and ascites (Sanders 1980; Runner et al. 2014), the GB thickness can increase significantly. Indeed, diseased GB body wall thickness was ranged in 3–5 mm (Sanders 1980; Mohammed et al. 2010; Runner et al. 2014). The GB wall thickness needs to be correlated to GB disease and the corresponding correlation should be involved into the model in future.

CLINICAL SIGNIFICANCE

As done in Chapter 9, six Young's moduli at three points on the wall of a GB were calculated by employing the Young's modulus formulas described in Chapter 9 based on the property constants listed in Table 10.1. The Young's modulus formulas at the three points are written as

$$\begin{cases} E_{\varphi}i = 2c^i(2\lambda_i^{\varphi}\dfrac{1}{\lambda_i^{\varphi 3}\lambda_i^{\varphi 2}} + 4k_1^i[\lambda_i^{\varphi 3} + \lambda_i^{\varphi}(I_4^i - 1) \\ \quad + 2k_2^i\lambda_i^{\varphi 3}(I_4^i - 1)^2]\exp[k_2^i(I_4^i - 1)^2] \\ E_{\theta}i = 2c^i(2\lambda_i^{\theta}\dfrac{1}{\lambda_i^{\theta 3}\lambda_i^{\theta 2}} + 4k_3^i[\lambda_i^{\theta 3} + \lambda_i^{\theta}(I_6^i - 1) \\ \quad + 2k_4^i\lambda_i^{\theta 3}(I_6^i - 1)^2]\exp[k_4^i(I_6^i - 1)^2] \end{cases} \tag{10.15}$$

where two-squared stretch ratios are $I_4^i = \lambda_i^{\varphi 2}$ and $I_6^i = \lambda_i^{\varphi 2}$, i is for the number of points, $i = 1, 2, 3$.

The extracted Young's modulus ratios $E_{\varphi 1}/E_{\theta 1}$, $E_{\varphi 2}/E_{\theta 2}$ and $E_{\varphi 3}/E_{\theta 3}$ are plotted in Figure 10.7 as a function of GB volume V. These Young's moduli ratios at the three points are dependent on GB volume, especially decrease with increasing GB volume. The ratios at points 1 and 2 differ each other, existing a heterogeneous, anisotropic and circumferentially stiff biomechanical property. At point 3, however, the strongest anisotropic property, i.e., the largest longitudinally stiff occurs because there is not circumferential fibre there.

The peak mean Young's modulus ratios of ten patients were extracted and used as GB pain index. The peak mean Young's modulus ratio is defined as the maximum arithmetic mean of the Young's modulus ratios at points 1 and 2 in the range of GB volume considered, i.e.,

$$E_{\varphi}/E_{\theta} = \text{Max}[0.5(E_{\varphi 1}/E_{\theta 1} + E_{\varphi 2}/E_{\theta 2})] \tag{10.16}$$

where Max [ ] represents to pick up the maximum E_{φ}/E_{θ} from a series of 0.5 ($E_{\varphi 1}/E_{\theta 1} + E_{\varphi 2}/E_{\theta 2}$), since it is as a function of GB volume V at 15 time instances.

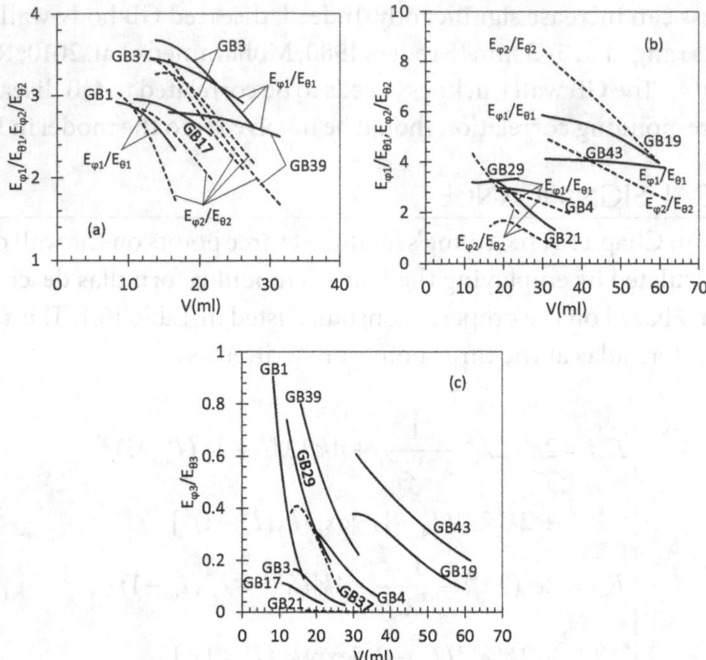

FIGURE 10.7 Young's modulus ratios $E_{\varphi1} / E_{\theta1}$, $E_{\varphi2} / E_{\theta2}$ and $E_{\varphi3} / E_{\theta3}$ are at points 1, 2 and 3 on GB walls are illustrated as a function of GB volume, the GB wall thickness is kept in 2.5 mm in the model.

TABLE 10.5 Acalculous GB pain prediction with the peak mean E_φ / E_θ and 3.0 threshold for 10 patients.

GB	Peak E_φ / E_θ	E_φ / E_θ threshold	Prediction	Observation	Agreement
1	2.98	3.0	Pain (−)	Pain (−)	Yes
3	3.47	3.0	Pain (+)	Pain (−)	No
4	2.92	3.0	Pain (−)	Pain (−)	Yes
17	2.90	3.0	Pain (−)	Pain (+)	No
19	7.21	3.0	Pain (+)	Pain (+)	Yes
21	2.44	3.0	Pain (−)	Pain (−)	Yes
29	3.78	3.0	Pain (+)	Pain (−)	No
37	3.45	3.0	Pain (+)	Pain (+)	Yes
39	3.12	3.0	Pain (+)	Pain (−)	No
43	4.54	3.0	Pain (+)	Pain (−)	No
Success rate					5/10 = 0.5

Pain (+) indicates that a GB is painful, Pain (−) means that a GB is not painful. The observation data were extracted from Table 6.1 in Chapter 6.

TABLE 10.6 Acalculous GB pain prediction with the peak mean E_φ / E_θ and 4.0 threshold for 10 patients.

GB	Peak E_φ / E_θ	E_φ / E_θ threshold	Prediction	Observation	Agreement
1	2.98	4.0	Pain (−)	Pain (−)	Yes
3	3.47	4.0	Pain (−)	Pain (−)	Yes
4	2.92	4.0	Pain (−)	Pain (−)	Yes
17	2.90	4.0	Pain (−)	Pain (+)	No
19	7.21	4.0	Pain (+)	Pain (+)	Yes
21	2.44	4.0	Pain (−)	Pain (−)	Yes
29	3.78	4.0	Pain (−)	Pain (−)	Yes
37	3.45	4.0	Pain (−)	Pain (+)	No
39	3.12	4.0	Pain (−)	Pain (−)	Yes
43	4.54	4.0	Pain (+)	Pain (−)	No
Success rate					7/10 = 0.7

Pain (+) indicates that GB is painful, Pain (−) means that GB is not painful. The observation data were extracted from Table 6.1 in Chapter 6.

The predicted 10 GB pain profiles are tubulated and compared with the clinical observation in Tables 10.5 and 10.6 for the E_φ / E_θ thresholds of 3.0 and 4.0, respectively. When the threshold varies from 3.0 to 4.0, the success rate rises from 50% to 70%. The peak mean Young's modulus ratio at points 1 and 2 might be applicable to GB pain prediction and will be helpful to GB pain diagnosis in future.

REFERENCES

Al-Muqbel, K. M., Bani Hani, M. N., Elheis, M. A., et al. 2010. Reproducibility of gallbladder ejection fraction measured by fatty meal cholescintigraphy. *Nuclear Medicine and Molecular Imaging* 44:246–251.

Bird, N., H. Wegstapel, R. Chess-Williams, et al. 1996. In vitro contractility of simulated human and non-stimulated gallbladder muscle. *Neurogastroenterology & Motility* 8:63–68.

Bocchi, L., and F. Rogai. 2011. Segmentation of ultrasound breast images: Optimization of algorithm parameters. In *Applications of Evolutionary Computation-LNCS 6624*, ed. C. Di Chio, S. Cagnoni, C. Cotta, et al., 163–172. Berlin: Springer-Verlag.

Crosby, J., B. H. Amundsen, T. Hørgum, et al. 2009. 3-D speckle tracking for assessment of regional left ventricular function. *Ultrasound in Medicine & Biology* 36:458–471.

Edvardsen, T., B. L. Gerber, J. Garot, et al. 2002. Quantitative assessment of intrinsic regional myocardial deformation by Doppler strain rate echocardiography in humans. *Circulation* 106:50–56.

Engel, J. M., E. A. Deitch, and W. Sikkema. 1980. Gallbladder wall thickness: Sonographic accuracy and relation to disease. *American Journal of Roentgenology* 134:907–909.

Genovese, K., L. Casaletto, J. D. Humphrey, et al. 2014. Digital image correlation-based point-wise inverse characterization of heterogeneous material properties of gallbladder in vitro. *Proceedings of Royal Society-Series A* 470:20140152.

Hoit, B. D. 2011. Strain and strain rate echocardiography and coronary artery disease. *Circulation* 4:179–190.

Khan, L. F., H. Naushaba, U. K. Paul, et al. 2012. Gross and histomorphological study of thickness of the gallbladder wall. *Journal Dhaka National Medical College & Hospital* 18:34–38.

Kleijn, S. A., M. F. Aly, C. B. Terwee, et al. 2011. Three-dimensional speckle tracking echocardiography for automatic assessment of global and regional left ventricular function based on area strain. *Journal of the American Society of Echocardiography* 24:314–321.

Li, W. G., N. C. Bird, and X. Y. Luo. 2017. A pointwise method for identifying biomechanical heterogeneity of the human gallbladder. *Frontiers in Physiology* 8:176.

Li, W. G., N. A. Hill, R. W. Ogden, et al. 2013. Anisotropic behaviour of human gallbladder walls. *Journal of the Mechanical Behavior of Biomedical Materials* 20:363–375.

Li, W. G., X. Y. Luo, N. A. Hill, et al. 2011. A mechanical model for CCK-induced acalculous gallbladder pain. *Annals of Biomedical Engineering* 39:786–800.

Li, W. G., X. Y. Luo, N. A. Hill, et al. 2012. A quasi-nonlinear analysis on anisotropic behaviour of human gallbladder wall. *ASME Journal of Biomechanical Engineering* 134:0101009.

Maffessanti, F., H. J. Nesser, L. Weinert, et al. 2009. Quantitative evaluation of regional left ventricular function using three-dimensional speckle tracking echocardiography in patients with and without heart disease. *American Journal of Cardiology* 104:1755–1762.

Marwick, T. H. 2006. Measurement of strain and strain rate by echocardiography. *Journal of American College of Cardiology* 47:1313–1327.

Marwick, T. H., R. L. Leano, J. Brown, et al. 2009. Myocardial strain measurement with 2-dimensional speckle-tracking echocardiography. *Journal of American College of Cardiology-Cardiovascular Imaging* 2:80–84.

Mohammed, S., A. Tahir, A. Ahidjo, et al. 2010. Sonographic gallbladder wall thickness in normal adult population in Nigeria. *South African Journal of Radiology* 14:84–87.

More, J. J., and D. C. Sorensen. 1983. Computing a trust region step. *SIAM Journal on Scientific and Statistical Computing* 4:553–572.

Oluseyi, K Y. 2018. Ultrasound determination of gall bladder size and wall thickness in normal adults in Abuja, North Central Nigeria. *Archives of International Surgery* 6:214–218.

Ozden, N., and J. K. DiBaise. 2003. Gallbladder ejection fraction and symptom outcome in patients with acalculous biliary-like pain. *Digestive Diseases and Sciences* 48:890–897.

Prasad, M. N., M. S. Brown, C. Ni, et al. 2008. Three-dimensional mapping of gallbladder wall thickness on computed tomography using Laplace's equation. *Academic Radiology* 15:1075–1081.

Runner, G. J., M. T. Corwin, B. Siewert, et al. 2014. Gallbladder wall thickening. *American Journal of Roentgenology* 202:W1–W12.

Sanders, R. 1980. The significance of sonographic gallbladder wall thickening. *Journal of Clinical Ultrasound* 8:143–146.

Su, Y. 2005. The mechanical properties of human gallbladder. BEng diss., University of Sheffield, Sheffield, UK.

Tanaka, H., H. Hara, S. Saba, et al. 2010. Usefulness of three-dimensional speckle tracking strain to quantify desynchrony and the site of latest mechanical activation. *American Journal of Cardiology* 105:235–242.

Timoshenko, S., and S. Woinowsky-Krieger. 1959. *Theory of Plates and Shells* (2nd edition). New York: McGraw-Hill Book Companies Inc.

Ugwu, A. C., and K. K. Agwu. 2010. Ultrasound quantification of gallbladder volume to establish baseline contraction indices in healthy adults: A pilot study. *The South African Radiographer* 48:9–12.

CHAPTER 11

Modelling the Viscoelastic Pressure-Volume Curve of the GB

GB PRESSURE-VOLUME CURVE WITH VISCOELASTICITY

In Chapters 5 and 6, it was learnt that ABP can be related to biomechanical factors. For patients with ABP, their GB smooth muscle is subject to an impaired contractile capacity in response to CCK (Merg et al. 2002) and the smooth muscle to fibrosis layer thickness ratio might be associated with abnormal EF (Lim et al. 2017). These two factors can alter the bile pressure in a GB and subsequently the stress level in the GB wall. Total (active plus passive) peak stress was estimated at the start of GB emptying phase in a CCK-ultrasound scan examination based on an ellipsoid model in the study by (Li et al. 2008, 2011a). The peak total stress can correlate to the ABP symptom well. These investigations are related to GB static biomechanical behaviour.

The human GB works with dynamic emptying-refilling cycle. A couple of *in vitro* experiments on GB dynamic biomechanical behaviour have been carried out by using saline infusion and withdrawal under passive and active states (Schoetz et al. 1981; Brotschi et al. 1984; Kaplan et al. 1984; Matsuki 1985). A typical pressure-volume curve of an intact GB in infusion-withdrawal experiments is shown in Figure 11.1.

The pressure path in the infusion is different from that in the withdrawal, forming a hysteresis loop. Clearly, an intact GB exhibits viscoelasticity

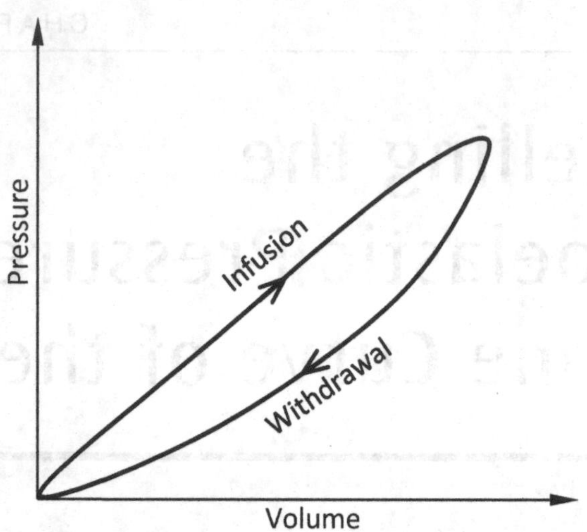

FIGURE 11.1 The hysteresis loop in pressure-volume curves of an intact GB in infusion-withdrawal cycle experiments.

like aorta (Goto and Kimoto 1966), artery (Gow and Taylor 1968), bladder (Remington and Alexander 1955), left ventricle (Rankin et al. 1977) and so on.

The hysteresis effect of an intact GB might have a relationship to ABP. As a first step, a nonlinear viscous model and a passive elastic model were proposed in terms of the discrete Voigt viscoelastic model in time domain by Li (2020) to identify the elastic and viscous responses in the experimental pressure-volume curves of an intact GB under passive and active conditions. The ratio of the work done on the viscous response to the elastic energy and the stiffness of the elastic pressure-volume curve were defined and extracted as well as discussed. In this chapter, the viscoelasticity modelling method for the pressure-volume curves of the intact GB was set forth briefly.

In the study by Bauer et al. (1979), the method was raised for separating elastic and viscous responses in the hysteresis loop in monitored arterial stress-radius curves in terms of the slope of radius-time relationships and a second-order viscous model. The method presented herein was similar to that by Bauer et al. (1979), but with two updated elements: (1) a new nonlinear viscous model and (2) a general biomechanical model for elastic pressure-volume curves of GBs for passive and active GBs.

THE EXPERIMENTS

In vitro experiments on pressure in the guinea pig GB were performed under passive and active conditions by Brotschi et al. (1984). During the experiments, a guinea pig was anaesthetised with urethane and laparotomy operated, a catheter sutured into the fundus of the GB was connected to an infusion-withdrawal pump and a pressure sensor as shown in Figure 11.2a. CCK was infused at 0.6 Ivy dog units/kg/min with a catheter inserted in the inferior vena cava over 20 min. The GB pressure-volume curve was recorded continuously with a computer in infusion (lasting 2 min) and withdrawal (2 min). The recorded GB pressure-time history

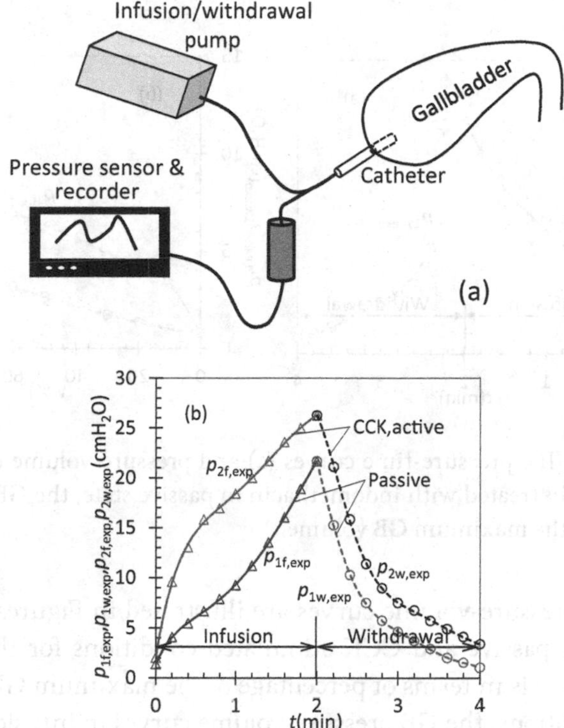

FIGURE 11.2 A sketch of experimental installation for GB pressure-volume measurements *in vitro* of a guinea pig (a), pressure-time curves (b), pressure volume curves (c) under passive and CCK stimulated conditions for the same GB, and pressure-time curves for the GB in passive state, the GB volume is expressed with percentage of the maximum GB volume [(a) is adapted from Brotschi et al. 1984].

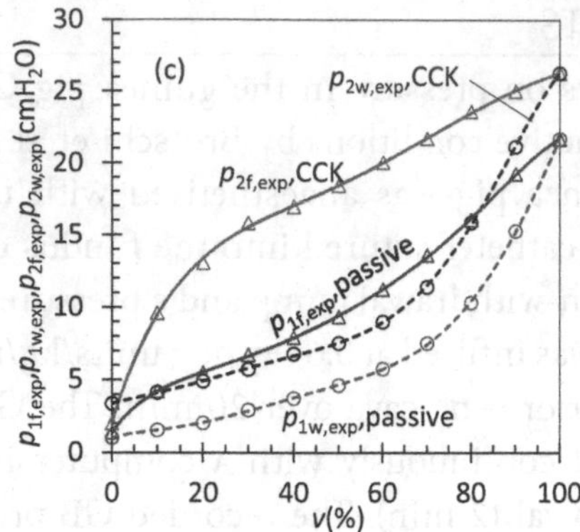

FIGURE 11.2 (Continued)

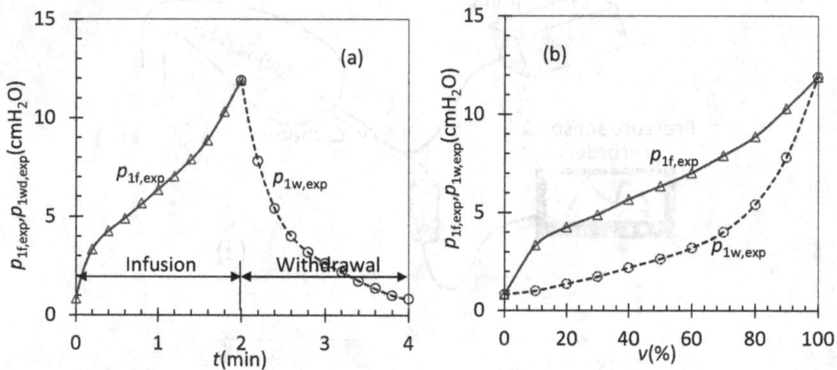

FIGURE 11.3 The pressure-time curves (a) and pressure-volume curves (b) for a guinea pig GB treated with indomethacin in passive state, the GB volume is in percentage of the maximum GB volume.

curves and pressure-volume curves are illustrated in Figures 11.2(b) and 11.2(c) under passive and CCK stimulated conditions for the same GB. The GB volume is in terms of percentage of the maximum GB volume. In the two conditions, the GB pressure-volume curves in infusion and withdrawal processes are in difference paths, exhibiting a hysteresis effect.

Indomethacin can diminish GB mobility or tone and its influence on guinea-pig gallstone formation and motility in control and cholesterol-fed was examined as well (Brotschi et al. 1984). The recorded pressure-time and pressure-volume curves of the GB treated with indomethacin in passive state are illustrated in Figure 11.3. Clearly, the shape of the curves is changed but the hysteresis effect still exists on the pressure-volume curves. This effect will be modelled by using nonlinear discrete viscoelastic models in the next sections.

VISCOELASTIC MODELS AVAILABLE

Existing viscoelastic models for soft tissues are subject to four classifications: (1) discrete model in time domain, (2) discrete model in frequency domain, (3) quasi-linear model or continuous model in time domain and (4) viscous potential function model in time domain.

In the discrete model in time domain, a soft tissue is represented mechanically with a spring and a dashpot in series (Maxwell model) or in parallel (Voigt model) or their combinations (van Duyl 1985) and time is independent variable. The model can involve a linear spring and a linear dashpot (Kondo and Susset 1973) or an exponential spring and a linear dashpot (Glantz 1974) or a linear spring and a nonlinear dashpot (Sanjeevi 1982) or a spring with isotropic exponential strain energy function and a linear dashpot (Vandenbrouck et al. 2010) or a spring with neo-Hookean strain energy function and a nonlinear dashpot with power function (Panda and Buist 2020).

In the discrete model in frequency domain, a soft tissue is still represented with a spring and a dashpot in series or in parallel or their combinations, but the governing equation of the model is transformed to the frequency domain with Laplace transform, and frequency is independent variable to determine the complex Young's modulus. Once the solution is obtained in the frequency domain, it can be mapped back to the time domain by employing the inverse Laplace transform (Westerhof and Noordergraaf 1970; Cox 1972; Goedhard and Knoop 1973).

In the quasi-linear model or continuous model in time domain or Fung model, the stress in a soft tissue equals to the product of an elastic stress and a reduced relaxation function (Woo et al. 1981; Provenzano et al. 2001; Davis and De Vita 2012; Ghahfarokhi et al. 2020). The total stress is calculated from a convolution integral between the reduced relaxation function and the rate of elastic stress. The reduced relaxation function is calculated by employing relaxation test data. The elastic stress function is estimated with total stress elongation test data at various strain rates.

In the viscous potential function model in time domain, the elastic stress is represented with a strain energy function, but the viscous stress is in terms of viscous potential function which is as a function of a few invariants related to viscoelasticity and strain rate, especially for short-term viscoelasticity (Pioletti et al. 1998; Limbert and Middleton 2004; Lu et al. 2010; Ahsanizadeh and Li 2015; Kulkarni et al. 2016; Yousefi et al. 2018), the property constants are determined with stress-strain curves at various strain rates.

VISCOELASTIC MODEL FOR GB WALLS

The GB is composed of three layers: mucosa, muscularis externa and adventitia without considering epithelium (serosa) (see Figure 8.12). The very tall, simple columnar epithelium of the mucosa forms a number of folds to absorb water in the bile effectively. The mucosa is full of vascular vessels but without lymphatics. The muscle layer includes the longitudinal smooth muscle fibres along the longitudinal direction near the lamina propria and circular smooth muscle fibres near the adventitia layer. These smooth muscle fibres contribute to contraction as the GB is under stimulation of CCK. The adventitia layer consists of collagen fibres such as I, III and IV types and fibronectin (Yamada et al. 1989) as well as elastin, which are responsible for GB passive expansion and shrinkage.

From the biomechanical point of view, the GB wall is regarded as three nonlinear springs, one contractile element, and one nonlinear dashpot in parallel phenomenologically and biomechanically, as shown in Figure 11.4. In the passive state, the elastic pressure response p_e is related

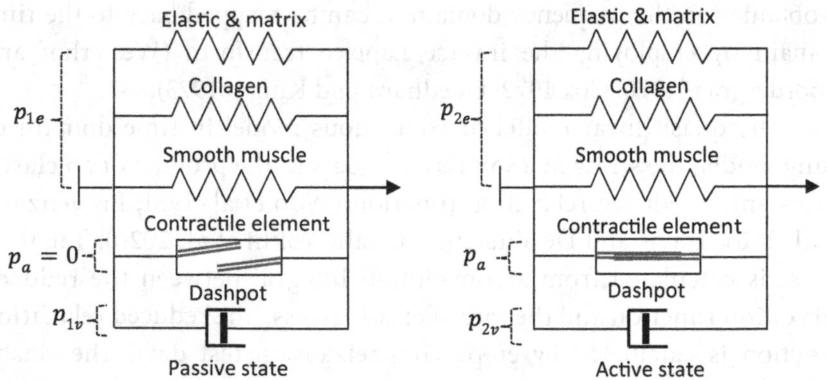

FIGURE 11.4 Biomechanical viscoelastic discrete models in time domain for the pressure-volume curve of GBs in passive and active states, $p_a = 0$ in the passive state.

to the apparent stiffness of elastin and matrix, collagen fibre and smooth muscle, that the viscous pressure response p_v exists is owing to the dashpot, the cross bridges in the muscle are not in engagement, so that the active response is zero, $p_a = 0$ (Li et al. 2011b). The collagen fibres only engage at high GB volume. In active stage, except the three passive springs and one dashpot, the active pressure response p_a comes into play as the cross bridges are in engagement. Due to effect of CCK, the model parameters for a GB wall in a passive state are different from those in an active state.

The pressure-volume curves shown in Figures 11.2 and 11.3 are decomposed into three components, namely, passive, active and viscoelastic components to match respective responses of elastic, contractile and viscoelastic elements in parallel in the GB wall tissue.

In the passive state experiment, the pressure p_{1f} in infusion process and p_{1w} in withdrawal process are divided into the pressures for elastic and viscous responses, and expressed as

$$\begin{cases} p_{1f} = p_{1fe}(V) + p_{1fv}(\mathrm{d}p_{1f} / \mathrm{d}t) \text{ infusion} \\ p_{1w} = p_{1we}(V) + p_{1wv}(\mathrm{d}p_{1w} / \mathrm{d}t) \text{ withdrawal} \end{cases} \tag{11.1}$$

where $p_{1fe}(V)$ and $p_{1we}(V)$ are the pressures for the elastic responses in infusion and withdrawal processes, $p_{1fv}(\mathrm{d}p_{if} / \mathrm{d}t)$ and $p_{1wv}(\mathrm{d}p_{iw} / \mathrm{d}t)$ are the pressures for the viscous responses in infusion and withdrawal processes.

First, $p_{1fe}(V)$ and $p_{1fv}(\mathrm{d}p_{1f} / \mathrm{d}t)$, $p_{1we}(V)$ and $p_{1wv}(\mathrm{d}p_{1w} / \mathrm{d}t)$ are determined based on the pressure-volume data in a passive state experiment. $p_{1fv}(\mathrm{d}p_{1f} / \mathrm{d}t)$ is supposed to be the product of nonlinear viscosity $\mu_1(\mathrm{d}p_{1f} / \mathrm{d}t, a_1, a_2, a_3, m_1)$ and $\mathrm{d}p_{1f} / \mathrm{d}t$, while $p_{1wv}(\mathrm{d}p_{1w} / \mathrm{d}t)$ is the product of nonlinear viscosity $\mu_1(\mathrm{d}p_{1w} / \mathrm{d}t, a_1, a_2, a_3, m_1)$ and $\mathrm{d}p_{1w} / \mathrm{d}t$. Note that $\mathrm{d}p_{1f} / \mathrm{d}t$ and $\mathrm{d}p_{1w} / \mathrm{d}t$ are known in the experiment, and the viscosity function μ_1 and its constants a_1, a_2, a_3, m_1 can be determined with the experimental pressure-volume curves. $p_{1fe}(V)$ and $p_{1we}(V)$ are calculated by the following expressions

$$\begin{cases} p_{1fe}(V_i) = p_{1f,\exp}(V_i) - \mu_1(\mathrm{d}p_{1f} / \mathrm{d}t, a_1, a_2, a_3, m_1)\mathrm{d}p_{1f} / \mathrm{d}t \text{ infusion} \\ p_{1we}(V_i) = p_{1w,\exp}(V_i) - \mu_1(\mathrm{d}p_{1w} / \mathrm{d}t, a_1, a_2, a_3, m_1)\mathrm{d}p_{1w} / \mathrm{d}t \text{ withdrawal} \end{cases} \tag{11.2}$$

where $p_{1fe}(V_i)$ and $p_{1we}(V_i)$ are elastic response in infusion and withdrawal processes at a GB specific volume V_i, $p_{1fe}(V_i)$ and $p_{1we}(V_i)$ are independent of loading paths. Hence, the proper constants a_1, a_2, a_3, m_1 should make the following objective function minimum

$$\text{fun}_1 = \Sigma_{i=1}^{N}[p_{1\text{fe}}(V_\text{i}) - p_{1\text{we}}(V_\text{i})]^2 \tag{11.3}$$

where N is the number of experimental data points.

Second, for the same intact GB in active state, the pressure decomposition for elastic and viscous responses is expressed as follows in infusion and withdrawal processes

$$\begin{cases} p_{2\text{f}} = p_{2\text{fe}}(V) + p_{2\text{fv}}(\text{d}p_{2\text{f}}/\text{d}t) \text{ infusion} \\ p_{2\text{w}} = p_{2\text{we}}(V) + p_{2\text{wv}}(\text{d}p_{2\text{w}}/\text{d}t) \text{ withdrawal} \end{cases} \tag{11.4}$$

where $p_{2\text{fv}}(\text{d}p_{2\text{f}}/\text{d}t)$ is a function of both $\text{d}p_{2\text{f}}/\text{d}t$ and nonlinear viscosity $\mu_2(\text{d}p_{2\text{f}}/\text{d}t, b_1, b_2, b_3, m_2)$, this is true for $\text{d}p_{2\text{wv}}(\text{d}p_{2\text{w}}/\text{d}t$. They are estimated by the following expressions

$$\begin{cases} p_{2\text{fe}}(V_\text{i}) = p_{2\text{f,exp}}(V_\text{i}) - \mu_2(\text{d}p_{2\text{f}}/\text{d}t, b_1, b_2, b_3, m_2)\text{d}p_{2\text{f}}/\text{d}t \text{ infusion} \\ p_{2\text{we}}(V_\text{i}) = p_{2\text{w,exp}}(V_\text{i}) - \mu_2(\text{d}p_{2\text{w}}/\text{d}t, b_1, b_2, b_3, m_2)\text{d}p_{2\text{w}}/\text{d}t \text{ withdrawal} \end{cases} \tag{11.5}$$

where $p_{2\text{fe}}(V_\text{i})$ and $p_{2\text{we}}(V_\text{i})$ are independent of loading path. Therefore, the proper constants b_1, b_2, b_3, m_2 should make the following objective function minimum as well

$$\text{fun}_2 = \Sigma_{i=1}^{N}[p_{2\text{fe}}(V_\text{i}) - p_{2\text{we}}(V_\text{i})]^2 \tag{11.6}$$

Thirdly, the pressure for active response $p_\text{a}(V_\text{i})$ at each experimental data point is calculated by

$$\begin{cases} p_\text{a}(V_\text{i}) = p_{2\text{e}}(V_\text{i}) - p_{1\text{e}}(V_\text{i}) \\ p_{1\text{e}}(V_\text{i}) = 0.5[p_{1\text{fe}}(V_\text{i}) + p_{1\text{we}}(V_\text{i})] \\ p_{2\text{e}}(V_\text{i}) = 0.5[p_{2\text{fe}}(V_\text{i}) + p_{2\text{we}}(V_\text{i})] \end{cases} \tag{11.7}$$

then, the mathematical models of pressure passive and active components can be established based on the data points of $p_{1\text{e}}(V_\text{i})$ and $p_\text{a}(V_\text{i})$. In this way, the dynamic behaviour of an intact GB can be resolved.

Finally, the extracted $p_{1e}(V_i)$, $p_a(V_i)$, $p_{1fv}(V_i)$ − $p_{1wv}(V_i)$ and $p_{2fv}(V_i)$ − $p_{2wv}(V_i)$ curves are analysed in terms of existing clinical observations to provide some insight into GB dynamic behaviour.

The viscous biomechanical model is essential for solving Eqs. (11.2) and (11.5). Thus, the following viscous biomechanical model is proposed for the nonlinear dashpot in passive state

$$\begin{cases} \mu_1\left(\frac{dp_{1f}}{dt}, a_1, a_2, a_3, m_1\right)\frac{dp_{1f}}{dt} = \left(a_1 + a_2 \left|\frac{dp_{1f}}{dt}\right|^{m_1} + a_3 \left|\frac{dp_{1f}}{dt}\right|^{2m_1}\right)\frac{dp_{1f}}{dt} \\ \mu_1\left(\frac{dp_{1w}}{dt}, a_1, a_2, a_3, m_1\right)\frac{dp_{1w}}{dt} = \left(a_1 + a_2 \left|\frac{dp_{1w}}{dt}\right|^{m_1} + a_3 \left|\frac{dp_{1w}}{dt}\right|^{2m_1}\right)\frac{dp_{1w}}{dt} \end{cases} \tag{11.8}$$

Likewise, in active state, the viscous biomechanical model reads

$$\begin{cases} \mu_2\left(\frac{dp_{2f}}{dt}, b_1, b_2, b_3, m_2\right)\frac{dp_{2f}}{dt} = \left(b_1 + b_2 \left|\frac{dp_{2f}}{dt}\right|^{m_2} + b_3 \left|\frac{dp_{2f}}{dt}\right|^{2m_2}\right)\frac{dp_{2f}}{dt} \\ \mu_2\left(\frac{dp_{2w}}{dt}, b_1, b_2, b_3, m_2\right)\frac{dp_{2w}}{dt} = \left(b_1 + b_2 \left|\frac{dp_{2w}}{dt}\right|^{m_2} + b_3 \left|\frac{dp_{2w}}{dt}\right|^{2m_2}\right)\frac{dp_{2w}}{dt} \end{cases} \tag{11.9}$$

where the property constants a_2 and b_2 are negative, but the rest constants are positive in Eqs. (11.8) and (11.9).

At three specific time moments, namely the start of infusion, t_1, the end of infusion, t_2, and the end of withdrawal, t_3, the slopes dp_{1f}/dt, dp_{1w}/dt, dp_{2f}/dt and dp_{2w}/dt should be zero, i.e.,

$$\frac{dp_{1f}}{dt} = \frac{dp_{1w}}{dt} = \frac{dp_{2f}}{dt} = \frac{dp_{2w}}{dt} = 0 \text{ at } t_1, t_2, t_3 \tag{11.10}$$

where $t_1 = 0$, $t_2 = 2$ and $t_3 = 4$ min based on Figure 11.2(b).

The scattered data points in Figure 11.2(b) are best fitted by using a fourth- or third-order polynomial, and then its derivative with respect to time is derived to obtain experimental slopes dp_{1f}/dt, dp_{1w}/dt, dp_{2f}/dt and dp_{2w}/dt. In passive state, their expressions are written as

$$\begin{cases} \frac{dp_{1t}}{dt} = -4\times 3.3052t^3 + 3\times 14.155t^2 - 2\times 18.983t + 15.722 \\ \frac{dp_{1w}}{dt} = -4\times 3.0114t^3 - 3\times 41.538t^2 + 2\times 215.06t - 499.03 \end{cases} \tag{11.11}$$

Whilst in active state, the expressions are deduced as follows:

$$\begin{cases} \dfrac{\mathrm{d}p_{2f}}{\mathrm{d}t} = -4\times 6.7026t^3 + 3\times 32.28t^2 + 2\times 54.1770t + 44.842 \\ \dfrac{\mathrm{d}p_{2w}}{\mathrm{d}t} = -3\times 4.4689t^2 + 2\times 47.664t - 127.54 \end{cases} \tag{11.12}$$

The condition expressed with Eq. (11.10) should be imposed in Eqs. (11.11) and (11.12), even though these slopes may not be exactly zero at $t_1 = 0$, $t_2 = 2$ and $t_3 = 4$ min.

The experimental pressure-time curves in infusion and withdrawal for the GB treated with indomethacin in passive state are illustrated in Figure 11.3a, the slopes of the curves are expressed as

$$\begin{cases} \dfrac{\mathrm{d}p_{1f}}{\mathrm{d}t} = -6\times 3.6236t^5 + 5\times 23.9049t^4 - 4\times 61.192t^3 + 3\times 77.8922t^2 \\ \qquad -2\times 51.5662t + 20.1179 \\ \dfrac{\mathrm{d}p_{1w}}{\mathrm{d}t} = 6\times 0.0644t^5 - 5\times 1.8649t^4 + 4\times 21.7908t^3 - 3\times 130.824t^2 \\ \qquad +2\times 427.1563t - 725.1213 \end{cases} \tag{11.13}$$

After the passive elastic pressure-volume scattered points are extracted, they should be formulated with a mathematical model. The following mathematical model is proposed in terms of dimensionless GB volume

$$p_e = p_{e,\exp}(0) + c_0 v + c_1 v^{n_1} + c_2 e^{c_2 v^{n_2}} \tag{11.14}$$

where v is dimensionless GB volume, $v = V/V_{max}$, V_{max} is the largest GB volume achievable in an experiment, V is a GB volume less than V_{max}, $p_{e,\exp}(0)$ is the GB pressure at $v = 0$ in the experiment, c_0, c_1, c_2, c_3, n_1 and n_2 are model constants and will be determined by employing a set of experimental pressure-volume data.

The biomechanical model and experimental data of a guinea-pig GB in Figure 11.2(b) and (c) were coded in MATLAB. First, the pressure-volume curves for the elastic and viscous responses in passive and active states are split by employing the *lsqnonlin* function based on Trust-Region-Reflective optimisation algorithm, then the elastic pressure-volume curves in both the states are best fitted by using Eq. (11.14).

VISCOELASTICITY CHARACTERISATION

The determined model constants and errors in the two procedures are listed in Table 11.1. The errors in passive and active states, χ_1 and χ_2 are defined as

$$\begin{cases} \chi_1 = \dfrac{\sqrt{\text{fun}_1}}{0.5\Sigma_{i=1}^{N}[p_{1\text{f},\exp^{(V_i)}} + p_{1\text{w},\exp^{(V_i)}}]} \times 100\% \\ \chi_2 = \dfrac{\sqrt{\text{fun}_2}}{0.5\Sigma_{i=1}^{N}[p_{2\text{f},\exp^{(V_i)}} + p_{2\text{w},\exp^{(V_i)}}]} \times 100\% \end{cases} \tag{11.15}$$

The split elastic and viscous pressure-volume curves are illustrated in Figure 11.5 along with the experimental data, and the elastic passive and active pressure-volume curves are shown, too. It is shown that the errors,

TABLE 11.1 Model constants optimised in pressure decomposition and pressure-volume curve fitting in passive and active states.

GB state	Viscous property		Elastic property		
	Constant and error	**Value**	**Constant**	**Value**	χ_p (%)
Passive	a_1	4.3667×10^{-1}	c_0	10.2740	0.41
	a_2	-8.9372×10^{-3}	c_1	9.0932	
	a_3	4.5843×10^{-5}	n_1	4.2455	
	m_1	1.2801	c_2	3.0288×10^{-7}	
	χ_1 (%)	1.63	c_3	15.2020	
			n_2	9.9131	
Active	b_1	6.3195×10^{1}	c_0	1.1972×10^{1}	0.69
	b_2	-1.0000×10^{2}	c_1	4.4048	
	b_3	3.8962×10^{1}	n_1	4.9091×10^{-10}	
	m_2	3.5578×10^{-2}	c_2	1.2123×10^{-1}	
	χ_2 (%)	1.61	c_3	4.1278	
			n_2	2.7313	
Indomethacin	a_1	8.4342×10^{-1}	c_0	4.8221	0.40
	a_2	-8.3111×10^{-2}	c_1	1.4452	
	a_3	2.1309×10^{-3}	n_1	4.8451×10^{-1}	
	m_1	1.1403	c_2	2.9176×10^{-3}	
	χ_1 (%)	1.16	c_3	7.4109	
			n_2	1.0950	

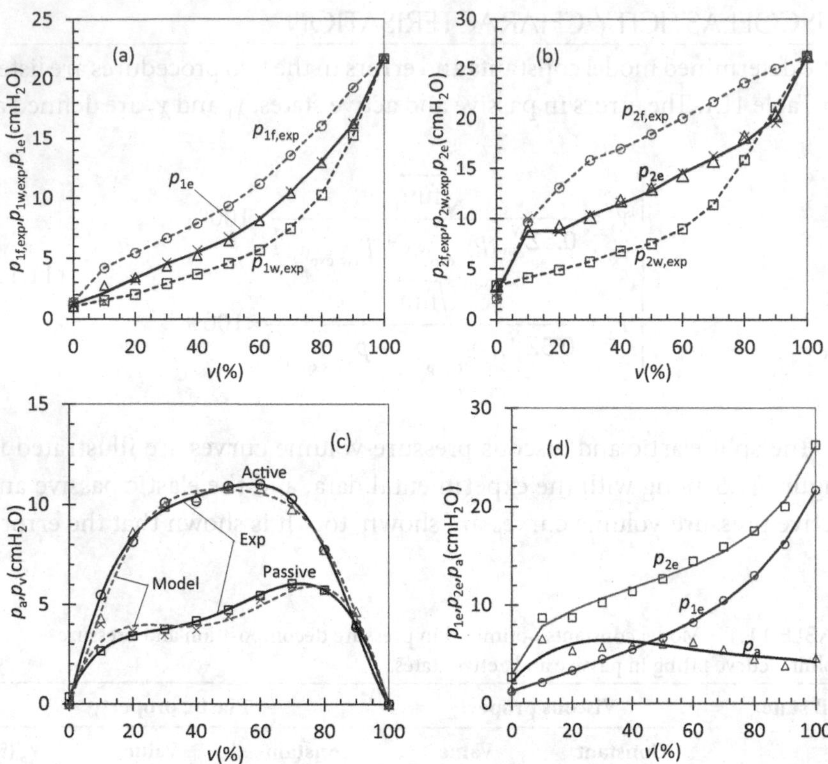

FIGURE 11.5 The discomposed pressure responses to dimensionless volume change in a guinea-pig GB in both passive and active states, (a) passive, (b) active, (c) viscous and (d) elastic passive pressure and active pressure, the symbols are for experimental data.

χ_1 and χ_2 are all less than 2%, suggesting the biomechanical models proposed are feasible and reasonable.

The extracted viscous pressure-volume curves are very similar to those from the experimental data. As a result, the error in the areas enclosed by the curves and the horizontal axis in Figure 11.5(c) is less than 1% between the experiments and the model predictions.

In Figure 11.5(d), the extracted elastic pressures, p_{1e} and p_{2e} can be best fitted by using the model expressed with Eq. (11.14), giving the error χ_p as small as 0.6% and 1.0%, respectively. The error χ_p is defined as

$$\chi_{\mathrm{p}} = \frac{\sqrt{\Sigma_{i=1}^{N}[p_{\mathrm{e}}(V_{\mathrm{i}}) - p_{\mathrm{e,exp}}(V_{\mathrm{i}})]^2}}{\Sigma_{i=1}^{N} p_{\mathrm{e,exp}}(V_{\mathrm{i}})} \times 100 \tag{11.16}$$

where $p_{e,exp}(V_i)$ and $p_e(V_i)$ are the experimental GB pressure and the pressure predicted with Eq. (11.14) at an experimental point i.

The active pressure response can be best fitted by employing a polynomial as done in (Gestrelius and Borgstron 1986; Schmitz and Bol 2011) for smooth muscle in active state as the following:

$$p_a = -3.3519\times10^{(-7)}v^4 + 9.2538\times10^{(-5)}v^3 - 9.11\times10^{(-3)}v^2 + 3.4058\times10^{(-1)}v + 2.2114 \quad (11.17)$$

where the correlation coefficient of the equation remains to be 0.84. The peak active state is located at v = 30%.

Likewise, the extracted elastic pressure-volume curve for a guinea-pig GB treated with indomethacin in passive state is illustrated in Figure 11.6a, and the pressure-volume curves for viscous response are demonstrated in Figure 11.6b, the determined model constants are listed in Table 11.1.

To clarify the difference in biomechanical properties between CCK stimulated GBs, normal passive GBs and the GBs treated with indomethacin in passive state, the ratio of the work done on the viscous response in the hysteresis loop to the elastic energy is proposed, which is defined as

$$\Gamma = \frac{W}{\text{En}}, W = \oint p_v \text{d}V, \text{En} = \int_0^{V_{\max}} p_e \text{d}V \quad (11.18)$$

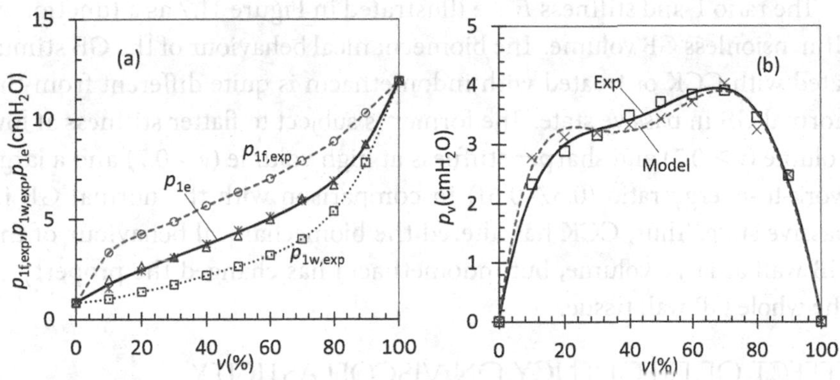

FIGURE 11.6 The extracted elastic pressure response (a) and viscous pressure response (b) in a guinea-pig GB treated with indomethacin in passive state, the symbols are for experimental data, the curves for model predictions.

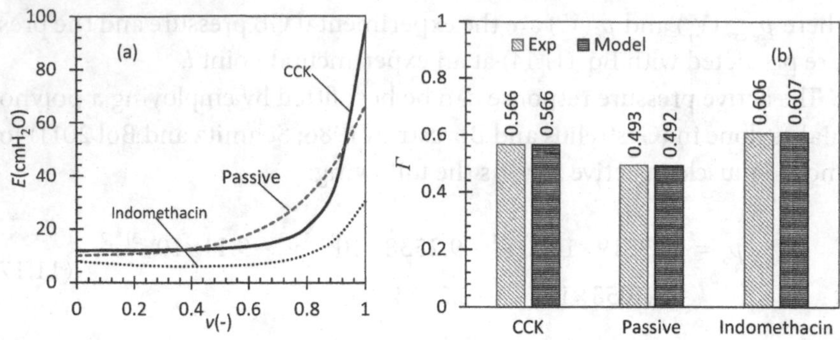

FIGURE 11.7 The elastic stiffness E and the ratio of the viscous loss energy to the elastic energy stored in GB wall Γ in active state stimulated with CCK, passive state in control and passive state treated with indomethacin.

where Γ is the work-to-energy ratio, En and W are the elastic energy stored and the work done by loading, p_v is the pressure for viscous response, p_e is the pressure for elastic response.

Additionally, the stiffness of the elastic pressure-volume curve expressed with Eq. (11.14) is extracted with the following expression

$$E = \frac{dp_e}{dv} = c_0 + c_1 n_1 v^{n_1 - 1} + c_2 c_3 n_2 v^{n_2 - 1} e^{c_2 v^{n_2}} \tag{11.19}$$

where E is the stiffness and the powers are $n_1, n_2 > 0$.

The ratio Γ and stiffness E are illustrated in Figure 11.7 as a function of dimensionless GB volume. The biomechanical behaviour of the GB stimulated with CCK or treated with indomethacin is quite different from the normal GB in passive state. The former is subject to flatter stiffness at low volume ($v \geq 0.7$) and sharper stiffness at high volume ($v > 0.7$) and a large work-to-energy ratio (0.57–0.61) in comparison with the normal GB in passive state. Thus, CCK has altered the biomechanical behaviour of the GB wall at large volume, but indomethacin has changed the property of the whole GB wall tissue.

EFFECT OF FREQUENCY ON VISCOELASTICITY

The slopes dp_{1f}/dt, dp_{1w}/dt, etc., were used as an independent variable in the viscous model. The pressure- and volume-time curves of a baboons GB are in almost identical phase angle and the frequency of infusion-withdrawal cycle is 1/20 Hz (Schoetz et al. 1981). It suggests that the slope of pressure-time curve or volume-time can be used as an independent variable in the viscoelastic model.

In the experimental data used in the chapter, the frequency of infusion-withdrawal cycle is 1/240 Hz, which is 1/12th of 1/20 Hz, hence the GB pressure and volume will be in the same phase angle during the cycle. In this context, that the slopes dp_{1f}/dt, dp_{1w}/dt, etc., were considered as independent variables in the viscous model is reasonable and feasible.

In GB infusion-withdrawal cycle experiments, the frequency of the infusion-withdrawal cycle can influence the viscoelastic property of intact GBs. The hysteresis loop size grows as the frequency is at 0.17 and 3.33 Hz for guinea pig GBs in passive state, respectively. But the loop size is the narrowest at the frequency of 1.67 Hz somehow (Matsuki 1985).

The mean timescales of emptying (withdrawal) are 10 and 18 min for the human healthy GBs and the GBs with gallstone, while the refilling (infusion) timescales are 60 min (Pomeranz and Shaffer 1985). Based on these timescales, the frequencies of emptying (withdrawal)-refilling (infusion) cycle can be as low as 1/4200 Hz for the healthy GBs and 1/4680 Hz for the GBs with gallstone. Therefore, there is a significant difference in the frequency of infusion-withdrawal cycle in experiment from physiology, but also the emptying is much faster than the refilling. The viscoelastic property in human intact GBs at such a low frequency of emptying-refilling cycle needs a confirmation in future.

CLINICAL SIGNIFICANCE

The GB pressure-volume curve based on *in vitro* infusion-withdrawal cycle sketched in Figure 11.1 differs from that in physiological state illustrated. The GB pressure-volume curve in the GB physiological refilling-emptying cycle includes refilling, isometric contraction, emptying and isometric relaxation (Li et al. 2013), see the solid lines in Figure 11.8, but that pressure-volume curve in infusion-withdrawal cycle experiments has not the isometric contraction and relaxation phases. Additionally, the hysteresis loop sense in the infusion-withdrawal cycle is opposite to that in the refilling-emptying cycle. These facts show that the infusion-withdrawal cycle experiment differs from the physiological refilling-emptying cycle and is essentially an *in vitro* or *in vivo* method for characterising the viscoelastic biomechanical property only of an intact GB,

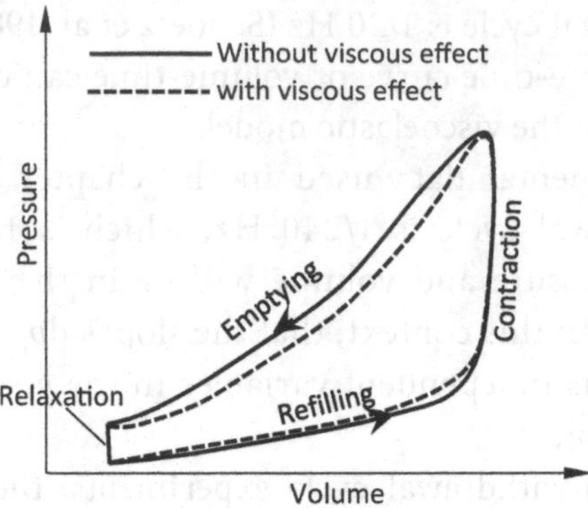

FIGURE 11.8 A comparison of GB physiological emptying-filling cycle between normal GB and GB with dominant viscosity.

then has nothing to do with the physiological refilling-emptying cycle of GBs.

Since the viscous effect in a GB wall usually exhibits a resistance to deformation and a delay in dynamic response, if the viscous effect in the GB wall is considered in the GB refilling-emptying cycle, the cycle should be like one in dashed lines shown in the dashed lines in Figure 11.8(b). If the effect is strong enough, it will impair the GB motor function.

CCK is a peptide hormone to stimulate GB muscle by attaching to CCK receptors imbedded in the smooth muscle of the GB (Yau et al. 1973; Krishnamurthy and Krisbnamurthy 1997). The change in the stiffness of the passive elastic pressure-volume curve of the GB stimulated with CCK occurs mainly at a large GB volume compared with the normal GB as shown in Figure 11.7(a). This effect apparently suggests CCK may result in significant alternation in the passive biomechanical property of the smooth muscle at large GB volume.

Indomethacin is a NSAID blocking the production of prostaglandins which not only influence smooth muscle contractility but also relieve inflammation. Indomethacin can reduce or abolish GB smooth muscle spontaneous rhythmical contractions by inhibiting endogenous prostaglandin synthesis (Brotschi et al. 1984; Doggrell and Scott 1980; Kotwall et al. 1984). The other studies showed that indomethacin can improve GB emptying of patients with gallstones (O'Donnell et al. 1992) or animal models with cholecystitis (Yau et al. 1973) and has no effect on healthy

GBs (Parkman et al. 2001; Murray et al. 1992). Based on those observations, it is concluded that the change in biomechanical property of GB smooth muscle is attributed to indomethacin.

In Figure 11.7a, however, indomethacin likely reduces the stiffness of the passive pressure-volume curve considerably at low and large GB volumes. Actually, indomethacin may alter the biomechanical property of the elastin and collagen fibre. Nonetheless, the corresponding experimental evidence is desirable.

Based on the present methods proposed here the elastic, active and viscous pressure-volume curves have been separated for the GB. If the GB geometrical parameters were known with volume change, then the corresponding 3D elastic, active and viscous stresses would be estimated by using the simple ellipsoid model of the GB in Chapters 5 and 6. Since the geometrical parameters of the experimental GB were not provided by Brotschi et al. (1984), these stress estimates cannot be workable in the chapter.

REFERENCES

Ahsanizadeh, S., and L. P. Li. 2015. Visco-hyperelastic constitutive modeling of soft tissues based on short and long-term internal variables. *BioMedical Engineering OnLine* 14:1–16. https://doi.org/10.1186/s12938-015-0023-7

Bauer, R. D., R. Busse, A. Schabert, et al. 1979. Separate determination of the pulsatile elastic and viscous developed in the arterial wall in vivo. *European Journal of Physiology* 380:221–226.

Brotschi, E. A., W. W. Lamore, and L. F. Williams. 1984. Effect of dietary cholesterol and indomethacin on cholelithiasis and gallbladder motility in guinea pig. *Digestive Diseases and Sciences* 29:1050–1056.

Cox, R. H. 1972. A model for the dynamic mechanical properties of arteries. *Journal of Biomechanics* 5:135–152.

Davis, F. M., and R. De Vita. 2012. A nonlinear constitutive model for stress relaxation in ligaments and tendons. *Annals of Biomedical Engineering* 40:2541–2550.

Doggrell, S. A., and G. W. Scott. 1980. The effect of time and indomethacin on contractile response of the guinea-pig gall bladder in vitro. *British Journal of Pharmacology* 71:429–434.

Gestrelius, S., and P. Borgstron. 1986. A dynamic model of smooth muscle contraction. *Biophysical Journal* 50:157–169.

Ghahfarokhi, Z. M., M. M. Zand, M. S. Tehrani, et al. 2020. A visco-hyperelastic constitutive model of short- and long-term viscous effects on isotropic soft tissues. *Proc IMechE Part C: Journal Mechanical Engineering Science* 234:3–17.

Glantz, S. A. 1974. A constitutive equation for the passive properties of muscle. *Journal of Biomechanics* 7:137–145.

Goedhard, W. J., and A. A. Knoop. 1973. A model of the arterial wall. *Journal of Biomechanics* 6:281–288.

Goto, M., and Y. Kimoto. 1966. Hysteresis and stress-relaxation of the blood vessels studied by a universal tensile testing instrument. *Japanese Journal of Physiology* 16:169–184.

Gow, B. S., and M. G. Taylor. 1968. Measurement of viscoelastic properties of arteries in the living dog. *Circulation Research* 23:111–122.

Kaplan, G. S., V. K. Bhutani, T. H. Shaffer, et al. 1984. Gallbladder mechanics in newborn piglets. *Pediatric Research* 18:1181–1184.

Kondo, A., and J. G. Susset. 1973. Physical properties of the urinary detrusor muscle: A mechanical model based upon the analysis of stress relaxation curve. *Journal of Biomechanics* 6:141–151.

Kotwall, C. A., A. S. Clanachan, H. P. Baer, et al. 1984. Effects of prostaglandins on motility of gallbladders removed patients with gallstones. *Archive of Surgery* 119:709–712.

Krishnamurthy, S., and G. T. Krishnamurthy. 1997. Biliary dyskinesia: Role of the sphincter of Oddi, gallbladder and cholecystokinin. *Journal of Nuclear Medicine* 38:1824–1830.

Kulkarni, S. G., X. L. Gao, S. E. Horner, et al. 2016. A transversely isotropic visco-hyperelastic constitutive model for soft tissues. *Mathematics and Mechanics of Solids* 21:747–770.

Li, W. G. 2020. Modelling of viscoelasticity in pressure-volume curve of an intact gallbladder. *Mechanics of Soft Materials* 2:1–16.

Li, W. G., N. A. Hill, and R. W. Ogden, et al. 2013. Anisotropic behaviour of human gallbladder walls. *Journal of Mechanical Behavior of Biomedical Materials* 20:363–375.

Li, W. G., X. Y. Luo, N. A. Hill, et al. 2008. Correlation of mechanical factors and gallbladder pain. *Computational & Mathematical Methods in Medicine* 9:27–45.

Li, W. G., X. Y. Luo, N. A. Hill, et al. 2011a. Cross-bridge apparent rate constants of human gallbladder smooth muscle. *Journal of Muscle Research and Cell Motility* 32:209–220.

Li, W. G., X. Y. Luo, N. A. Hill, et al. 2011b. A mechanical model for CCK-induced acalculous gallstone pain. *Annals of Biomedical Engineering* 39:786–800.

Lim, J. U., K. R. Joo, K. Y. Won, et al. 2017. Predictor of abnormal gallbladder ejection fraction in patients with atypical biliary pain: Histopathological point of view. *Medicine (Baltimore)* 96:e9269.

Limbert, G., and J. Middleton. 2004. A transversely isotropic viscohyperelastic material Application to the modeling of biological soft connective tissues. *International Journal of Solids and Structures* 41:4237–4260.

Lu, Y. T., H. X. Zhu, S. Richmond, et al. 2010. Visco-hyperelastic model for skeletal muscle tissue under high strain rates. *Journal of Biomechanics* 43:2629–2632.

Matsuki, Y. 1985. Dynamic stiffness of the isolated guinea pig gall-bladder during contraction induced by cholecystokinin. *Japanese Journal of Smooth Muscle Research* 21:427–438.

Merg, A. R., S. E. Kalinowski, M. M. Hinkhous, et al. 2002. Mechanisms of impaired gallbladder contractile response in chronic acalculous cholecystitis. *Journal of Gastrointestinal Surgery* 6:432–437.

Murray, F. E., S. J. Stinchcombe, and C. J. Hawkey. 1992. Effect of indomethacin and misoprostol on fast gallbladder volume and meal-induced gallbladder contractility in humans. *Digestive Diseases and Sciences* 37:1228–1231.

O'Donnell, L. J., P. Wilson, P. Guest, et al. 1992. Indomethacin and postprandial gallbladder emptying. *Lancet* 339:269–271.

Panda, S. K., and M. L. Buist. 2020. A viscoelastic framework for inflation testing of gastrointestinal tissue. *Journal of Mechanical Behavior of Biomedical Materials* 103:103569.

Parkman, H. P., A. N. James, R. M. Thomas, et al. 2001. Effect of indomethacin on gallbladder inflammation and contractility during acute cholecystitis. *Journal of Surgical Research* 96:135–142.

Pioletti, D. P., L. R. Rakotomanana, J. F. Benvenuti, et al. 1998. Viscoelastic constitutive law in large deformations: Application to human knee ligaments and tendons. *Journal of Biomechanics* 31:753–757.

Pomeranz, I. S., and E. A. Shaffer. 1985. Abnormal gallbladder emptying in a subgroup of patients with gallstones. *Gastroenterology* 88:787–791.

Provenzano, P., R. Lakes, T. Keenan, et al. 2001. Nonlinear ligament viscoelasticity. *Annals of Biomedical Engineering* 29:908–914.

Rankin, J. S., C. E. Arentzen, P. A. McHale, et al. 1977. Viscoelastic properties of the diastolic left ventricle in the conscious dog. *Circulation Research* 41:37–45.

Remington, J. W., and R. S. Alexander. 1955. Stretch of the bladder as an approach to vascular distensibility. *American Journal of Physiology* 181:240–248.

Sanjeevi, R. 1982. A viscoelastic model for the mechanical properties of biological materials. *Journal of Biomechanics* 15:107–109.

Schmitz, A., and M. Bol. 2011. On a phenomenological model for active smooth muscle contraction. *Journal of Biomechanics* 44:2090–2095.

Schoetz, D. J., W. W. Lamore, W. E. Wise, et al. 1981. Mechanical properties of primate gallbladder description by a dynamic method. *American Journal of Physiology-Gastrointestinal Liver Physiology* 41:G376–G381.

Vandenbrouck, A., H. Laurent, N. A. Hocine, et al. 2010. A hyperelasto-visco-hysteresis model for an elastomeric behaviour: Experimental and numerical investigations. *Computational Materials Science* 48:495–503.

van Duyl, W. A. 1985. A model for both the passive and active properties of urinary bladder tissue related to bladder function. *Neurourology and Urodynamics* 4:275–283.

Westerhof, N., and A. Noordergraaf. 1970. Arterial viscoelasticity: A generalized model. *Journal of Biomechanics* 3:357–379.

Woo, S. L., M. A. Gomez, and W. H. Akeson. 1981. The time and history-dependent viscoelastic properties of the canine medical collateral ligament. *ASME Journal of Biomechanical Engineering* 103:293–297.

Yamada, K., Y. Matsumura, H. Suzuki, et al. 1989. The histochemistry of collagens and fibronectin in the gallbladders of the guinea pig and mouse. *Acta Histochemica et Cytochemica* 22:675–683.

Yau, W. M., G. M. Makhouf, L. E. Edwards, et al. 1973. Mode of actin of cholecystokinin and related peptides on gallbladder muscle. *Gastroenterology* 65:451–456.

Yousefi, A. K., M. A. Nazari, P. Perrier, et al. 2018. A visco-hyperelastic constitutive model and its application in bovine tongue tissue. *Journal of Biomechanics* 71:190–198.

CHAPTER 12

Constitutive Law of GB Walls with Damage Effects

BIOMECHANICAL PROPERTY OF GB WALLS

For the passive biomechanical property of GB walls, there are a few measurements in the literature. GB pressure-volume curves were measured *in vitro* by inflating the GB with saline in passive and active states (Miura and Saito 1967; Ryan and Cohen 1976; Schoetz et al. 1981; Brotschi et al. 1984; Matsuki 1985a,b; Borly et al. 1996). It was shown that GB pressure-volume curves presented viscoelastic property (Miura and Saito 1967; Schoetz et al. 1981; Brotschi et al. 1984; Matsuki 1985a,b). The viscoelasticity on the pressure-volume curves of an intact GB has been clarified analytically in Chapter 11.

The compliance of GB was figured out based on experimental data (Schoetz et al. 1981). Porcine GB walls were measured under compression loads on a material testing machine and engineering stress-strain curves were established in passive compressed state (Rosen et al. 2009). An organ inflating experimental set-up was built and a lamb GB shape was recorded *in vitro* optically when the GB was pressurised with phosphate-buffered solution in the study by Genovese et al. (2014). Based on the membrane mechanics model, the passive biomechanical property constants were optimised numerically with FEA.

Porcine GBs were harvested from a slaughterhouse and experienced indentation experiments along the circumferential and longitudinal directions when the GBs chamber was full of bile. The GBs were cut into specimens in both the directions, and the specimens were elongated in passive state on a uniaxial material testing machine (Xiong et al. 2013). The GB wall material property constants were determined based on the indentation experimental data and strain energy function proposed by Fung et al. (1979).

The passive uniaxial biomechanical property of human healthy GB walls was measured on a material testing machine by Karimi et al. (2017) based on a few specimens harvested from the GBs of corpses in hospital. The engineering stress-strain curves were provided.

GB passive material constitutive law plays an important role in GB refilling and emptying phases, perhaps with a connection with GB disease diagnosis. In the chapter, two constitutive laws with damage were established based on the uniaxial tensile test data on the human (Karimi et al. 2017) and porcine (Xiong et al. 2013) GB walls in passive state. The details for this approach can be found in the study by Li (2019).

THE EXPERIMENTAL DATA

Uniaxial tensile tests on five pairs of the circumferential and longitudinal specimens of porcine GB walls were conducted by Xiong et al. (2013), and the engineering stresses were presented as a function of stretch. The circumferential samples were stiffer than the longitudinal one in stress-stretch curves. The experimental Cauchy stress-stretch curves are illustrated in Figure 12.1.

Sixteen GBs were excised from the cadavers of human subjects that were 69.3 ± 9.8 years old, and GB wall specimens were cut along the axial and transversal directions and tested on DBBP-50 material testing machine by Karimi et al. (2017). The axial direction is the longitudinal direction, while the transversal direction is the circumferential direction in common sense. The stress level seems to be one order larger than the porcine GB wall in tension measured by Xiong et al. (2013) and in compression tested by Rosen et al. (2008). Therefore, the experimental stress levels have to be reduced by 1/10 to make them comparable with those in the porcine GB wall. The experimental Cauchy-stretch curves are presented in Figure 12.1. Two sets of the Cauchy-stretch curves in the figure will be employed to produce passive constitutive laws of GB walls.

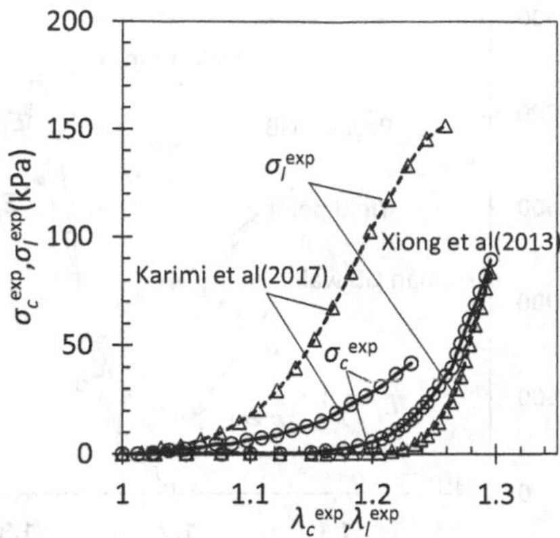

FIGURE 12.1 Two sets of experimental Cauchy stress-stretch curves of porcine and human GB walls.

DAMAGE IN SPECIMENS

Damage is related to change in instant Young's modulus of a specimen during its uniaxial tensile tests (Lemaitre 1984; Voyiadjis and Kattan 2013, 2017; Fett et al. 2018). To make sure whether there is the damage effect in the experimental data shown in Figure 12.1, the scattered data points were best fitted by using a sixth-order polynomial, the Young's moduli such as circumferential modulus E_c^{exp} and longitudinal modulus E_l^{exp} were worked out by calculating the instant slopes of the curves, which are defined as

$$\begin{cases} E_c^{exp} = d\sigma_c^{exp} / d\lambda_c^{exp} \\ E_l^{exp} = d\sigma_l^{exp} / d\lambda_l^{exp} \end{cases} \tag{12.1}$$

where σ_c^{exp} and λ_c^{exp} are the experimental circumferential Cauchy stress and stretch, σ_l^{exp} and λ_l^{exp} are the experimental longitudinal Cauchy stress and stretch. Since the Young's moduli are very small and less change when λ_c^{exp} and λ_l^{exp} are smaller than certain values, just the parts with substantial change in instant Young's moduli are fitted and demonstrated in Figure 12.2.

For the human GB wall, the longitudinal Young's modulus E_l^{exp} is always larger than the circumferential E_c^{exp}, indicating the longitudinal specimen is stiffer than the circumferential one. For the porcine GB wall, however, the longitudinal specimen is not stiffer than the circumferential specimen until the stretch of 1.25.

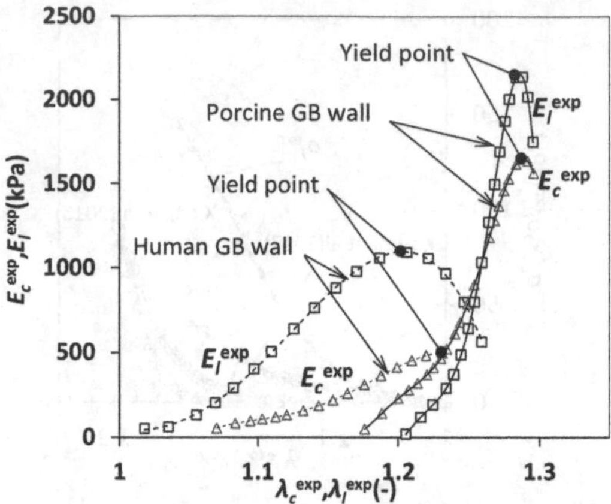

FIGURE 12.2 The longitudinal and circumferential Young's moduli determined by fitting the uniaxial test data in Figure 12.1 with sixth-order polynomial.

These Young's moduli grow with increasing stretch until the peak value for both the GB walls. Beyond that point, the modulus starts declining, suggesting the yield point existence in the curves. Therefore, there is a damage effect in the circumferential and longitudinal specimens.

CONSTITUTIVE LAW WITHOUT DAMAGE EFFECTS

First, the constitutive law is established without considering the damage effect above in GB wall tissues. A constitutive law for a passive human GB wall without damage in the wall tissue was proposed by Li et al. (2013) based on the law for passive arterial walls in Holzapfel et al. (2000). The strain energy function based constitutive law by Li et al. (2013) is written as

$$\psi = c(I_1 - 3) + \frac{\kappa_1}{2\kappa_2}[e^{k_2(I_4-1)^2} - 1] + \frac{\kappa_3}{2\kappa_4}[e^{\kappa_4(I_6-1)^2} - 1 \tag{12.2}$$

where c is the matrix material stiffness, k_1 and k_2 are the initial stiffness and its change rate with stretch of the circumferential fibres, k_3 and k_4 are the initial stiffness and its change rate with stretch of the longitudinal fibres, I_1 is the trace of the Cauchy-Green deformation tensor, $I_1 = \lambda_c^2 + \lambda_l^2 + \lambda_h^2, \lambda_c, \lambda_l$ and λ_h are the stretches in the circumferential, longitudinal and thickness directions, respectively, I_4 is the squared λ_c, $I_4 = \lambda_c^2$ and I_6 is the squared λ_l, $I_6 = \lambda_l^2$.

Five model parameters c, k_1, k_2, k_3 and k_4 can be determined based on the experimental stress-stretch curves shown in Figure 12.1 by using *lsqnonlin* function in MATLAB in terms of 'Trust-Region-Reflective' optimisation algorithm to minimise the value of the following objective function

$$\text{fun}(c,k_1,k_2,k_3,k_4)=\Sigma_{i=1}^{N_c}(\sigma_{ci}^{\text{mod}}-\sigma_{ci}^{\text{exp}})^2+\Sigma_{j=1}^{N_l}(\sigma_{lj}^{\text{mod}}-\sigma_{lj}^{\text{exp}})^2 \quad (12.3)$$

where σ_{ci}^{mod} and σ_{lj}^{mod} are the circumferential and longitudinal Cauchy stresses calculated by using the strain energy function in Eq. (12.2) at the i-th experimental stretch $\lambda_{ci}^{\text{exp}}$ in the uniaxial tensile test of a circumferential sample and at the j-th experimental stretch $\lambda_{lj}^{\text{exp}}$ in the similar test of a longitudinal sample, respectively; N_c and N_l are the total numbers of experimental points in the uniaxial tensile tests on the circumferential and longitudinal samples; σ_{ci}^{exp} and σ_{lj}^{exp} are the measured circumferential and longitudinal Cauchy stresses at the i-th and j-th experimental points, and presented in Figure 12.1.

σ_{ci}^{mod} and σ_{lj}^{mod} are calculated by the following equations when the incompressible condition $\lambda_c\lambda_l\lambda_h=1$ is held in the GB wall. For the uniaxial tensile test of circumferential samples, σ_{ci}^{mod} reads as

$$\sigma_{ci}^{\text{mod}}=\lambda_{ci}^{\text{exp}}\frac{\partial\psi}{\partial I_1}\frac{\partial I_1}{\partial\lambda_c}\bigg|_{\lambda_c=\lambda_{ci}^{\text{exp}}}+\lambda_{ci}^{\text{exp}}\frac{\partial\psi}{\partial I_4}\frac{\partial I_4}{\partial\lambda_c}\bigg|_{\lambda_c=\lambda_{ci}^{\text{exp}}}, \quad (12.4)$$
$$I_1=\lambda_c^2+\lambda_l^2+\lambda_h^2,\lambda_l=\lambda_h=1/\lambda_c$$

and, for the uniaxial tensile test of longitudinal samples, σ_{lj}^{mod} is expressed as

$$\sigma_{lj}^{\text{mod}}=\lambda_{lj}^{\text{exp}}\frac{\partial\psi}{\partial I_1}\frac{\partial I_1}{\partial\lambda_l}\bigg|_{\lambda_l=\lambda_{lj}^{\text{exp}}}+\lambda_{lj}^{\text{exp}}\frac{\partial\psi}{\partial I_6}\frac{\partial I_6}{\partial\lambda_l}\bigg|_{\lambda_l=\lambda_{lj}^{\text{exp}}}, \quad (12.5)$$
$$I_1=\lambda_c^2+\lambda_l^2+\lambda_h^2,\lambda_c=\lambda_h=1/\lambda_l$$

The standard deviation error in the Cauchy stress is calculated to evaluate the curve fitting quality quantitatively, the expression for the error is written as

$$\chi_\sigma=\frac{1}{\sigma_{\text{mean}}^{\text{exp}}}\sqrt{\frac{\sum_{i=1}^{N_c}\left(\sigma_{ci}^{\text{mod}}-\sigma_{ci}^{\text{exp}}\right)^2+\sum_{j=1}^{N_l}\left(\sigma_{lj}^{\text{mod}}-\sigma_{lj}^{\text{exp}}\right)^2}{N_c+N_l}}\times100\% \quad (12.6)$$

TABLE 12.1 The determined parameters from uniaxial tensile test data based on the model without damage.

Damage effect	Parameter	GB wall	
		Human	**Porcine**
Excluded	c (kPa)	0.0182	0.0037
	k_1 (kPa)	9.1076	1.8718
	k_2 (-)	4.2018	6.7299
	k_3 (kPa)	47.0016	0.2681
	k_4 (-)	1.9348	10.8841
	χ_σ (%)	15.9825	11.6071

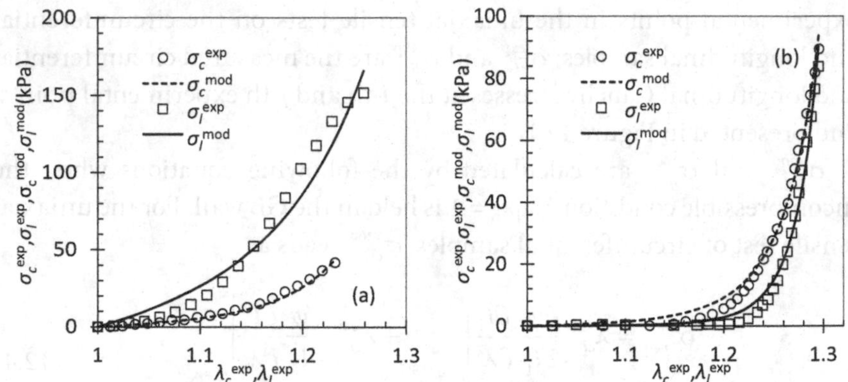

FIGURE 12.3 The uniaxial test data and predicted Cauchy stress-stretch curves by using model Eq. (12.2) based on the determined parameters in Table 12.1, (a) human GB wall and (b) porcine GB wall.

where $\sigma_{\text{mean}}^{\text{exp}}$ is the mean Cauchy stress in both kinds of uniaxial tensile test, i.e., $\sigma_{\text{mean}}^{\text{exp}} = \left(\sum_{i=1}^{N_c} \sigma_{ci}^{\text{exp}} + \sum_{j=1}^{N_l} \sigma_{lj}^{\text{exp}} \right) \Big/ (N_c + N_l)$.

The equations above were programmed in MATLAB, the determined five parameters are listed in Table 12.1 and a comparison is made in Figure 12.3 between the measured and predicted stresses at the same stretch values. For both the human and the porcine GB walls, the predicted and measured Cauchy stress-stretch share different slopes clearly. As a result, the errors in the stress curves are as high as 16.0% and 11.6%, suggesting a poor performance of the model presented by Eq. (12.2).

CONSTITUTIVE LAW WITH DAMAGE

The results in the last section suggest that the damage effect in the experimental stress-stretch curves must be included to reduce the error in the constitutive law. In doing so phenomenologically, based on the work by Li and Luo (2016) and Li (2018), the constitutive model presented by Eq. (12.2) is expanded by adding three extra terms and rewritten as

$$\psi^{\mathrm{dam}} = c\left[(I_1-3)-\frac{(I_1-3)^{m_\mathrm{d}+1}}{(m_\mathrm{d}+1)(\zeta-3)^{m_\mathrm{d}}}\right]+\frac{k_1}{2k_2}$$
$$\left\{\exp\left[k_2(I_4-1)^2\right]-1-\frac{2k_2(I_4-1)^{n_\mathrm{d}+2}}{(n_\mathrm{d}+2)(\eta^2-1)^{n_\mathrm{d}}}\right\} \tag{12.7}$$
$$+\frac{k_3}{2k_4}\left\{\exp\left[k_4(I_6-1)^2\right]-1-\frac{2k_4(I_6-1)^{n_\mathrm{d}+2}}{(n_\mathrm{d}+2)(\eta^2-1)^{n_\mathrm{d}}}\right\}$$

where m_d, n_d, η and ζ are the phenomenological parameters to describe the damage in the GB wall, m_d and ζ are relevant to the matrix damage; m_d specifies the sharpness of the stress-stretch curve when the damage occurs, and ζ indicates the value of I_1 when the matrix damage occurs, n_d and η are the corresponding parameters for the fibre damage; n_d is the counterpart of m_d, and η demonstrates the fibre stretch λ_f at which the fibres damage occurs. If these parameters are chosen to be $\eta = \zeta = +\infty$ and $m_\mathrm{d} = n_\mathrm{d} = 1$, then the constitutive model Eq. (12.2) is restored.

The uniaxial tensile test data in Figure 12.1 were read into a MATLAB code to perform optimisation processes to determine nine model constants c, k_1, k_2, k_3, k_4, m_d, n_d, η and ζ simultaneously by using the same algorithm above. The estimated parameters are present in Table 12.2, and the predicted stress-stretch curves with them are plotted in Figure 12.4 along with the corresponding experimental data for comparison.

Based on Tables 12.1 and 12.2, after the damage effect is involved in both specimens, the fitting errors in the stress are reduced to 4.6% and 8.2% from 16.0% and 11.6%, respectively, for both the GB walls.

For the porcine GB wall, the stresses are very flat as the stretch is less than 1.2 or so, suggesting just the matrix material engages in tension. However, when the stretch is in the range of 1.2–1.3, the stress level grows

TABLE 12.2 The determined parameters from uniaxial tensile test data of human and porcine GB walls based on the model with damage in the matrix and fibres.

Damage effect	Parameter	GB wall	
		Human	Porcine
Included damage effects in matrix and fibres	c (kPa)	0.4770	0.0129
	k_1 (kPa)	5.8311	0.6444
	k_2 (-)	8.9293	10.2208
	k_3 (kPa)	23.8486	0.1964
	k_4 (-)	8.5417	12.0102
	η (-)	1.1825	1.2447
	n_{d} (-)	7.0763	18.6265
	m_{d} (-)	10.0174	18.5992
	ζ (-)	5.8792	4.3933
	χ_σ (%)	4.5836	8.1985

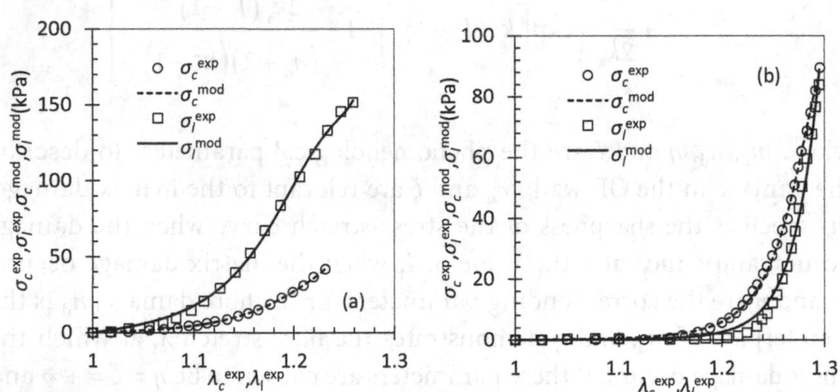

FIGURE 12.4 The uniaxial test data and predicted Cauchy stress-stretch curves by using model Eq. (12.7) based on the determined parameters in Table 12.2, (a) human GB wall and (b) porcine GB wall.

markedly with increasing stretch, implying fibres are recruited extensively. This effect results in a difficulty in constitutive behaviour modelling of the porcine GB wall. Additionally, compared with the human GB wall, constants c, k_1 and k_2 are smaller, but k_3 and k_4 are larger in value.

DAMAGE VARIABLES

For convenience of damage variable estimation, the strain energy function in Eq. (12.7) is divided into three parts: the first part is the strain energy function for the matrix, $\psi_{\mathrm{m}}^{\mathrm{dam}}$, the second part is the strain energy function for the circumferential fibres, $\psi_{\mathrm{fc}}^{\mathrm{dam}}$ and the third part is the strain energy function for the longitudinal fibres, $\psi_{\mathrm{fl}}^{\mathrm{dam}}$, then the updated equation is rewritten as

$$
\begin{aligned}
\psi^{\mathrm{dam}} &= \psi_{\mathrm{m}}^{\mathrm{dam}} + \psi_{\mathrm{fc}}^{\mathrm{dam}} + \psi_{\mathrm{fl}}^{\mathrm{dam}} \\
\psi_{\mathrm{m}}^{\mathrm{dam}} &= c\left[(I_1-3) - \frac{(I_1-3)^{m_{\mathrm{d}}+1}}{(m_{\mathrm{d}}+1)(\zeta-3)^{m_{\mathrm{d}}}}\right] \\
\psi_{\mathrm{fc}}^{\mathrm{dam}} &= \frac{k_1}{2k_2}\left\{\exp\left[k_2(I_4-1)^2\right] - 1 - \frac{2k_2(I_4-1)^{n_{\mathrm{d}}+2}}{(n_{\mathrm{d}}+2)(\eta^2-1)^{n_{\mathrm{d}}}}\right\} \\
\psi_{\mathrm{fl}}^{\mathrm{dam}} &= \frac{k_3}{2k_4}\left\{\exp\left[k_4(I_6-1)^2\right] - 1 - \frac{2k_4(I_6-1)^{n_{\mathrm{d}}+2}}{(n_{\mathrm{d}}+2)(\eta^2-1)^{n_{\mathrm{d}}}}\right\}
\end{aligned}
\tag{12.8}
$$

The Cauchy stress components in the tissue can be divided into three parts accordingly: the first part is the stress in the matrix, $\sigma_{\mathrm{m}}^{\mathrm{dam}}$, due to $\psi_{\mathrm{m}}^{\mathrm{dam}}$, the second part is the stress in the circumferential fibres, $\sigma_{\mathrm{fc}}^{\mathrm{dam}}$ and the third part is the stress in the longitudinal fibres, $\sigma_{\mathrm{fl}}^{\mathrm{dam}}$, i.e.,

$$
\begin{aligned}
\sigma_{\mathrm{c}}^{\mathrm{dam}} &= \sigma_{\mathrm{mc}}^{\mathrm{dam}} + \sigma_{\mathrm{fc}}^{\mathrm{dam}}, \sigma_{\mathrm{l}}^{\mathrm{dam}} = \sigma_{\mathrm{ml}}^{\mathrm{dam}} + \sigma_{\mathrm{fl}}^{\mathrm{dam}} \\
\sigma_{\mathrm{mc}}^{\mathrm{dam}} &= \lambda_{\mathrm{c}}\frac{\partial\psi_{\mathrm{m}}^{\mathrm{dam}}}{\partial I_1}\frac{\partial I_1}{\partial\lambda_{\mathrm{c}}} = 2\left(\lambda_{\mathrm{c}}^2 - 1/\lambda_{\mathrm{c}}\right)\left[c - \left(\frac{I_1-3}{\zeta-3}\right)^m\right] \\
\sigma_{\mathrm{fc}}^{\mathrm{dam}} &= \lambda_{\mathrm{c}}\frac{\partial\psi_{\mathrm{fc}}^{\mathrm{dam}}}{\partial I_4}\frac{\partial I_4}{\partial\lambda_{\mathrm{c}}} = 2\lambda_{\mathrm{c}}^2 k_1(I_4-1)\left[\mathrm{e}^{k_2(I_4-1)^2} - \left(\frac{I_4-1}{\xi^2-1}\right)^n\right] \\
\sigma_{\mathrm{ml}}^{\mathrm{dam}} &= \lambda_{\mathrm{l}}\frac{\partial\psi_{\mathrm{m}}^{\mathrm{dam}}}{\partial I_1}\frac{\partial I_1}{\partial\lambda_{\mathrm{l}}} = 2\left(\lambda_{\mathrm{l}}^2 - 1/\lambda_{\mathrm{l}}\right)\left[c - \left(\frac{I_1-3}{\zeta-3}\right)^m\right] \\
\sigma_{\mathrm{fl}}^{\mathrm{dam}} &= \lambda_{\mathrm{l}}\frac{\partial\psi_{\mathrm{fl}}^{\mathrm{dam}}}{\partial I_6}\frac{\partial I_6}{\partial\lambda_{\mathrm{l}}} = 2\lambda_{\mathrm{l}}^2 k_3(I_6-1)\left[\mathrm{e}^{k_4(I_6-1)^2} - \left(\frac{I_6-1}{\xi^2-1}\right)^n\right]
\end{aligned}
\tag{12.9}
$$

Taking the derivatives of $\sigma_{\mathrm{c}}^{\mathrm{dam}}$, $\sigma_{\mathrm{mc}}^{\mathrm{dam}}$ and $\sigma_{\mathrm{fc}}^{\mathrm{dam}}$ with respect to λ_{c} and the derivatives of $\sigma_{\mathrm{l}}^{\mathrm{dam}}$, $\sigma_{\mathrm{ml}}^{\mathrm{dam}}$ and $\sigma_{\mathrm{fl}}^{\mathrm{dam}}$ with respect to λ_{l}, the Young's moduli can be calculated by using the following equations

$$E_c^{dam} = \partial\sigma_c^{dam}/\partial\lambda_c, E_{mc}^{dam} = \partial\sigma_{mc}^{dam}/\partial\lambda_c, E_{fc}^{dam} = \partial\sigma_{fc}^{dam}/\partial\lambda_c$$
$$E_l^{dam} = \partial\sigma_l^{dam}/\partial\lambda_l, E_{ml}^{dam} = \partial\sigma_{ml}^{dam}/\partial\lambda_l, E_{fl}^{dam} = \partial\sigma_{fl}^{dam}/\partial\lambda_l \quad (12.10)$$

where E_c^{dam}, E_{mc}^{dam} and E_{fc}^{dam} are the circumferential Young's moduli in total, in the matrix and in the fibres; similarly, E_l^{dam}, E_{ml}^{dam} and E_{fl}^{dam} are the longitudinal Young's moduli in total, in the matrix and in the fibres. These moduli are calculated by using second-order difference scheme based on the predicted stress-stretch curves with 100 scattered points.

The model property constants in virgin/undamaged state are the model parameters c, k_1, k_2, k_3 and k_4 presented in Table 12.2. In this case, the first part is the strain energy function for the matrix ψ_m^{vir}, the second part is the strain energy function for the circumferential fibres, ψ_{fc}^{vir}, and the third part is the strain energy function for the longitudinal fibres, ψ_{fl}^{vir}, in the virgin sate, then the equation in the virgin state is rewritten as

$$\psi^{vir} = \psi_m^{vir} + \psi_{fc}^{vir} + \psi_{fl}^{vir}$$
$$\psi_m^{vir} = c(I_1 - 3), \psi_{fc}^{vir} = \frac{k_1}{2k_2}[e^{k_2(I_4-1)^2} - 1], \psi_{fl}^{vir} = \frac{k_3}{2k_4}[e^{k_4(I_6-1)^2} - 1] \quad (12.11)$$

Similarly, the Cauchy stress components in the tissue are divided into three parts: the first part is the stress in the matrix σ_m^{vir} due to ψ_m^{vir}, the second part is the stress in the circumferential fibres σ_{fc}^{vir} and the third part is the stress in the longitudinal fibres σ_{fl}^{vir}, i.e.,

$$\sigma_c^{vir} = \sigma_{mc}^{vir} + \sigma_{fc}^{vir}, \sigma_l^{vir} = \sigma_{ml}^{vir} + \sigma_{fl}^{vir}$$
$$\sigma_{mc}^{vir} = \lambda_c \frac{\partial\psi_m^{vir}}{\partial I_1}\frac{\partial I_1}{\partial\lambda_c}, \sigma_{fc}^{vir} = \lambda_c \frac{\partial\psi_{fc}^{vir}}{\partial I_4}\frac{\partial I_4}{\partial\lambda_c} \quad (12.12)$$
$$\sigma_{ml}^{vir} = \lambda_l \frac{\partial\psi_m^{vir}}{\partial I_1}\frac{\partial I_1}{\partial\lambda_l}, \sigma_{fl}^{vir} = \lambda_l \frac{\partial\psi_{fl}^{vir}}{\partial I_6}\frac{\partial I_6}{\partial\lambda_l}$$

Similarly, the circumferential Young's moduli in total, in the matrix and in the fibres E_c^{dam}, E_{mc}^{dam} and E_{fc}^{dam} as well as the longitudinal Young's moduli in total, in the matrix and in the fibre E_l^{dam}, E_{ml}^{dam} and E_{fl}^{dam} are expressed as

$$E_c^{vir} = \partial\sigma_c^{vir}/\partial\lambda_c, E_{mc}^{vir} = \partial\sigma_{mc}^{vir}/\partial\lambda_c, E_{fc}^{vir} = \partial\sigma_{fc}^{vir}/\partial\lambda_c$$
$$E_l^{vir} = \partial\sigma_l^{vir}/\partial\lambda_l, E_{ml}^{vir} = \partial\sigma_{ml}^{vir}/\partial\lambda_l, E_{fl}^{vir} = \partial\sigma_{fl}^{vir}/\partial\lambda_l \quad (12.13)$$

For linear materials, the damage variable is related to the ratio of the Young's modulus in the damaged state to the modulus in the virgin state, (Lemaitre 1984; Becker and Gross 1987; Chaboche 1987; Lemaitre and Dufailly 1987; Voyiadjis and Kattan 2013, 2017; Fett et al. 2018), and written as

$$d = 1 - E^{\mathrm{dam}} / E^{\mathrm{vir}} \tag{12.14}$$

For anisotropic, nonlinear GB walls, it is postulated that this definition is held for each pair of Young's moduli in Eq. (12.10) in the damaged state and those in Eq. (12.13) in the virgin state. Then the damage variables for GB walls are decided by the following expressions

$$\begin{aligned} d_{\mathrm{c}} &= 1 - E_{\mathrm{c}}^{\mathrm{dam}} / E_{\mathrm{c}}^{\mathrm{vir}}, d_{\mathrm{mc}} = 1 - E_{\mathrm{mc}}^{\mathrm{dam}} / E_{\mathrm{mc}}^{\mathrm{vir}}, d_{\mathrm{fc}} = 1 - E_{\mathrm{fc}}^{\mathrm{dam}} / E_{\mathrm{fc}}^{\mathrm{vir}} \\ d_{\mathrm{l}} &= 1 - E_{\mathrm{l}}^{\mathrm{dam}} / E_{\mathrm{l}}^{\mathrm{vir}}, d_{\mathrm{ml}} = 1 - E_{\mathrm{ml}}^{\mathrm{dam}} / E_{\mathrm{ml}}^{\mathrm{vir}}, d_{\mathrm{fl}} = 1 - E_{\mathrm{fl}}^{\mathrm{dam}} / E_{\mathrm{fl}}^{\mathrm{vir}} \end{aligned} \tag{12.15}$$

where d_{c} represents the global damage degree of GB walls in the circumferential direction, d_{mc} and d_{fc} describe the damage degree in the matrix and fibres in the walls; accordingly, d_{l}, d_{mc} and d_{fc} reflect the damage degree in total, in the matrix and in the fibres in the longitudinal direction.

The Cauchy stresses, Young's moduli in total, in the matrix and fibres are illustrated in Figures 12.5 and 12.6 for the human and porcine GB walls. In comparison with the stresses and Young's moduli in the fibres, the stresses and Young's moduli in the matrix are so small that they can be neglected for two kinds of GB walls.

In the matrix, the stresses and Young's moduli fail to demonstrate any difference in value in the damaged state from those in the virgin state. In the fibres, however, the stresses and Young's moduli are reduced significantly in the damaged state from the virgin state at a stretch more than 1.15 for the human GB wall and 1.25 for the porcine GB wall. This fact suggests that the damage does occur in the fibres at a high stretch rather in the matrix.

The damage variables estimated by Eq. (12.15) are presented in Figure 12.7 for the human and porcine GB walls. It is clear that two damage variables, d_{mc} and d_{ml}, are nearly zero, implying there is no damage effect in the matrix basically. In the fibres, two damage variables, d_{fc} and d_{fl} rise markedly with increasing stretch, suggesting a substantially developed damage there. In consequence, the damage in the fibres attributes to the structure failure in the GB walls, i.e., $d_{\mathrm{c}} \approx d_{\mathrm{fc}}$ and $d_{\mathrm{l}} \approx d_{\mathrm{fl}}$.

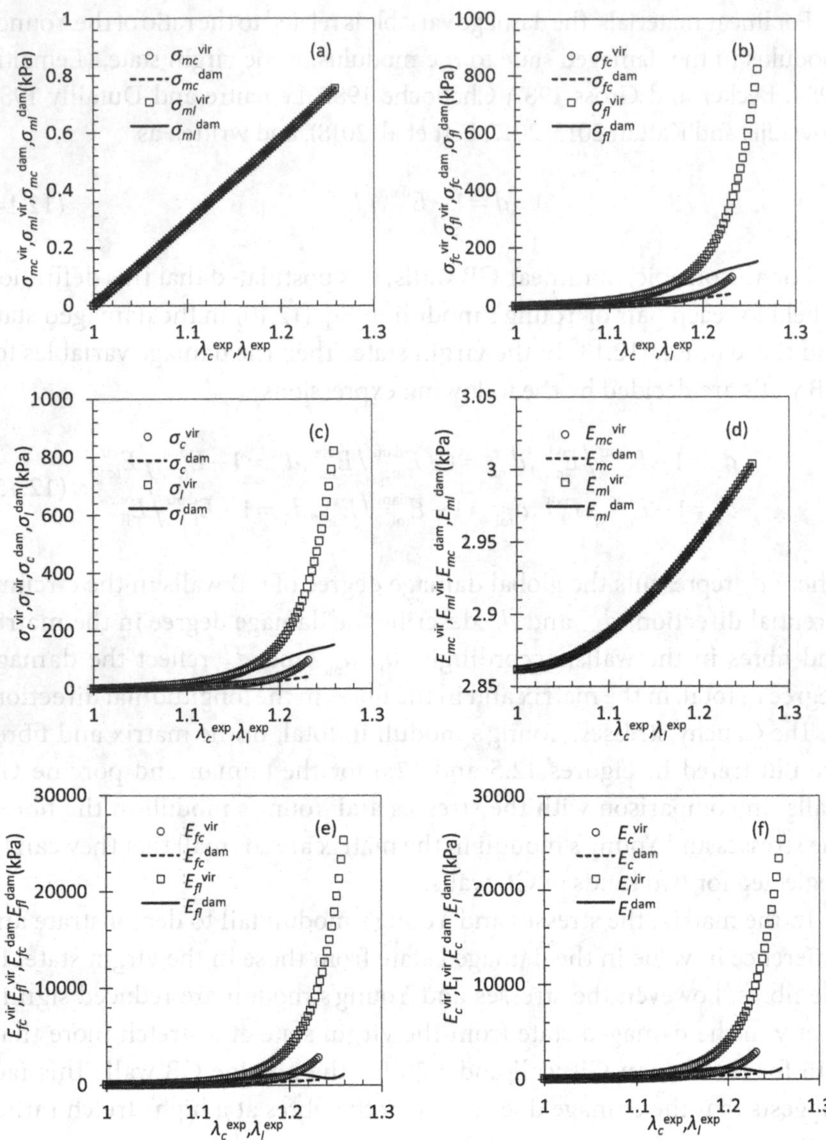

FIGURE 12.5 The Cauchy stresses and Young's moduli in total, in the matrix and fibres for the human GB wall, (a)–(c) for stresses, (d)–(f) for Young's moduli.

Based on the values of the damage variables, d_c and d_l, the damage effect in the longitudinal direction is more dominant than in the circumferential direction for the human GB wall. However, the dominant damage situation is in the circumferential direction for the porcine GB wall.

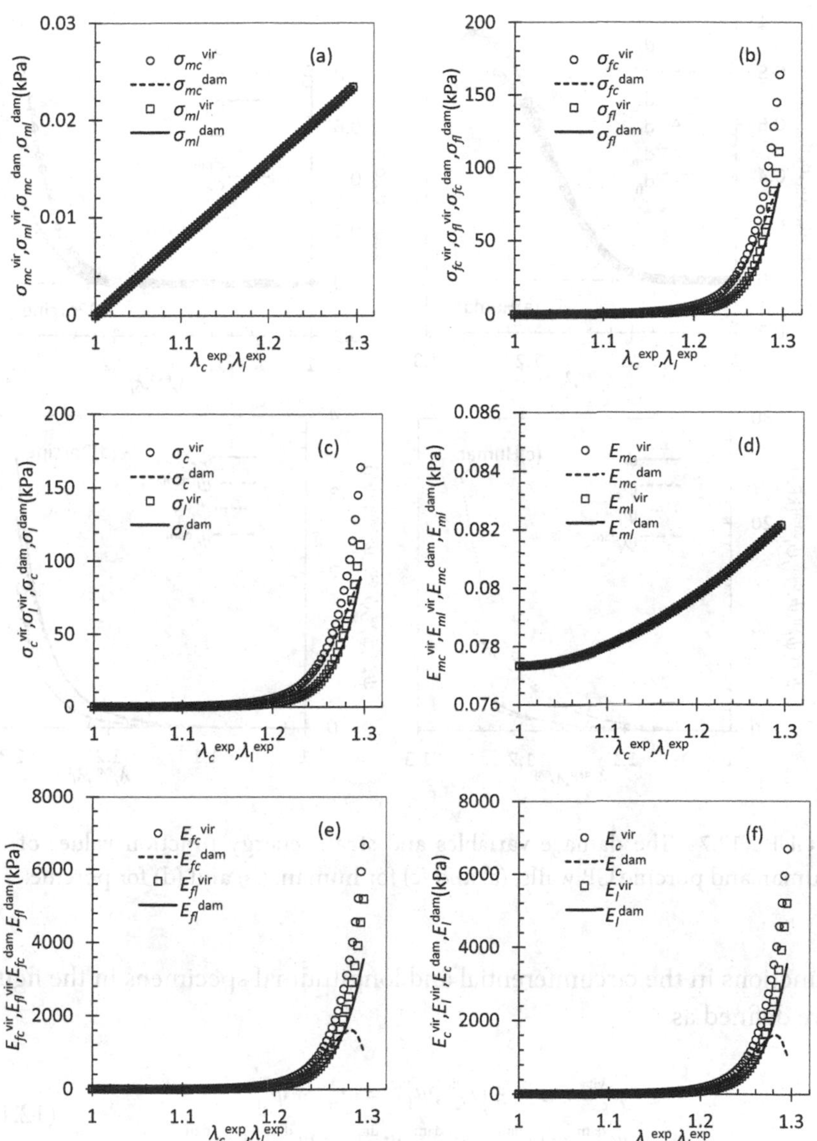

FIGURE 12.6 The Cauchy stresses and Young's moduli in total, in the matrix and fibres for the porcine GB wall, (a)–(c) for stresses, (d)–(f) for Young's moduli.

Since the tissue damage needs energy to generate cracks, the strain energy of GB walls with damage is always lower than the strain energy in the virgin state, see Figure 12.7(c) and 12.7(d). The strain energy

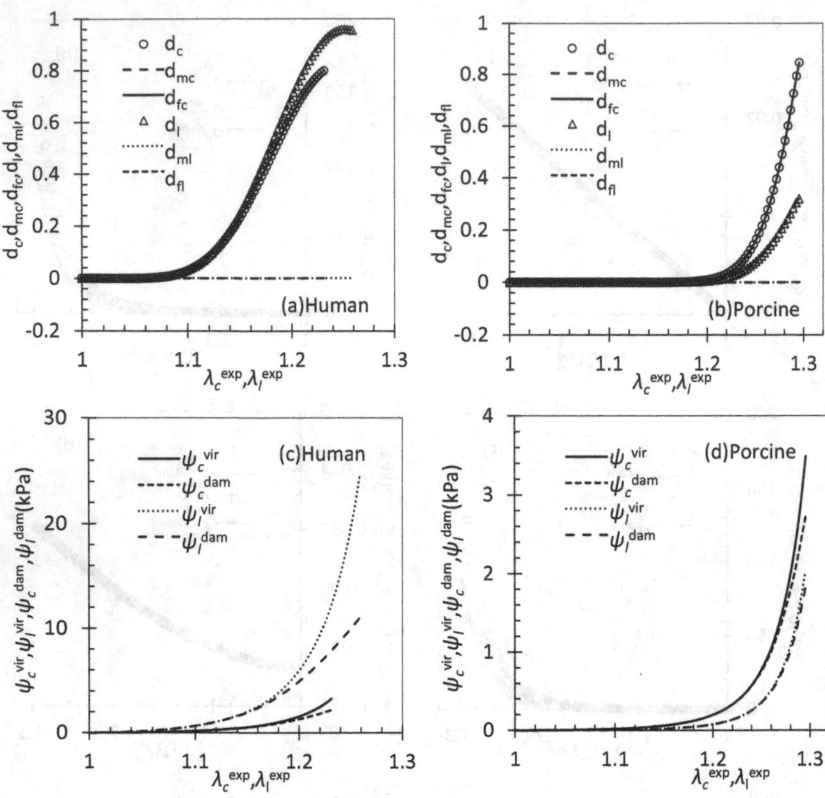

FIGURE 12.7 The damage variables and strain energy function values of the human and porcine GB walls, (a) and (c) for human, (b) and (d) for porcine.

functions in the circumferential and longitudinal specimens in the figure are defined as

$$\begin{aligned} \psi_{\mathrm{c}}^{\mathrm{vir}} &= \psi_{\mathrm{m}}^{\mathrm{vir}} + \psi_{\mathrm{fc}}^{\mathrm{vir}}, \psi_{\mathrm{l}}^{\mathrm{vir}} = \psi_{\mathrm{m}}^{\mathrm{vir}} + \psi_{\mathrm{fl}}^{\mathrm{vir}} \\ \psi_{\mathrm{c}}^{\mathrm{dam}} &= \psi_{\mathrm{m}}^{\mathrm{dam}} + \psi_{\mathrm{fc}}^{\mathrm{dam}}, \psi_{\mathrm{l}}^{\mathrm{dam}} = \psi_{\mathrm{m}}^{\mathrm{dam}} + \psi_{\mathrm{fl}}^{\mathrm{dam}} \end{aligned} \tag{12.16}$$

From the physics point of view, the value of the strain energy function with damage should be equal to or larger than zero. The zero strain energy occurring after being stretched corresponds to the complete structure failure. This means that two stretch components in Eq. (12.8) are not infinite and should have limits. Beyond the limits, the value of the strain energy function is less than zero, which is meaningless in physics.

MECHANISM OF GB WALL DAMAGE

Based on the results above, the damage in the matrix is negligible in comparison with the fibres for two kinds of GB walls. Thus, the strain energy function with damage in both the matrix and the fibres presented with Eq. (12.7) should be updated by removing the term for the damage in the matrix as follows

$$\psi^{\text{dam}} = c(I_1 - 3) + \frac{k_1}{2k_2}\left\{\exp\left[k_2(I_4-1)^2\right] - 1 - \frac{2k_2(I_4-1)^{n_\text{d}+2}}{(n_\text{d}+2)(\eta^2-1)^{n_\text{d}}}\right\}$$
$$+\frac{k_3}{2k_4}\left\{\exp\left[k_4(I_6-1)^2\right] - 1 - \frac{2k_4(I_6-1)^{n_\text{d}+2}}{(n_\text{d}+2)(\eta^2-1)^{n_\text{d}}}\right\} \quad (12.17)$$

The determined model parameters are listed in Table 12.3. As expected, the values of the parameters and error remain unchanged. Obviously, the updated strain energy function in Eq. (12.17) is proper for two sets of uniaxial tensile test adopted.

For linear materials, under the hypothesis of strain equivalence, scalar damage variable is related to the Young's moduli in undamaged/virgin and damaged states (Lemaitre 1984; Voyiadjis and Kattan 2013, 2017; Fett

TABLE 12.3 The decided parameters from uniaxial tensile test data of human and porcine GB walls without damage in the matrix but in the fibres.

Damage effect	Parameter	GB wall	
		Human	Porcine
Included damage effects in fibres only	c (kPa)	0.4770	0.0129
	k_1 (kPa)	5.8311	0.6445
	k_2 (-)	8.9293	10.2208
	k_3 (kPa)	23.8486	0.1964
	k_4 (-)	8.5417	12.0102
	η (-)	1.1825	1.2447
	n_d (-)	7.0763	18.6269
	m_d (-)	N/A	N/A
	ζ (-)	N/A	N/A
	χ_σ (%)	4.5836	8.1985

et al. 2018). Once a brittle material is damaged, its Young's modulus must be degraded in comparison with its undamaged state, showing Young's modulus degradation or stress softening effect.

For nonlinear anisotropic GB walls, the same approach was applied to judge the damage happens in the GB walls in the chapter. Firstly, the Cauchy stress-stretch experimental data points of the circumferential and longitudinal specimens were best fitted by using the sixth-order polynomial; subsequently, the local/instant Young's moduli of the fitted curves were calculated across the experimental ranges of stretch with the expressions of $E_c^{exp} = d\sigma_c^{exp} / d\lambda_c^{exp}$ and $E_l^{exp} = d\sigma_l^{exp} / d\lambda_l^{exp}$. And then, the yield points where the maximum Young's modulus occurred were identified by satisfying the conditions of $dE_c^{exp} / d\lambda_c^{exp} = d^2\sigma_c^{exp} / d\lambda_c^{exp2} = 0$ and $dE_l^{exp} / d\lambda_l^{exp} = d^2\sigma_l^{exp} / d\lambda_l^{exp2} = 0$, respectively. When a stretch is beyond these points, the instant Young's modulus is degraded, indicating the material is damaged. Finally, the corresponding terms are involved in the strain energy function to take the damage effect into account.

After pre-conditioning, GB walls exhibited a hyperelastic behaviour without any plastic characteristics (Xiong et al. 2013; Karimi et al. 2017), suggesting the GB walls with brittle damage. In this sense, the damage variable for isotropic, homogeneous and brittle materials (Lemaitre 1984; Becker and Gross 1987; Voyiadjis and Kattan 2013, 2017; Fett et al. 2018) was extended into anisotropic homogeneous GB walls by introducing a scalar damage variable to the matrix and two families of fibres individually.

Based on the values of these damage variables, the damage states of the matrix and two families of fibres inside a GB wall were examined. For two pairs of GB wall samples used, the damage variable of the matrix is zero, but the variables of two families of fibres are greater than zero. Nonetheless, the damage occurs in the fibres rather than in the matrix.

PROGRAMS FOR MODEL CONSTANT OPTIMISATION

In Chapters 8–11, the model constants were optimised in custom codes with the *lsqnonlin* function in MATLAB. In this section, the optimisation process for the model constants is outlined by using the constitutive law with damages. This procedure is similar to the optimisation process in Chapters 8–11.

The damage model described with Eqs. (12.2)–(12.16) was programmed in MATLAB by using a main program and a user function. Firstly, the experimental data of two uniaxial tensile tests presented with the curves in Figure 12.1 are read into the main program after the curves were digitalised by employing a digitiser. The lower and upper bounds of nine model constants are specified. To guarantee a global optimisation result, the lower bound should be small enough while the upper bound should be large enough. Table 12.4 lists the lower and upper bounds applied in the parameter optimisation process. For the model without damage effect, the lower and upper bounds of ξ and ζ are 10^8, and those of m_d and n_d are 1 to remove their effect on the model and restore the model represented by Eq. (12.2) without damage, but the bounds of the rest parameter are the same as those in the model with damage.

The *lsqnonlin* function in MATLAB was chosen to carry out the parameter optimisation by minimising the objective function Eq. (12.3). In the *lsqnonlin* function, 'Trust-Region-Reflective' optimisation algorithm is implanted. In the algorithm, the objective function is approximated with a model function i.e. a quadratic function. Trust region is a subset of

TABLE 12.4 The lower and upper bounds of nine parameters used in their optimisation process.

Parameter	Bounds		Model type
	Lower	Upper	
c (kPa)	0	10	With damage
k_1 (kPa)	0	50	
k_2 (-)	0	50	
k_3 (kPa)	0	50	
k_4(-)	0	50	
ξ (-)	1	1.5	
n_d (-)	0.1	20	
m_d (-)	0.1	20	
ζ (-)	3	6	
c (kPa)	0	10	Without damage
k_1 (kPa)	0	50	
k_2 (-)	0	50	
k_3 (kPa)	0	50	
k_4 (-)	0	50	
ξ (-)	10^8	10^8	
n_d (-)	1	1	
m_d (-)	1	1	
ζ (-)	10^8	10^8	

the region of the objective function. The minimum objective function is achieved in the trust region. In the trust region algorithm, the search step and size of trust region are decided and updated according to the ratio of the real change of the objective function to the predicted change in the objective function by the model function to ensure sufficient reduction of the objective function. Such procedures can result in the trust region being out of one bound. Thus, the search direction should be reflected to the interior region constrained by the bounds with the law of reflection in optics on that bound. Compared with Newton method and Levenberg-Marquardt algorithm, the Trust-Region-Reflective algorithm can ensure the optimisation iteration remaining in the strict feasible region and its convergence rate is in the second-order (Li 1993).

Nine internal optimisation variables in the *lsqnonlin* function $[x_1, x_2, x_3, \ldots, x_9]$ were selected to represent nine parameters $[c, k_1, k_2, k_3, k_4, \xi, n_d, m_d, \zeta]$ in the physical domain. However, the variables of $[x_1, x_2, x_3, \ldots, x_9]$ in the computational domain of the *lsqnonlin* function is subject to the same lower bound 0 and upper bound 1, but also the step sizes for searching the optimum solution are identical to all the variable. Thus, a transformation relationship between $[x_1, x_2, x_3, \ldots, x_9]$ in the computational domain and $[c, k_1, k_2, k_3, k_4, \xi, n_d, m_d, \zeta]$ in the physical domain is needed. Here, a linear relationship is employed and written as

$$\begin{cases} c = c_{\min} + x_1 \times (c_{\max} - c_{\min}) \\ k_1 = k_{1\min} + x_2 \times (k_{1\max} - k_{1\min}) \\ k_2 = k_{2\min} + x_3 \times (k_{2\max} - k_{2\min}) \\ k_3 = k_{3\min} + x_4 \times (k_{3\max} - k_{3\min}) \\ k_4 = k_{4\min} + x_5 \times (k_{4\max} - k_{4\min}) \\ \xi = \xi_{\min} + x_6 \times (\xi_{\max} - \xi_{\min}) \\ n_d = n_{d\min} + x_7 \times (n_{d\max} - n_{d\min}) \\ m_d = m_{d\min} + x_8 \times (m_{d\max} - m_{d\min}) \\ \zeta = \zeta_{\min} + x_9 \times (\zeta_{\max} - \zeta_{\min}) \end{cases} \tag{12.18}$$

where the lower and upper bounds of nine parameters, such as $c_{\min}$, $c_{\max}$, $k_{1\min}$, $k_{1\max}$, and so on, have been listed in Table 12.14. Accordingly, the step sizes in the computational domain are related to those in the counterpart in the physical domain by the following from Eq. (12.18)

$$\begin{cases} \Delta c = \Delta x_1 \times (c_{\max} - c_{\min}) \\ \Delta k_1 = \Delta x_2 \times (k_{1\max} - k_{1\min}) \\ \Delta k_2 = \Delta x_3 \times (k_{2\max} - k_{2\min}) \\ \Delta k_3 = \Delta x_4 \times (k_{3\max} - k_{3\min}) \\ \Delta k_4 = \Delta x_5 \times (k_{4\max} - k_{4\min}) \\ \Delta \xi = \Delta x_6 \times (\xi_{\max} - \xi_{\min}) \\ \Delta n_{\mathrm{d}} = \Delta x_7 \times (n_{\mathrm{dmax}} - n_{\mathrm{dmin}}) \\ \Delta m_{\mathrm{d}} = \Delta x_8 \times (m_{\mathrm{dmax}} - m_{\mathrm{dmin}}) \\ \Delta \zeta = \Delta x_9 \times (\zeta_{\max} - \zeta_{\min}) \end{cases} \tag{12.19}$$

Based on Eq. (12.19), even though the step sizes of the internal variables $[x_1, x_2, x_3, \ldots, x_9]$ are the same, i.e. $\Delta x_1 = \Delta x_2 = \Delta x_3 = \ldots = \Delta x_9$ in the *lsqnonlin* function, the step sizes such as Δc, Δk_1, Δk_2, . . . , $\Delta \zeta$ in the physical domain still vary across the variables.

Actually, k_1 (x_2) and k_2 (x_4) vary little and affect the optimisation results negligibly, but c (x_1) changes significantly during the optimisation process. Hence, k_1 (x_2) and k_2 (x_4) have to be updated by c (x_1) after they were calculated with Eq. (12.18) in the following expressions

$$\begin{cases} k_1 \times c \Rightarrow k_1 \\ k_3 \times c \Rightarrow k_3 \end{cases} \tag{12.20}$$

where k_1 and k_3 on the left-hand side have been determined by Eq. (12.18).

Additionally, the initial nine parameters $[c_0, k_{10}, k_{20}, k_{30}, k_{40}, \xi_0, n_{\mathrm{d}0}, m_{\mathrm{d}0}, \zeta_0]$ are generated randomly in the bounds by using *rand* function of MATLAB in terms of $[x_{10}, x_{20}, x_{30}, \ldots, x_{90}]$ to make sure a global optimisation process, i.e.,

$$\begin{cases} c_0 = c_{\min} + x_{10} \times (c_{\max} - c_{\min}) \\ k_{10} = k_{1\min} + x_{20} \times (k_{1\max} - k_{1\min}) \\ k_{20} = k_{2\min} + x_{30} \times (k_{2\max} - k_{2\min}) \\ k_{30} = k_{3\min} + x_{40} \times (k_{3\max} - k_{3\min}) \\ k_{40} = k_{4\min} + x_{50} \times (k_{4\max} - k_{4\min}) \\ \xi_0 = \xi_{\min} + x_{60} \times (\xi_{\max} - \xi_{\min}) \\ n_{\mathrm{d}0} = n_{\mathrm{dmin}} + x_{70} \times (n_{\mathrm{dmax}} - n_{\mathrm{dmin}}) \\ m_{\mathrm{d}0} = m_{\mathrm{dmin}} + x_{80} \times (m_{\mathrm{dmax}} - m_{\mathrm{dmin}}) \\ \zeta_0 = \zeta_{\min} + x_{90} \times (\zeta_{\max} - \zeta_{\min}) \end{cases} \tag{12.21}$$

where x_{10} =rand(1,1), x_{20} =rand(1,1), x_{30} =rand(1,1), . . . , x_{90} =rand(1,1).

The option in the *lsqnonlin* function is as follows: MaxIter=4000, TolFun=10^{-8}, TolX=10^{-8}, Diffminchange=10^{-4}, Diffmaxchange=10^{-2} and MaxFunEvals=50000 where MaxIter is the maximum number of iterations allowed, TolFun is the termination tolerance on the objective function value, TolX is the termination tolerance on $[x_1, x_2, x_3, \ldots, x_9]$, Diffminchange and Diffmaxchange are minimum and maximum changes in variables for finite difference derivatives of the objective function, respectively; MaxFunEvals is the maximum number of the objective function evaluations allowed.

The temporary nine parameters, stresses and objective function value at the experimental stretches are calculated in the user function. The user function is called repeatedly by the *lsqnonlin* function until a convergent optimisation process arrives. The stress-stretch curves, strain energy function values, Young's moduli, damage variables and relevant plots are figured out in the main program based on the determined nine parameters.

REFERENCES

Becker, W., and D. Gross. 1987. A one-dimensional micromechanical model of elastic-microplastic damage evolution. *Acta Mechanica* 70:221–233.

Borly, L., L. Hojgaard, S. Gronvall, et al. 1996. Human gallbladder pressure and volume: Validation of a new direct method for measurements of gallbladder pressure in patients with acute cholecystitis. *Clinical Physiology and Functional Imaging* 16:145–156.

Brotschi, E. A., W. W. Lamorte, and L. F. Williams. 1984. Effect of dietary cholesterol and indomethacin on cholelithiasis and gallbladder motility in guinea pig. *Digestive Diseases and Sciences* 29:1050–1056.

Chaboche, J. L. 1987. Continuum damage mechanics: Present state and future trends. *Nuclear Engineering and Design* 105:19–33.

Fett, T., K. G. Schell, M. J. Hoffmann, et al. 2018. Effect of damage by hydroxyl generation on strength of silica fibers. *Journal of American Ceramic Society* 101:2724–2726.

Fung, Y. C., K. Fronek, and P. Patitucci. 1979. Pseudoelasticity of arteries and the choice of its mathematical expression. *American Journal of Physiology* 237:H620–H631.

Genovese, K., L. Casaletto, and J. D. Humphrey, et al. 2014. Digital image correlation-based point-wise inverse characterization of heterogeneous material properties of gallbladder in vitro. *Proceedings of Royal Society-Series A* 470:20140152.

Holzapfel, G. A., T. C. Gasser, and R. W. Ogden. 2000. A new constitutive framework for arterial wall mechanics and a comparative study of material models. *Journal of Elasticity* 61:1–48.

Karimi, A., A. Shojaei, and P. Tehrani. 2017. Measurement of the mechanical properties of the human gallbladder. *Journal of Medical Engineering & Technology* 41:541–545.

Lemaitre, J. 1984. How to use damage mechanics. *Nuclear Engineering and Design* 80:233–245.

Lemaitre, J., and J. Dufailly. 1987. Damage measurements. *Engineering Fracture Mechanics* 28:643–661.

Li, W. G. 2018. Constitutive laws with damage effect for the human great saphenous vein. *Journal of the Mechanical Behavior of Biomedical Materials* 81:202–213.

Li, W. G. 2019. Constitutive law of healthy gallbladder walls in passive state with damage effect. *Biomedical Engineering Letters* 9:189–201.

Li, W. G., N. A. Hill, and R. W. Ogden, et al. 2013. Anisotropic behaviour of human gallbladder walls. *Journal of the Mechanical Behavior of Biomedical Materials* 20:363–375.

Li, W. G., and X. Y. Luo. 2016. An invariant-based damage model for human and animal skins. *Annals of Biomedical Engineering* 44:3109–3122.

Li, Y. 1993. Centering, trust region, reflective techniques for nonlinear minimization subject to bounds. Cornell Theory Center Technical report-CTC93TR152, Cornell University.

Matsuki, Y. 1985a. Spontaneous contractions and the visco-elastic properties of the isolated guinea-pig gall-bladder. *Japanese Journal of Smooth Muscle Research* 21:71–78.

Matsuki, Y. 1985b. Dynamic stiffness of the isolated Guinea-pig gallbladder during contraction induced by cholecystokinin. *Japanese Journal of Smooth Muscle Research* 21:427–438.

Miura, K., and S. Saito. 1967. Visco-elastic properties of the gallbladder in rabbit and guinea-pig. *Journal of the Showa Medical Association* 27:135–138.

Rosen, J. J. D. Brown, and S. De, et al. 2008. Biomechanical properties of abdominal organs in vivo and postmortem under compression loads. *ASME Journal of Biomechanical Engineering* 130:021020.

Ryan, J., and S. Cohen. 1976. Gallbladder pressure-volume response to gastrointestinal hormones. *American Journal of Physiology* 230:1461–1465.

Schoetz, D. J., W. W. LaMorte, and W. E. Wise, et al. 1981. Mechanical properties of primate gallbladder: Description by a dynamic method. *American Journal of Physiology-Gastrointestinal and Liver Physiology* 241:G376–G381.

Voyiadjis, G. Z., and P. I. Kattan. 2013. On the theory of elastic undamageable materials. *ASME Journal of Engineering Materials and Technology* 135:021002.

Voyiadjis, G. Z., and P. I. Kattan. 2017. Decomposition of elastic stiffness degradation in continuum damage mechanics. *ASME Journal of Engineering Materials and Technology* 139:021005.

Xiong, L., C. K. Chui, and C. L. Teo. 2013. Reality based modelling and simulation of gallbladder shape deformation using variational methods. *ASME International Journal of Computer Assisted Radiology and Surgery* 8:857–865.

CHAPTER 13

GB 3D Models from Ultrasound Images

GB 3D GEOMETRICAL MODELS

In the previous chapters, the human GB was assumed to be an ellipsoid model and the corresponding GB volume and stress were determined. Even though this method is easily applied in GB disease diagnosis, differences in GB shape and stress pattern and level between ellipsoid model and actual GB are unknown.

Ultrasound/ultrasonography is a sensitive non-invasive screening tool on the diagnosis of biliary tract and GB disorders since 1970s (Bartrum et al. 1977; Kishk et al. 1987; Fitzgerald and Toi 1987; Nelson and Pretorius 1998; Portincsa et al. 2003; Xu et al. 2003; Irshad et al. 2011; Frank and Kurian 2016; Serra et al. 2016). Simultaneously, ultrasound was adopted to study GB motor function or motility by determining its real-time volume during the emptying phase to help the GB diseases diagnosis (Everson et al. 1980; Dodds et al. 1985; Hopman et al. 1985; Stolk et al. 1990; Wedmann et al. 1991; Andersen et al. 1993; Pauletzki et al. 1996; Hashimoto et al. 1999; Stads et al. 2007).

Even though 3D ultrasound can measure GB real-time volume accurately, getting 3D GB geometrical models for solid mechanics analysis is unavailable (Pauletzki et al. 1996; Hashimoto et al. 1999; Yoon et al. 2005; Stads et al. 2007). To extract real-time GB geometrical models from ultrasound images, image registration is required. Because of artefacts and speckle in ultrasound images, automatic registration of GB ultrasound images in 3D ultrasound system is still in developing (Rohling et al. 1998).

Automatic segmentation methods were proposed for 2D ultrasound static long-axis/sagittal view images of GB based on pixel grey level (Bodzioch and Ogiela 2009a, 2009b; Ogiela and Bodzioch 2011) or brightness (Ciecholewski and Chocholowicz 2013). However, there are not any 3D GB geometrical models generated in them.

In this chapter, a method for generating human 3D GB geometrical models is introduced based on two static ultrasound 2D images of the human GB scanned in the long-axis/sagittal and short-axis/transverse directions, respectively. The detailed method is referred to by Li (2020).

ULTRASOUND IMAGES OF THE HUMAN GB

Ultrasound static 2D images of the human GB scanned in the long-axis and short-axis directions were obtained from a hospital for seven patients. The images at the start of emptying phase are shown in Figure 13.1 to reflect a variety of GB shapes. These images were taken with an ordinary clinical ultrasound system in routine GB disease inspection. Three linear dimensions D_1, D_2 and D_3 are listed in the images to determine the image scales.

STEPS OF THE IMAGE PROCESSING METHOD

In the images, there are two views: one is in the long-axis plane where one longitudinal GB cross section is present; the other is in the short-axis plane where one cross section with the maximum cross-sectional area is specified. These two will be utilised to generate GB 3D geometrical models. There are three separate steps in the method, as demonstrated in Figure 13.2.

In the first step, GB ultrasound images are segmented manually in MATLAB with a custom program to obtain scattered points of a GB wall edge. Two edges are generated by fitting the points with piecewise cubic spline function in MATLAB. The closed short-axis cross section is lofted, and the GB volume is calculated. Finally, the data files of short-axis cross sections and guidelines are made for SolidWorks.

In the second step, GB short-axis cross sections and guideline profiles are established in SolidWorks, and a surfacical is lofted with these profiles, the surface is enclosed with a small spherical cap on the bottom of GB fundus. Geometrical model information exchange files in IGES and Parasolid formats are prepared for Abaqus to perform FEA under a specific bile pressure and boundary condition.

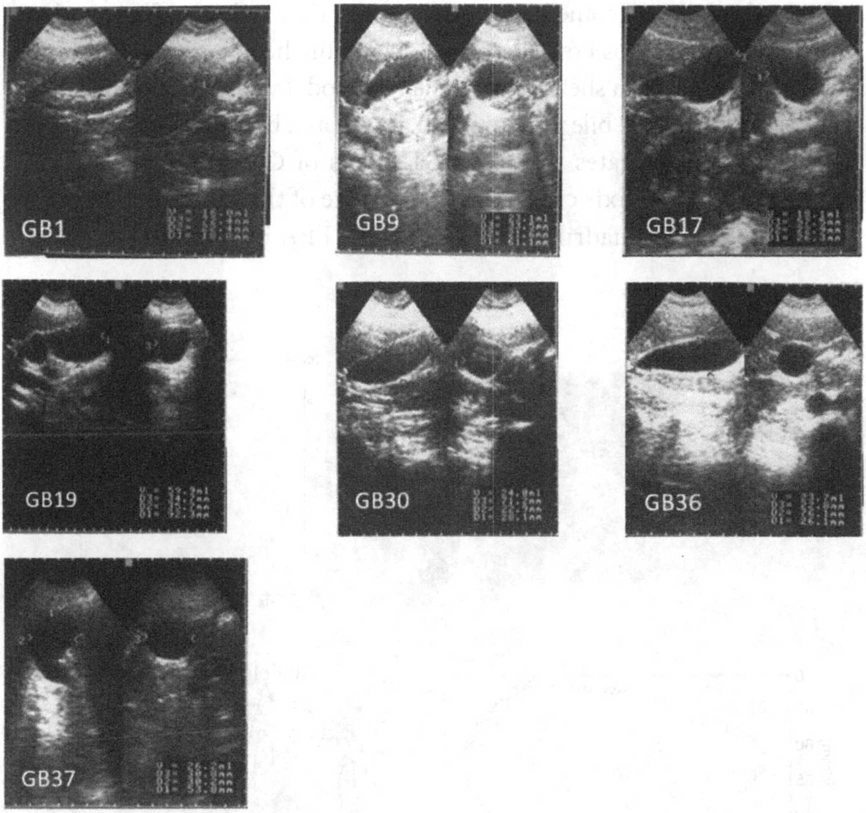

FIGURE 13.1 Static ultrasound 2D images of seven human GBs at the start of emptying phase, three linear dimensions D_1, D_2 and D_3 are given.

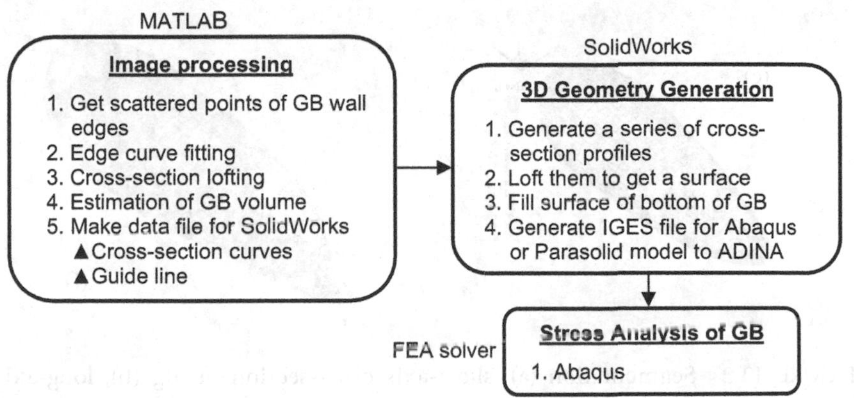

FIGURE 13.2 Flowchart of image processing strategy for GB geometrical models.

In the third step, a geometrical model information exchange file is read into Abaqus 6.11. A mesh is created and a FEA is launched by means of homogeneous, isotropic, thin shell solid mechanics model with a certain thickness and a specific internal bile pressure load as well as a boundary condition.

Figure 13.3 illustrates the scattered points of GB36 wall edges, fitted short-axis and long-axis cross sections, surface of the GB wall, GB surface presentation and a quadrilateral mesh for the FEA in Abaqus.

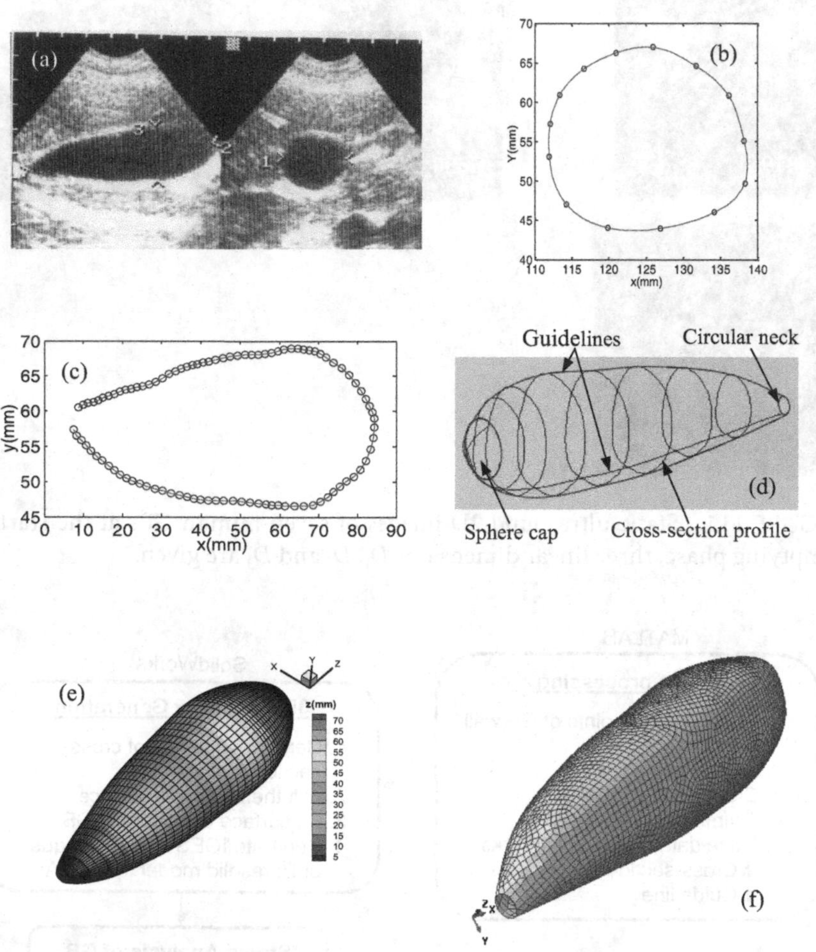

FIGURE 13.3 Segmentation (a), short-axis cross-section fitting (b), long-axis cross-section fitting (c), cross-section lofting in SolidWorks (d), generated GB wall surface (e) and quadrilateral mesh for FEA for GB36 (f).

DETAILS IN IMAGE PROCESSING

GB cross-section curve fitting, lofting and volume calculation are the essential in GB 3D geometrical model generation. The details of the program for accomplishing these functions are as follows:

(1) Determine image scales based on the marks and number figures in the images in Figure 13.1.

(2) Read data files of two edges, scattered points are made in the previous segmentation.

(3) Calculate the geometrical centre coordinates of the short-axis cross-sectional curve (x_c, y_c), determine the polar-coordinate system (θ-r) and perform the curve fitting with MATLAB spline function $r = csape(\theta, r, \text{'periodic'})$, which is cubic spline interpolation with end conditions, the coordinates of the edge are ($x_c + r \cos\theta$, $y_c + r \sin\theta$).

(4) Choose the middle point of GB neck (y_{neck}, z_{neck}) and specify it as the reference point of the long-axis cross-section edge scattered points, determine the other reference point (y_{bmax}, z_{bmax}) at which the maximum length is achieved from the middle point (y_{neck}, z_{neck}), rotate the cross-section edge points in an angle of -atan[(y_{bmax} - y_{neck})/z_{bmax} - z_{neck})] to allow the cross section to be laid horizontally in new coordinates (y_{rot}, z_{rot}), compute the arc length s_c of the scattered points from the first one to the last one in the neck, define MATLAB spline function

$$\begin{cases} z = csape(s_c, z_{rot}, q) \\ y = csape(s_c, z_{rot}, q) \\ q = [1 + (z_{rot2} - z_{rot1})^3 / 6]^{-1} \end{cases} \tag{13.1}$$

where q is the end condition parameter, z_{rot1} and z_{rot2} are the horizontal coordinates of the first and second scattered points after rotation, then the rotated scattered points are interpolated, the reference point (y_{bmax}, z_{bmax}) is searched again, and the angle of rotation is updated, and the points are rotated accordingly, the MATLAB spline function is redefined, and a subsequent interpolation is launched once more; such a cycle is repeated for five times until the sagittal cross section is in horizontal exactly.

(5) Interpolate more dense scattered points with the functions defined in (4), divide the whole the long-axis cross-section edge into two parts: i.e., upper and lower parts, determine the location where there is the maximum height in the long-axis cross-section, rescale the existing short-axis cross-section height and insert this cross section into the long-axis cross section orthogonally.

(6) Specify the short-axis cross section at the GB neck and the short-axis cross section near the GB fundus bottom is a circle, and then loft the rest short-axis cross sections from three existing short-axis cross sections based on the ratio of local short-axis cross-section height to the maximum short-axis cross-section height;

(7) Calculate the volume of a GB sample by summarising the volumes of each truncated cone formed by two neighbouring short-axis cross sections, this method is slightly more complex than the sum-of-cylinders method (Everson et al. 1980; Dodds et al. 1985; Stolk et al. 1990; Wedmann et al. 1991; Andersen et al. 1993; Pauletzki et al. 1996).

(8) Write the Cartesian coordinates (x_{surf}, y_{surf}, z_{surf}) of all the short-axis cross-section edges (lofted plus existing) into separate data files for SolidWorks, and the Cartesian coordinates of the long-axis cross-section edges in the horizontal and vertical planes are also written into data files for SolidWorks as guidelines.

Note that the cubic spline function in MATLAB was chosen to fit the scattered points in the long-axis and short-axis cross sections for the edges. The function can represent the complex edge shape of a human GB wall. In the study by Liao et al. (2004), the Fourier series with five terms was utilised to fit the scattered points in the short-axis cross section. Initially, this method was employed to perform the fitting task. However, the fitted curve does not pass through a few points picked up from the image, as seen Figure 13.4 for GB 36. The fitting curve generated by the cubic spline function passes through the points and preserves the edge geometrical feature.

GB SHAPE

GB wall surfaces of the rest six GB samples, which were generated by means of the method just mentioned, are illustrated in Figure 13.5. GBs 1 and 17 present a pear shape, while GBs 9, 30 and 36 in Figure 13.3 exhibit a slender structure, but GBs 19 and 37 are in an odd shape.

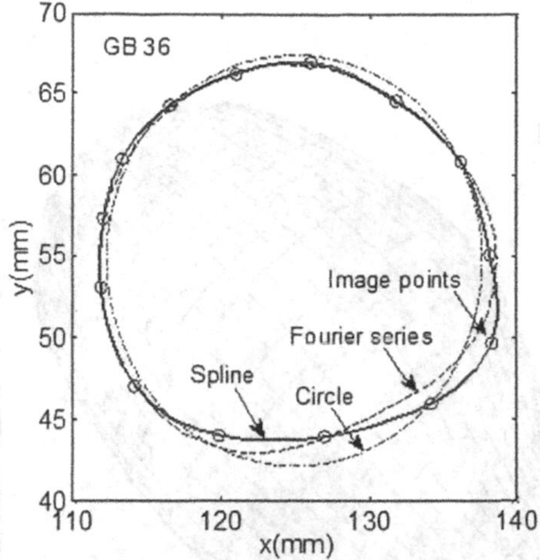

FIGURE 13.4 A comparison of fitting curves established by Fourier series (Liao et al. 2004), circle and cubic spline function in MATLAB for the scattered points in the short-axis/transverse cross section of GB 36.

These GB shapes have been observed in GB 3D volumes rendered CT cholangiographic images shown by Fidler et al. (2013).

GB VOLUME

The volumes of seven GB samples are calculated based on the ellipsoid model with three principal axis lengths measured in each 3D model dimensions from the image and the 3D model itself, respectively. The ellipsoid model (Dodds et al. 1985) and its original 3D model for the image of GB 36 are illustrated in Figure 13.6(a) and (b) as an example. The ellipsoid model volume is calculated by the expression (Dodds et al. 1985):

$$V_{\mathrm{el}} = \frac{\pi D_1 D_2 D_3}{6} \tag{13.2}$$

The formula of the sum-of-cylinders method for GB volume can be found in Everson et al. (1980), Dodds et al. (1985) and Pauletzki et al. (1996). A comparison of GB volume is made between three methods in Figure 13.6(c) and Table 13.1. The GB volume was estimated by means of the sum-of-cylinders method (Everson et al. 1980) and presented in the figure, too.

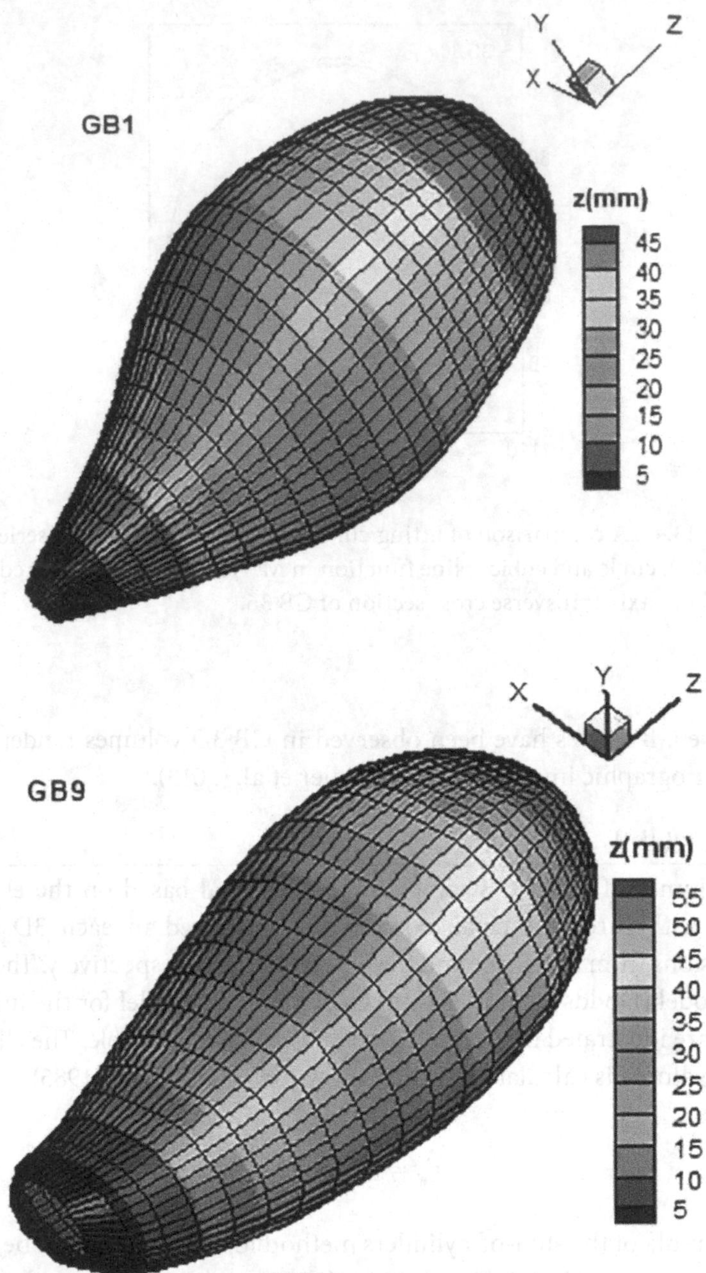

FIGURE 13.5 GB wall surfaces of six GB samples, generated by a custom MATLAB code.

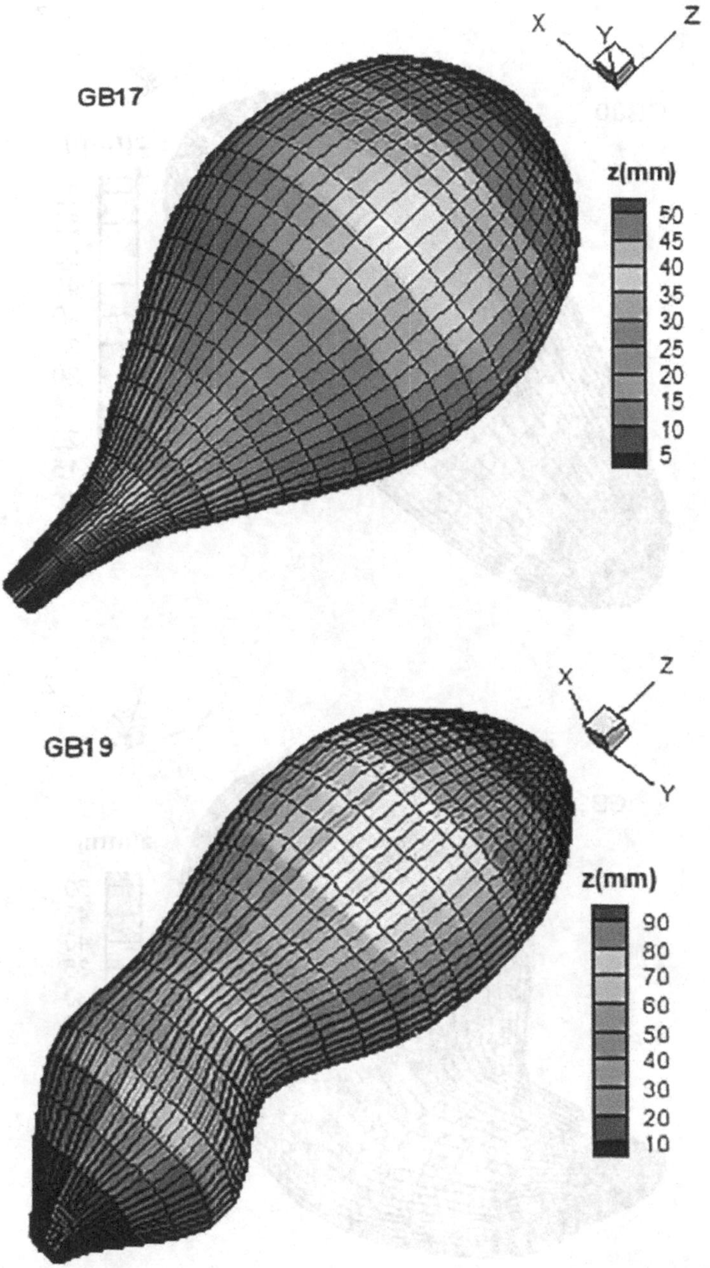

FIGURE 13.5 GB wall surfaces of six GB samples, generated by a custom MATLAB code.

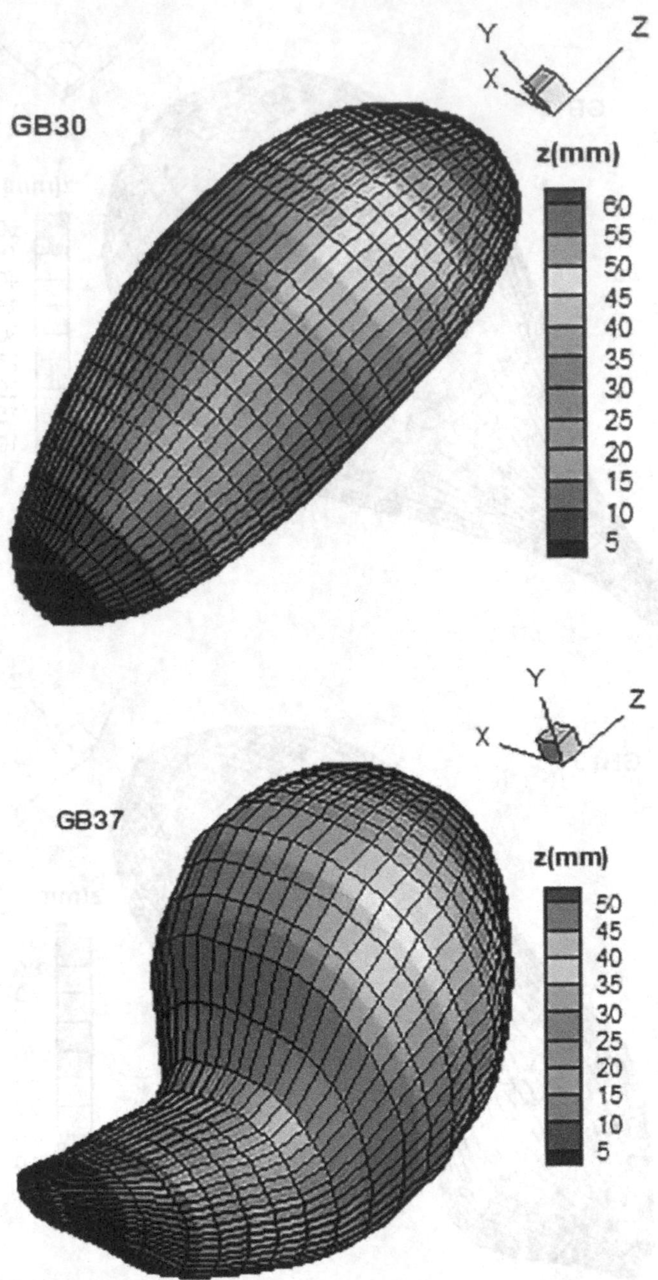

FIGURE 13.5 GB wall surfaces of six GB samples, generated by a custom MATLAB code.

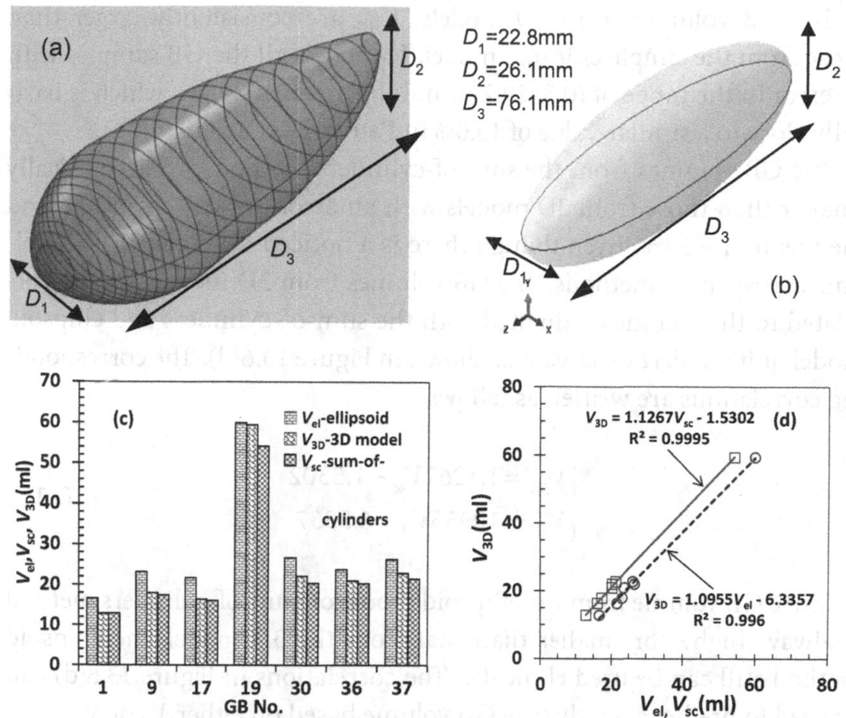

FIGURE 13.6 GB 3D model from the image of GB36 (a), its corresponding ellipsoid model with three principal axis lengths D_1, D_2 and D_3 (b), GB volumes based on ellipsoid model, sum-of-cylinders method and 3D model from images of seven GB samples (c), and regressed relationships of V_{3D} to V_{el} and V_{sc}, respectively (d).

TABLE 13.1 Comparison of GB volumes estimated with three methods.

GB	1	9	17	19	30	36	37
V_{el} (mL)	16.57	23.15	21.66	59.87	26.66	23.71	26.20
V_{sc} (mL)	12.87	17.35	16.06	54.15	20.32	20.28	21.45
V_{3D} (mL)	12.82	17.93	16.36	59.41	22.07	20.98	22.79
$(V_{el}/V_{3D}-1)\times 100\%$	29.2	29.1	32.4	0.8	20.8	13.0	15.0
$(V_{sc}/V_{3D}-1)\times 100\%$	0.4	-3.2	-1.8	-8.9	-7.9	-3.3	-5.9

V_{el} and V_{sc} are the GB volumes predicted by the ellipsoid method and sum-of-cylinders method, respectively; V_{3D} is the volume estimated by the present GB 3D geometrical models

The GB volumes from 3D models, V_{3D}, are consistently larger than those from the simple ellipsoid model, V_{el}, across all the GB samples with an error in the range of (0.8–29.2)% and the mean of 20.1%, which is basically close to a similar value of 15.0% in Pauletzki et al. (1996).

The GB volumes from the sum-of-cylinders method, V_{sc}, are basically smaller than those from 3D models with an error of (−8.9 to +0.4)% and the mean of −4.4%. Even though there is a noticed difference in GB volume across three methods, The GB volumes from 3D models can be correlated to the volumes estimated with the sum-of-cylinders and ellipsoid model quite well, respectively, as shown in Figure 13.6(d). The corresponding correlations are written as follows

$$\begin{cases} V_{3D} = 1.1267V_{sc} - 1.5302 \\ V_{3D} = 10955V_{el} - 6.3357 \end{cases} \tag{13.3}$$

Since GB volume from the ellipsoid model or sum-of-cylinders method is always higher or smaller than that from the 3D model, the ellipsoid method still can be used clinically. The correlations in Figure 13.6(d) can be used to predict a nearly true GB volume based on either V_{el} or V_{sc}.

FEA

Seven GB 3D geometrical models were put into Abaqus 6.11 standard to carry out a linear FEA with the aim to identify whether the geometrical models can be applicable in FEA and if the stress pattern and level are different from those based on the analytical solution and ellipsoid GB model in Chapters 5 and 6. The material property constants and parameter settings in FEA are presented in Table 13.2.

The Young's modulus is 500 kPa approximately based on the experimental data summarised in Li (2019), and the Poisson ratio of 0.49 is chosen for an incompressible GB wall. The GB wall thickness was assumed to be uniform and in 2.5 mm (Li et al. 2011). Thin shell elements used in FEA are included by type S3 (3-node triangular general-purpose shell) and S4 (4-node quadrilateral general-purpose shell). The pressure load was applied on the GB inside surface, and the load magnitudes are patient-specific and referred to Li et al. (2011). The boundary condition was imposed in the GB neck edge by fixing six freedoms. The effects of mesh size and Young's modulus on stress level can be found in Li (2020).

TABLE 13.2 Property constants of the GB wall and parameter settings in FEA.

GB		1	9	17	19	30	36	37
Young's modulus (kPa)					500			
Poisson's ratio					0.49			
GB wall thickness (mm)					2.5			
Software		Abaqus 6.11, standard						
Element type		Thin shell element S3 and S4						
Number of elements	S3	4	4	137	176	164	122	148
	S4	1621	3754	4515	5125	5933	5193	6031
Internal pressure (kPa)		2.0328	2.2265	2.2065	3.5124	2.3777	2.2770	2.3619
Boundary condition		Six freedoms at GB neck inlet edge are fixed						

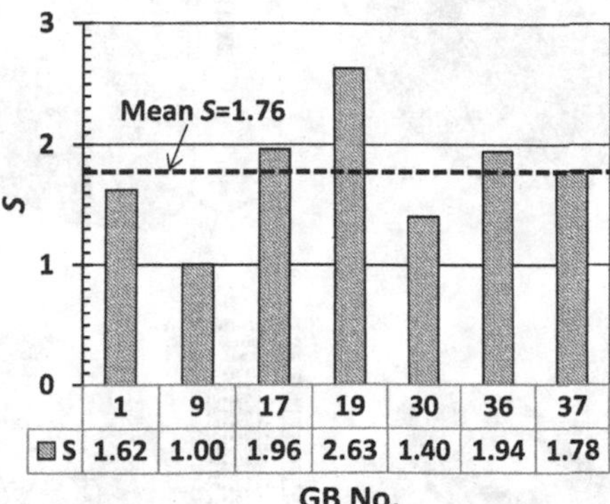

FIGURE 13.7 The first principal in-plane stress ratio of seven GB samples, it is defined as the ratio of the first principal in-plane stress in a GB 3D geometrical model, $\sigma_{1,3D}$, to the counterpart in its ellipsoid model, $\sigma_{1,e}$, $S = \sigma_{1,3D}/\sigma_{1,e}$.

The ratio of the first principal in-plane stress in a GB 3D geometrical model to the counterpart in its ellipsoid model, i.e., $S = \sigma_{1,3D} / \sigma_{1,e}$, is shown in Figure 13.7 to quantity the difference in stress level between two geometrical models. The first principal in-plane stress in the ellipsoid model was calculated based on the method described in Li et al. (2011). Clearly, under the same loading condition and material property constants, the stress level in the 3D geometrical model is higher than that in the ellipsoid model across the samples. The mean ratio is 1.76 with a 0.11 standard deviation.

The contours of the first principal in-plane stress across seven GB 3D geometrical models are illustrated in Figure 13.8. Furthermore, the

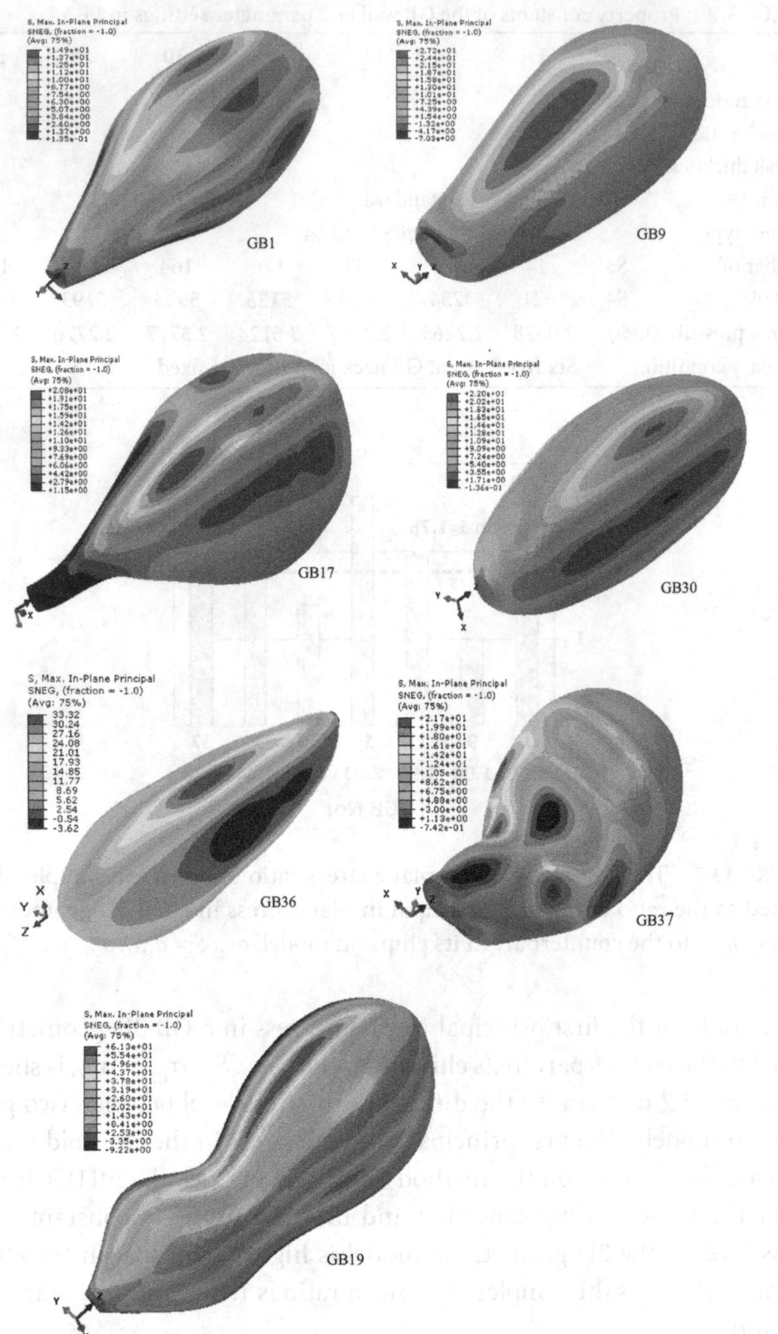

FIGURE 13.8 Contours of the first principal in-plane stress in seven GB 3D geometrical models.

location for the maximum stress is coincident with the location for largest strain, as shown in Figure 13.9.

Since the curvature of the cross-sectional edge changes significantly in the short-axis plane, see Figure 13.3(b), the stress exhibits a bumpy

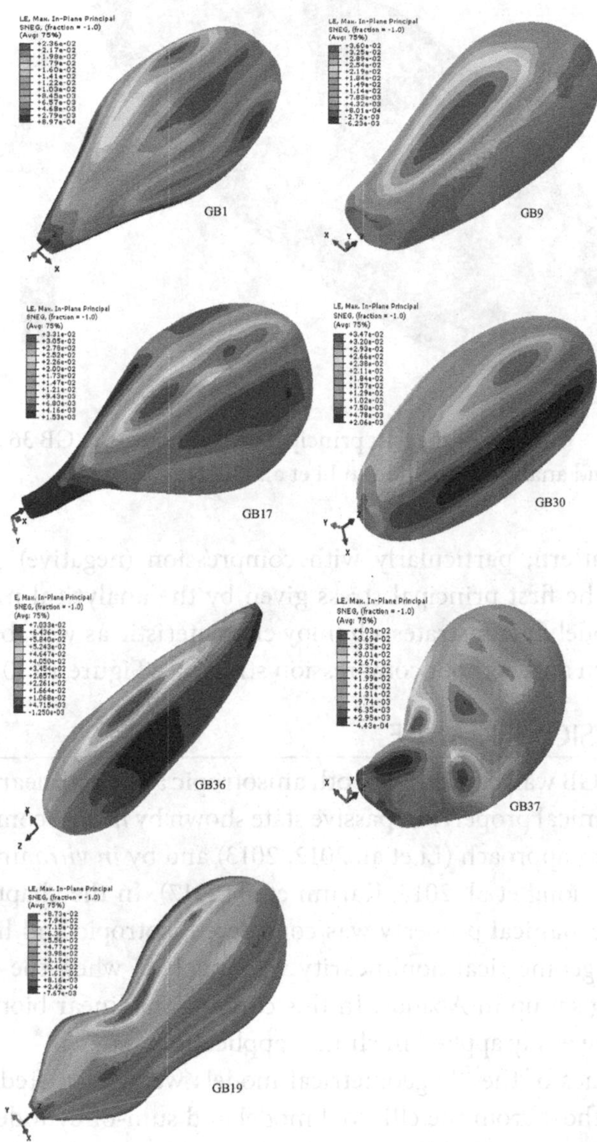

FIGURE 13.9 Contours of the first principal in-plane strain in seven GB 3D geometrical models.

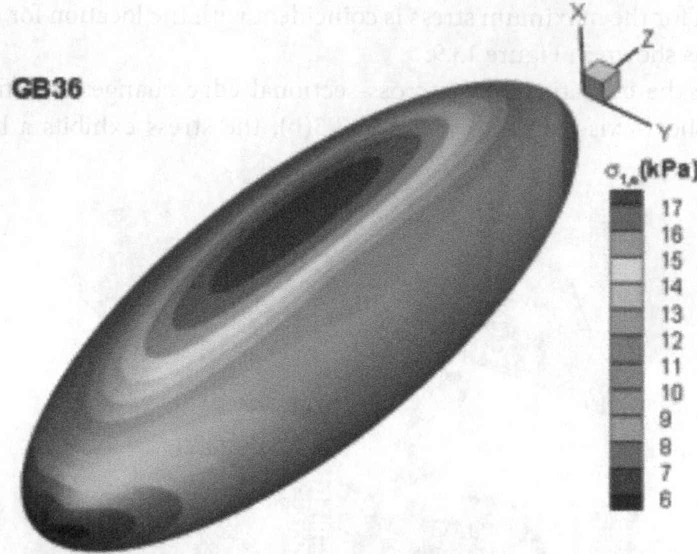

FIGURE 13.10 Contour of the first principal in-plane stress in GB 36 due to ellipsoid model and analytical method in Li et al. (2011).

variation pattern, particularly with compression (negative) stress. The contour of the first principal stress given by the analytical method and ellipsoid model demonstrates a bumpy characteristic as well, but in a different pattern and without compression stress, see Figure 13.10 for GB 36.

CLINICAL SIGNIFICANCE

The human GB wall can exhibit both anisotropic and nonlinear behaviour in biomechanical property in passive state shown by *in vivo* computational biomechanics approach (Li et al. 2012, 2013) and by *in vitro* uniaxial tensional test (Xiong et al. 2013; Karimi et al. 2017). In the chapter, the GB wall's biomechanical property was considered isotropic and linear, even though the geometrical nonlinearity was involved when the mechanics model being set up in Abaqus. In this context, the linear biomechanical model can be easily applied in clinical application.

GB volumes of the 3D geometrical models were calculated and compared with those from the ellipsoid model and sum-of-cylinders method (Everson et al. 1980; Dodds et al. 1985; Pauletzki et al. 1996). The GB volumes from 3D geometrical models can correlate to those from the ellipsoid model and sum-of-cylinders method. The latter two methods still can be used as a clinical tool. If the volumes are corrected with Eq. (13.3), more accurate GB volumes will be obtained.

The GB ellipsoid model provides us with a simple clinical mechanics model, however, its volume and stress level estimations can be lower by a certain factor than those based on GB 3D geometrical models from ultrasound images. If the first principal in-plane stress is amplified by 1.76-fold, the ellipsoid model can provide a realistic stress level for in the GB wall during the emptying phase.

3D spiral/helical CT (Kwon et al. 1995; Polkowski et al. 1999; Caoili et al. 2000; Chopra et al. 2000; Hirao et al. 2000; Ichii et al. 2002; Persson et al. 2005) and magnetic resonance imaging (Vyas et al. 2002; Elsayes et al. 2007; Catalano et al. 2008; Brugger et al. 2010; Tan and Lim 2013) have been involved in the diagnosis of GB diseases, as they can illustrate GB 3D images but also show a 3D pattern of the whole biliary duct. Therefore, the method in the chapter faces a new challenge.

REFERENCES

Andersen, I. B., H. Monrad, S. Gronvall, et al. 1993. In vitro and in vivo accuracy of sonographic gallbladder volume determinations. *Journal of Clinical Ultrasound* 21:157–162.

Bartrum, R. J., H. C. Crow, and S. R. Foote. 1977. Ultrasonic and radiographic cholecystography. *New England Journal of Medicine* 296:538–541.

Bodzioch, S., and M. R. Ogiela. 2009a. Effective filtration techniques for gallbladder ultrasound images with variable contrast. *Journal of Signal Processing Systems* 54:127–144.

Bodzioch, S., and M. R. Ogiela. 2009b. New approach to gallbladder ultrasonic images analysis and lesions recognition. *Computerized Medical Imaging and Graphics* 33:154–170.

Brugger, P. C., M. Weber, and D. Prayer. 2010. Magnetic resonance imaging of the fetal gallbladder and bile. *European Radiology* 20:2862–2869.

Caoili, E. M., E. K. Paulson, L. E. Heyneman, et al. 2000. Helical CT cholangiography with three-dimensional volume rendering using an oral biliary contrast agent: Feasibility of a novel technique. *American Journal of Roentgenology* 174:487–492.

Catalano, O. A., D. V. Sahani, S. P. Kalva, et al. 2008. MR imaging of the gallbladder: A pictorial essay. *Radio Graphics* 28:131–155.

Chopra, S., K. N. Chintapalli, K. Ramakrishna, et al. 2000. Helical CT cholangiography with oral cholecystographic contrast materials. *Radiology* 214:596–601.

Ciecholewski, M., and J. Chocholowicz. 2013. Gallbladder shape extraction from ultrasound images using active contour models. *Computers in Biology and Medicine* 43:2238–2255.

Dodds, W. J., W. J. Groh, R. M. Darweesh, et al. 1985. Sonographic measurement of gallbladder volume. *American Journal of Roentgenology* 145:1009–1011.

Elsayes, K. M., E. P. Oliveira, V. R. Narra, et al. 2007. Magnetic resonance imaging of the gallbladder: Spectrum of abnormalities. *Acta Radiologica* 48:476–482.

Everson, G. Y., D. Z. Braverman, M. L. Johnson, et al. 1980. A critical evaluation of real-time ultrasonography for the study of gallbladder volume and contraction. *Gastroenterology* 79:40–46.

Fidler, J. L., J. M. Knudsen, and D. A. Collins. 2013. Prospective assessment of dynamic CT and MR cholangiography in functional biliary pain. *American Journal of Roentgenology* 201:W271–W282.

Fitzgerald, E. J., and A. Toi. 1987. Fitfalls in the ultrasonographic diagnosis of gallbladder diseases. *Postgraduate Medical Journal* 63:525–532.

Frank, S. J., and J. Kurian. 2016. Three-dimensional sonography of biliary tract disorders. *Journal of Ultrasound in Medicine* 35:791–804.

Hashimoto, S., H. Goto, Y. Hirooka, et al. 1999. An evaluation of three-dimensional ultrasonography for the measurement of gallbladder volume. *American Journal of Gastroenterology* 94:3492–3496.

Hirao, K., A. Mayazaki, T. Fujimoto, et al. 2000. Evaluation of aberrant bile ducts before laparoscopic cholecystectomy: Helical CT cholangiography versus MR cholangiography. *American Journal of Roentgenology* 175:713–720.

Hopman, W. P., W. F. Brouwer, G. Rosenbusch, et al. 1985. A computerized method for rapid quantification of gallbladder volume from real-time sonograms. *Radiology* 154:236–237.

Ichii, H., M. Takada, R. Kashiwagi, et al. 2002. Three-dimensional reconstruction of biliary tract using spiral computed tomography for laparoscopic cholecystectomy. *World Journal of Surgery* 26:608–611.

Irshad, A., S. J. Ackerman, K. Spicer, et al. 2011. Ultrasound evaluation of gallbladder dyskinesia: Comparison of scintigraphy and dynamic 3D and 4D ultrasound techniques. *American Journal of Roentgenology* 197:1103–1110.

Karimi, A., A. Shojaei, and P. Tehrani. 2017. Measurement of the mechanical properties of the human gallbladder. *Journal of Medical Engineering & Technology* 41:541–545.

Kishk, S. M., R. M. Darweesh, W. J. Dodds, et al. 1987. Sonographic evaluation of resting gallbladder volume and postprandial emptying in patients with gallstones. *American Journal of Roentgenology* 148:875–879.

Kwon, A. H., S. Uetsuji, O. Yamada, et al. 1995. Three-dimensional reconstruction of the biliary tract using spiral computed tomography. *British Journal of Surgery* 82:260–263.

Li, W. G. 2019. Constitutive law of healthy gallbladder walls in passive state with damage effect. *Biomedical Engineering Letters* 9:189–201.

Li, W. G. 2020. Ultrasound image based human gallbladder 3D modelling along with volume and stress level assessment. *Journal of Medical and Biological Engineering* 40:112–127.

Li, W. G., N. A. Hill, R. W. Ogden, et al. 2013. Anistropic behaviour of human gallbladder walls. *Journal of the Mechanical Behavior of Biomedical Materials* 20:363–375.

Li, W. G., X. Y. Luo, N. A. Hill, et al. 2011. A mechanical model for CCK-induced acalculous gallstone pain. *Annals of Biomedical Engineering* 39:786–800.

Li, W. G., X. Y. Luo, N. A. Hill, et al. 2012. A quasi-nonlinear analysis on anisotropic behaviour of human gallbladder wall. *ASME Journal of Biomechanical Engineering* 134:0101009.

Liao, D. H., B. U. Duch, H. Stodkilde-Jorgensen, et al. 2004. Tension and stress calculations in a 3-D Fourier model of gall bladder geometry obtained from MR images. *Annals of Biomedical Engineering* 32:744–755.

Nelson, T. R., and D. H. Pretorius. 1998. Three-dimensional ultrasound imaging. *Ultrasound in Medicine & Biology* 24:1243–1270.

Ogiela, M. R., and S. Bodzioch. 2011. Computer analysis of gallbladder ultrasonic images towards recognition of pathological lesions. *Opto-Electronics* 19:155–168.

Pauletzki, J., M. Sackman, J. Holl, et al. 1996. Evaluation of gallbladder volume and emptying with a novel three-dimensional ultrasound system: Comparison with the sum-of-cylinders and the ellipsoid methods. *Journal of Clinical Ultrasound* 24:277–285.

Persson, A., N. Dahlstrom, O. Smedby, et al. 2005. Volume rendering of three-dimensional drip infusion CT cholangiography in patients with suspected obstructive biliary disease: A retrospective study. *British Journal of Radiology* 78:1078–1085.

Polkowski, M., J. Palucki, J. Regula, et al. 1999. Helical computed tomographic cholangiography versus endosonography for suspected bile duct stones: A prospective blinded study in non-jaundiced patients. *Gut* 45:744–749.

Portincasa, P., A. Moschetta, A. Colecchia, et al. 2003. Measurements of gallbladder motor function by ultrasonography: Towards standardization. *Digestive and Liver Disease* 35:S56–S61.

Rohling, R. N., A. H. Gee, and L. Berman. 1998. Automatic registration of 3-D ultrasound images. *Ultrasound in Medicine and Biology* 24:841–854.

Serra, C., F. Pallotti, M. Bortolotti, et al. 2016. A new reliable method for evaluating gallbladder dynamics. *Journal of Ultrasound in Medicine* 35:297–304.

Stads, S., N. G. Venneman, R. C. Scheffer, et al. 2007. Evaluation of gallbladder motility: Comparison of two-dimensional and three-dimensional ultrasonography. *Annals of Hepatology* 6:164–169.

Stolk, M. F., K. J. van Erpecum, H. G. van Berge, et al. 1990. Gallbladder volume and contraction measured by sum-of-cylinders methods compared with ellipsoid and area-length methods. *Acta Radiology* 31:591–596.

Tan, C. H., and K. S. Lim. 2013. MRI of gallbladder cancer. *Diagnostic and Interventional Radiology* 19:312–319.

Vyas, P. K., T. L. Vesy, O. Konez, et al. 2002. Estimation of gallbladder ejection fraction utilizing cholecystokinin-stimulated magnetic resonance cholangiography and comparison with hepatobiliary scintigraphy. *Journal of Magnetic Resonance Imaging* 15:75–81.

Wedmann, B., G. Schmidt, M. Wegener, et al. 1991. Sonographic evaluation of gallbladder kinetics: In vitro and in vivo comparison of different methods to assess gallbladder emptying. *Journal of Clinical Ultrasound* 19:341–349.

Xiong, L., C. K. Chui, and C. L. Teo. 2013. Reality based modelling and simulation of gallbladder shape deformation using variational methods. *International Journal of Computer Assisted Radiology and Surgery* 8:857–865.

Xu, H. X., X. Y. Yin, M. D. Lu, et al. 2003. Comparison of three- and two-dimensional sonography in gallbladder diseases. *Journal of Ultrasound in Medicine* 22:181–191.

Yoon, H. J., P. N. Kim, A. Y. Kim, et al. 2005. Three-dimensional sonographic evaluation of gallbladder contractility: Comparison with cholescintigraphy. *Journal of Clinical Ultrasound* 34:123–127.

CHAPTER 14

Fluid Mechanics of Bile Flow in Biliary Drainage Catheters

PERCUTANEOUS TRANSHEPATIC BILIARY DRAINAGE

Obstructive jaundice is a special situation of jaundice when the bile is stopped flowing into the duodenum and remains in the blood due to gallstones in the CBD or extrinsic compression by tumours external to the CBD. The obstructive jaundice is a serious condition associated with high mortality rates and should be treated instantly by using percutaneous transhepatic biliary drainage (PTBD). PTBD is a procedure based on which the blocked bile is discharged into a drainage bag outside of the human body [external drainage in Figure 14.1(a)] or the duodenum through the CBD [internal drainage in Figure 14.1(b)] by using a catheter (Hoevels et al. 1978; Ring et al. 1979). The internal drainage, the bile can flow in the internal catheter or flow into the bag through the external drainage catheter, depending on the resistance in the two catheters.

Even though PTBD is palliative, but it can improve quality of life for patients with benign and malignant biliary diseases with success rates of (82–99)% (Nakayama et al. 1978; Ferrucci et al. 1980; Mueller et al. 1982; Mendez et al. 1984; Hamlin et al. 1986; Joseph et al. 1986; Sirinek and Levine 1989; Weber et al. 2009; Knap et al. 2016). PTBD has been become

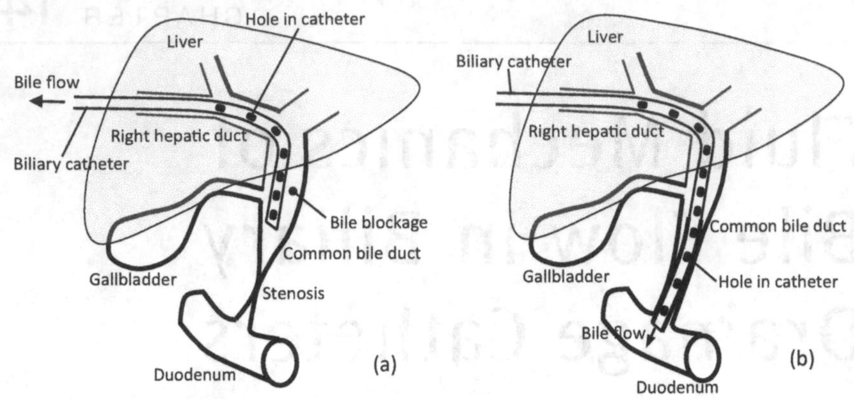

FIGURE 14.1 The external (a) and internal (b) percutaneous transhepatic biliary drainage.

an effective method for relief of biliary obstruction associated with both benign and malignant conditions.

Even though PTBD in patients is replaced every 4 or 6 weeks, unfortunately, PTBD is subject to complications after patients with malignant biliary disease undergo placement of drainage catheter (Mendez et al. 1984; Hamlin et al. 1986; Joseph et al. 1986; Weber et al. 2009; Clark et al. 1981; Yee and Ho 1987). The complications can be cholangitis, catheter dislodgement, leaking around catheter, obstructed catheter, haemobilia, hypersecretion of bile, biliopleural fistula, bile duct perforation and pneumothorax (Carrasco et al. 1984), their occurrence percentages in 179 patients are illustrated in Figure 14.2. The total percentage of the complications related to catheter is as high as 43%. This means that the catheter performance plays a vitally important role in PTBD technique.

The catheter dislodgement and leaking complications connect with catheter design and soft tissue biomechanical property, while the catheter obstruction is associated with catheter design and bile fluid mechanics inside. Even though the catheter obstruction has made a 12.3% contribution to the total complication occurrence, it can cause catheter malfunction and eventually result in PTBD failure. From this point of view, attention should be paid to design and fluid mechanics of catheters in PTBD.

Presently, investigations into design and fluid mechanics of catheters in PTBD are lax. Kerlan et al. (1984), for the first time, measured bile flow

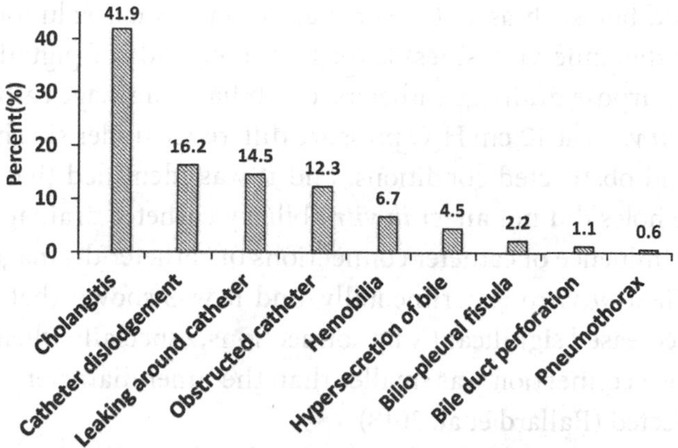

FIGURE 14.2 The percentages of complications after PTBD occurred in 179 patients with malignant biliary obstruction, the percentages were recalculated based on the data in the table in Carrasco et al. (1984).

rates through a series of different catheters at various pressure differences across the catheters *in vitro*, but their raw data remained unprocessed. Bret et al. (1986) clinically applied large size silicone catheters with 12, 15 and 18 French outer diameters (1 mm=3 French), 2, 3, 4 mm inner diameters and 3×5, 4×7, 5×9 mm side holes into 30 patients with obstructive jaundice due to stenoses and tumours in bile ducts. It was shown that PTBD was effective in treating benign and malignant bile duct strictures in the long term, but frequently minor problems, mostly catheter-related, did persist (Born et al. 1998).

A few *in vitro* experiments on fluid flow in percutaneous drainage catheters have been conducted by measuring drainage time at various catheter sizes and fluid viscosities (Park et al. 1993). It was demonstrated that a more viscous fluid required larger catheter to secure a rapid drainage. The flow rates of commercial multipurpose pigtail drainage catheters were measured *in vitro* at 30 mmHg pressure difference. Since their inner diameter sizes were comparable, their flow rates were similar in values (Macha et al. 2006). The effects of number and location of drainage catheter side holes on liquid flow rate were measured *in vitro* by employing unilateral and bilateral side hole models, the catheters with bilateral side holes had a higher flow rate than those with unilateral side holes, adding more side holes could not improve flow rate once the number of the holes beyond a critical number of holes (Ballard et al. 2015). The flow rates of simulated bile such as water, three additional water solutions of guar gum (four dynamic viscosities) through three kinds of pigtail catheters (two multipurpose drainage catheters, one biliary drainage catheter) were measured *in vitro* at 12 cm H_2O pressure difference under side hole unobstructed and obstructed conditions, and it was identified that the number of side holes did not affect *in vitro* biliary catheter drainage (Li et al. 2017). The influence of catheter connections of catheter drainage flow rate was identified *in vitro* experimentally, and it was shown that flow rates could be decreased significantly by connections, especially when the inner diameter of a connection was smaller than the inner diameter of the catheters connected (Ballard et al. 2018).

A method was proposed to predict the clogging effect drainage catheters based on *in vivo* rabbit experiments by monitoring intracatheter pressure (Lee et al. 2003). The commercial catheter for PTBD was ligated with a nylon thread just proximal to the first side hole to prevent the catheter obstruction caused from jejunobiliary reflux of the intestinal contents in internal PTBD (Hamada et al. 2013). Fracture of the PTBD catheter could occur and cause bile peritonitis (Nghiem 1984). Three new techniques were developed to retrieve fractured and intrahepatically dislodged PTBD catheters (Liu et al. 2017).

Based on the existing results on PTBD catheters mentioned above, bile fluid mechanics associated with clinical performance of catheters has been documented a little so far, and there are no empirical relationships for bile flow through biliary drainage catheters to assess their clinical performance in the literature currently. In the chapter, the raw experimental data of a series of catheters on bile flow rate and pressure difference were analysed based on the elements of fluid mechanics to make the dead experimental data alive. An equivalent friction factor through the catheters was proposed and determined by using these observed data. An empirical relationship of bile flow rate through a catheter was established accordingly and applied to predict the bile flow rates through the catheters with various inner diameters under measured normal and abnormal biliary pressures and two bile viscosities in a CBD.

BILE FLOW MEASUREMENTS

Catheters for PTBD are a flexile, plastic central-hollowed tube with taping head and side holes, as shown in Figure 14.1. *In vitro* bile flow measurements through catheters are very rare in the literature so far. Currently, just one relatively complete data set for such flow measurement was accomplished by Kerlan et al. (1984). The experimental set-up and method were referred to Kerlan et al. (1984), which is full access paper downloaded feely. In experiments, a tested catheter was connected to an Erlenmeyer flask and a measuring cylinder. The flask was pressurised and depressurised by adding and removing air with a syringe to maintain a constant pressure difference across the catheter and establish a bile flow in it. This pressure difference was measured by using a U-tube manometer. Bile level is held to be constant with another syringe by infusing bile. The Erlenmeyer flask and cylinder were submerged in a water bath to allow bile temperature to be at 37°C. The bile flow rate was measured by the net bile volume in the cylinder and the time elapsed.

Freshly aspirated human hepatic bile with a dynamic viscosity of 0.01Poise (0.001Pa.s) and a density of 1000 kg/m³ serves as experimental fluid (Kerlan et al. 1984). The experimental pressure differences, catheter sample lengths and diameters and measured bile flow rates are listed in Table 14.1 (Kerlan et al. 1984). These raw experimental data are going to be utilised to establish an empirical relationship between flow resistance factor/equivalent friction factor and Reynolds number and a correlation between bile flow rate and pressure gradient.

EQUIVALENT FRICTION FACTOR

The hydraulic losses in the catheter shown in Figure 14.1 include the friction loss over the wet surface, the incidence loss at the inlet of catheter, the secondary flow loss in the 90° bend and the diffusion loss across the side holes. At first, these losses are supposed to contribute an equivalent friction factor to simplify the problem; then, according to the skin friction factor formula for ducts (White 2011), the equivalent friction factor λ will be calculated based on a known pressure difference Δp, bile flow rate Q, effective length of the catheter L and inner diameter of the catheter d in the following manner

$$f = \frac{\Delta p}{\dfrac{V^2 L}{2gd}} \tag{14.1}$$

where g is the gravity acceleration, g = 9.81 m/s², V is the mean bile flow velocity. Because of $V = Q/(\pi d^2/4)$ (White 2011), Eq. (14.1) can be rewritten as

TABLE 14.1 Measured flow rates through selected drainage catheters under specified pressure differences.

Material	d (French)	d (mm)	L (cm)	Flow rate Q (ml/s) under pressure difference (cmH_2O)			
				1	2	6	9
Polyethylene	6.3	1.4	65	0.023	0.056	0.088	0.153
	7.1	1.5	65	0.028	0.068	0.110	0.165
	8.3	1.8	50	0.114	0.184	0.296	0.428
	10	2.2	50	0.210	0.385	0.495	0.701
Polyurethane	10	2.0	50	0.094	0.182	0.300	0.450
	12	2.7	50	0.256	0.456	0.769	1.110
Polyvinyl-chloride (PVC)	10	2.2	60	0.111	0.250	0.410	0.620
	10	2.2	50	0.153	0.333	0.500	0.800
Silicon elastomer	16	2.8	65	0.294	0.549	0.694	1.000
	16	2.8	50	0.310	0.595	0.833	1.330
Teflon	12	3.0	65	0.303	0.400	0.690	0.952
	12	3.0	31	1.300	1.540	2.000	2.500

Effective length of catheter L is the distance from catheter hub to first side hole, 1French = 1/3 mm

$$f = \frac{g\pi^2 d^5}{8Q^2}\frac{\Delta p}{L} \tag{14.2}$$

As a result, the corresponding *f-Re* scattered data points based on the experimental data in Table 14.1 are present in Figure 14.3, where $Re = 4Q/\pi d v$, v is the bile kinematic viscosity, v = 1 mm²/s (Kerlan et al. 1984). Because of $Re \le 1000$, the bile flow in the experimental catheters was in laminar regime. For comparison, the analytical friction factor $64/Re$, for the fully developed laminar flow in a circular pipe in White (2011) is plotted, too.

In the figure, some experimental points are below the $64/Re$ curve as $Re \le 200$, and some points are above the curve, suggesting the experimental f shares a different slope with the analytical friction factor curve. The experimental f has been best fitted by a power function of Re, the empirical formula reads as

$$f = \frac{12.265}{Re^{0.6540}}, R^2 = 0.7389 \tag{14.3}$$

where R is the correlation coefficient. A comparison of the experimental f with the fitted curve is made in Figure 14.4.

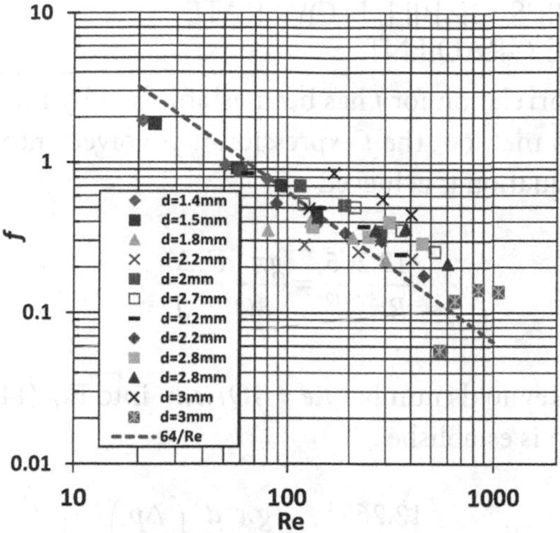

FIGURE 14.3 The scattered equivalent friction factors based on the experimental data in Table 14.1 and the analytical friction factor 64/*Re* are plotted as a function Reynolds number.

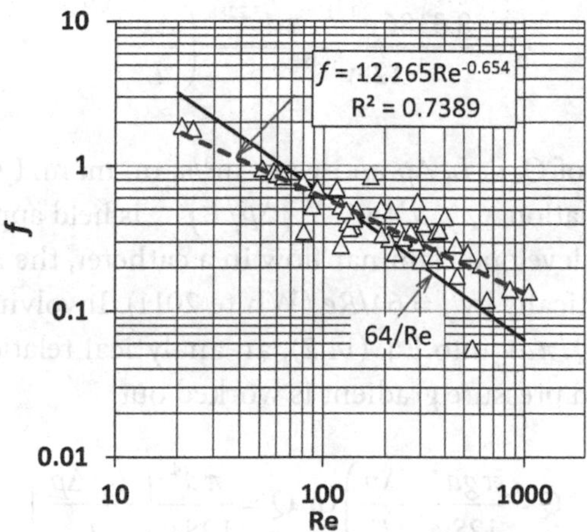

FIGURE 14.4 The fitted empirical correlation of equivalent friction factor in terms of Reynolds number and its comparison with the analytical friction factor for a catheter, the scattered data points are the same as those in Figure 14.3, the inner diameter is no longer indicated.

RELATIONSHIPS OF BILE FLOW RATE TO PRESSURE GRADIENT

An empirical correlation for f has been established by Eq. (14.3) based on fluid mechanics method, the f expression is involved into Eq. (14.2), and the following equation is achieved

$$\frac{12.265}{Re^{0.6540}} = \frac{g\pi^2 d^5}{8Q^2}\frac{\Delta p}{L} \tag{14.4}$$

Putting the Reynolds number $Re = 4Q/\pi dv$ into Eq. (14.4), and the following equation is established

$$\frac{12.265}{\left(\dfrac{4Q}{\pi dv}\right)^{0.6540}} = \frac{g\pi^2 d^5}{8Q^2}\left(\frac{\Delta p}{L}\right) \tag{14.5}$$

From Eq. (14.5), the bile flow rate Q through a catheter with the length L driven by a pressure gradient $\Delta p/L$ can be solved and expressed by

$$Q = \frac{8.3186 g^{0.7429} \pi d^{3.2288}}{128 v^{0.4859}}\left(\frac{\Delta p}{L}\right)^{0.7429} \tag{14.6}$$

where the units of Q, d, L, Δp and v are m^3/s, m, m, mH_2O, m^2/s, respectively. In the equation, $Q \propto d^{3.2}\ v^{-0.49}\left(\Delta p/L\right)^{0.74}$ is held approximately.

For the fully developed laminar flow in a catheter, the friction factor is expressed analytically by $f = 64/Re$ (White 2011). Involving the Reynolds number $Re = 4Q/\pi dv$ into Eq. (14.2), an analytical relationship between bile flow rate and pressure gradient is worked out

$$Q = \frac{\pi g d^4}{128 v}\left(\frac{\Delta p}{L}\right) \text{ or } Q = \frac{\pi d^4}{128\mu}\left(\frac{\rho g \Delta p}{L}\right) \tag{14.7}$$

where the dynamic viscosity μ is related to the kinematic viscosity with $\mu = \rho v$, ρ is the bile density, the unit of Δp is mH_2O. In these expressions, $Q \propto d^4\ v^{-1}\ (\Delta p/L)^1$, suggesting parameters d, v and $\Delta p/L$ exhibit a stronger effect on Q in comparison with those in Eq. (14.6) originated from experimental observations.

In forgoing section, an empirical friction factor is determined first, then it replaces the f in Eq. (14.2), and Re is expressed in terms of bile flow rate; finally, a relationship between bile flow rate and pressure gradient is sought as expressed with Eq. (14.6). In fact, we can derive an empirical relationship between bile flow rate and pressure gradient directly based on analytical Eq. (14.7). This method has been used in the determination of an empirical expression between bile flow rate and pressure gradient across animal biliary tree *in vitro* in Rodkiewicz et al. (1979). This method will be tried on the experimental data on the catheters (Kerlan et al. 1984) here.

A few perfusion experiments were performed on the biliary tree (hepatic, cystic and CBDs) of six fasting mongrel dogs (Rodkiewicz et al. 1979) by using saline and bovine bile at different bile perfusion flow rates, respectively. The pressure differences across the tree were recorded. It was identified that the experimental scattered points of $(\Delta p/L, 128\nu Q/\pi g d^4)$ and $(\rho g \Delta p/L, 128\mu Q/\pi d^4)$ can be best fitted with a linear relationship in a log-log plot. Then the bile flow rate can be obtained in terms of pressure gradient, like Eq. (14.7).

The experimental data points of $\Delta p/L$ - $128\nu Q/\pi g d^4$ and $\rho g \Delta p/L$ - $128\mu Q/\pi d^4$ are plotted and fitted, respectively, for the experimental data in the catheters in Kerlan et al. (1984). The experimental scattered data points and the corresponding regression formulas are illustrated in Figure 14.5.

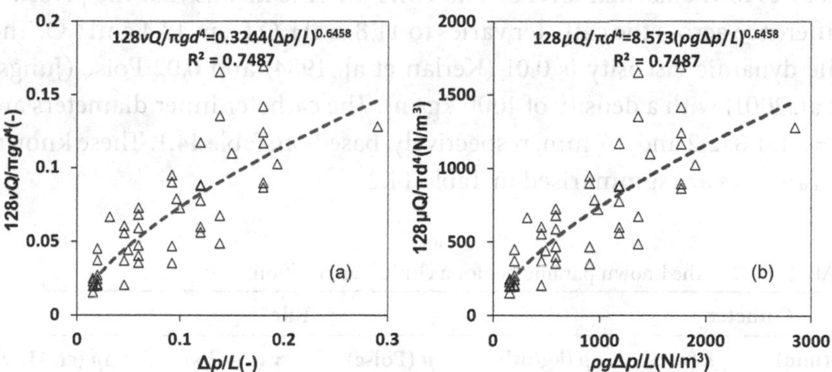

FIGURE 14.5 The scattered data points of $(\Delta p/L, 128\nu Q/\pi g d^4)$ and $(\rho g \Delta p/L, 128\mu Q/\pi d^4)$ as well as the corresponding regression formulas, the scattered data points are the same as those in Figure 14.3, the inner diameter is no longer indicated.

Based on these formulas, the bile flow rate Q through a catheter under a known pressure gradient $\Delta p/L$ is given by

$$Q=\frac{0.3234\pi gd^4}{128\nu}\left(\frac{\Delta p}{L}\right)^{0.6458} \quad \text{or}$$
$$Q=\frac{8.573\pi d^4}{128\mu}\left(\frac{\rho g\Delta p}{L}\right)^{0.6458},\ R^2=0.7487 \tag{14.8}$$

where the units of Q, d, L, Δp, ρ, ν and μ are m^3/s, m, m, mH_2O, kg/m^3, m^2/s and Pa·s, respectively.

A COMPARISON OF TWO RELATIONSHIPS

From the same set of experimental data, two relationships have been obtained for bile flow rate in terms of pressure gradient across a catheter expressed by Eqs. (14.6) and (14.8). Two relationships may result in a different bile flow rate under the same clinical condition. To confirm this effect, a computational example is provided here.

The biliary mean resting pressure in normal human CBD is 11.8 cmH_2O, but in the duct with obstructive jaundice, it is 18.4 cmH_2O (White et al. 1972; van Sonnenberg et al. 1983). It is assumed that a 50-cm-long catheter is connected to a CBD with obstructive jaundice at 18.4 cmH_2O initial pressure, after drainage persists for a certain long of time, the biliary pressure restores to the normal level of 11.8 cmH_2O. This means that the pressure difference across the catheter varies to 11.8 cmH_2O from 18.4 cmH_2O. The bile dynamic viscosity is 0.01 (Kerlan et al. 1984) and 0.02 Poise (Jungst et al. 2001) with a density of 1000 kg/m^3. The catheter inner diameters are d =1.4, 1.8, 2.2 and 2.7 mm, respectively, based on Table 14.1. These known parameters are summarised in Table 14.2.

TABLE 14.2 The known parameters for a clinical application.

Catheter		Bile			
d (mm)	L (cm)	ρ (kg/m^3)	μ (Poise)	ν (mm^2/s)	Δp (cmH_2O)
1.4	50	1000	0.01, 0.02	1, 2	11.8–18.4
1.8	50	1000	0.01, 0.02	1, 2	11.8–18.4
2.2	50	1000	0.01, 0.02	1, 2	11.8–18.4
2.7	50	1000	0.01, 0.02	1, 2	11.8–18.4

First, Eqs. (14.6) and (14.8) are used to predict the bile flow rates under the experimental conditions such as 1, 2, 6 and 9 cmH_2O pressure differences and 0.01 Poise viscosity as shown in Table 14.1 and at four inner diameters in Table 14.2. The two equations result in nearly the same bile flow rate profiles as shown in Figure 14.6(a). This is not surprised because they have originated from the same experimental data set and applied under nearly the same condition in terms of pressure difference, viscosity and catheter inner diameter as in the experiments (Kerlan et al. 1984).

Second, two equations are employed to estimate the bile flow rates at 0.02 Poise viscosity, while the rest condition remain the same as those

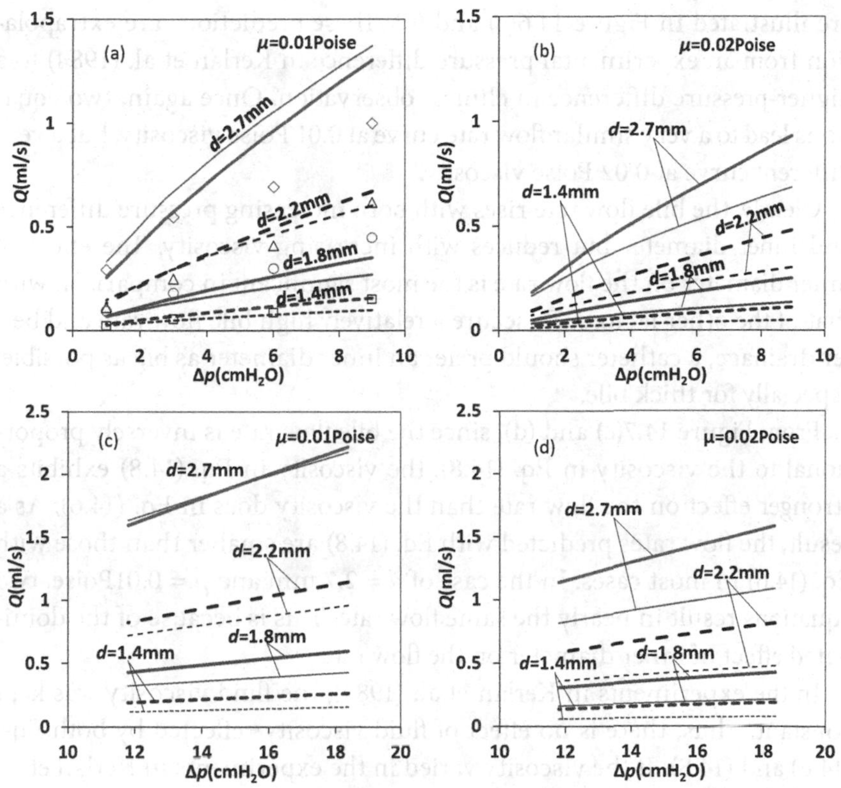

FIGURE 14.6 The predicted bile flow rate through four catheters in terms of bile pressure difference across the catheters at two viscosities, the thick lines are for Eq. (14.6), but the thin lines for Eq. (14.8); in (a) and (b), the pressure difference is in the range of the experiments in Table 14.1 (Kerlan et al. 1984); while in (c) and (d), the pressure difference is based on clinical observation (White et al. 1972; van Sonnenberg et al. 1983).

for Figure 14.6(a). In the experiments of Kerlan et al. (1984), the tested liquid viscosity was kept being 0.01 Poise. The prediction at 0.02 Poise viscosity is an extrapolation from the results at 0.01 Poise viscosity. The flow rates from Eq. (14.6) are larger than those from Eq. (14.8), as demonstrated in Figure 14.6(b) because of $Q \propto \nu^{-0.49}$ in Eq. (14.6) rather than $Q \propto \nu^{-1}$ in Eq. (14.8). These suggest that the flow rates predicted with two equations at a viscosity more than 0.01 Poise are not accurate as those at 0.01 Poise.

Finally, two equations are utilised to calculate $Q - \Delta p$ curves at four inner diameters and two viscosities in Table 14.2 and under the pressure differences higher than those in Table 14.1. The predicted $Q - \Delta p$ curves are illustrated in Figure 14.6(c) and (d). These predictions are extrapolation from an experimental pressure difference in Kerlan et al. (1984) to a higher-pressure difference in clinical observation. Once again, two equations lead to a very similar flow rate curve at 0.01 Poise viscosity, but a very different curve at 0.02 Poise viscosity.

Clearly, the bile flow rate rises with both increasing pressure difference and inner diameter but reduces with increasing viscosity. The effect of inner diameter on the flow rate is the most significant in comparison with that of the other factors. To secure a relatively high bile flow rate and better drainage, a catheter should prefer an inner diameter as big as possible, especially for thick bile.

From Figure 14.7(c) and (d), since the bile flow rate is inversely proportional to the viscosity in Eq. (14.8), the viscosity in Eq. (14.8) exhibits a stronger effect on the flow rate than the viscosity does in Eq. (14.6). As a result, the flow rates predicted with Eq. (14.8) are smaller than those with Eq. (14.6) in most cases. In the case of $d = 2.7$ mm and $\mu = 0.01$Poise, two equations result in nearly the same flow rate. This is because of the dominated effect of inner diameter on the flow rate.

In the experiments in Kerlan et al. (1984), the fluid viscosity was kept constant. Thus, there is no effect of fluid viscosity reflected by both Eqs. (14.6) and (14.8). If the viscosity varied in the experiments in Kerlan et al. (1984), Eqs. (14.6) and (14.8) should lead to a nearly identical bile flow under the same clinical condition. To validate two empirical relationships of Eqs. (14.6) and (14.8), more experimental data on *in vitro* bile flow measurements in catheters are desirable with more viscous liquids and under pressure differences higher than 9 cmH_2O.

MINOR HYDRAULIC LOSSES AND BILE VISCOSITY EFFECT

In the equivalent friction factor, there are minor hydraulic losses, namely entry loss at the catheter inlet, secondary flow loss in the bend of a catheter and expansion loss through the side holes in the catheter. It is not easily to measure and estimate these minor losses. Here, using the ratio of the equivalent friction factor to the theoretical friction factor for the fully developed laminar flow in a circular pipe, i.e., $f/(64/Re)$ is used to estimate these minor losses. As a result, the scattered data points and a regression equation are illustrated in Figure 14.7. Clearly, ratio $f/(64/Re)$ augments with increasing Re, particularly, if $Re > 100$, then $f/(64/Re) > 1$, indicating the dominant minor loss. When $Re < 100$, the ratio is less than one. This effect may be due to some errors in the experiments or the thickening effect of non-Newtonian bile at low flow rate.

Note that, in clinical practice, the bile flows into the side holes of a catheter rather than out of the holes as shown in the experiments as shown in Figure 14.3. The expansion loss in two scenarios may be different each other. This issue needs to be confirmed experimentally in the future. CFD studies on the minor hydraulic losses in biliary drainage catheters are also worthy of being attempted.

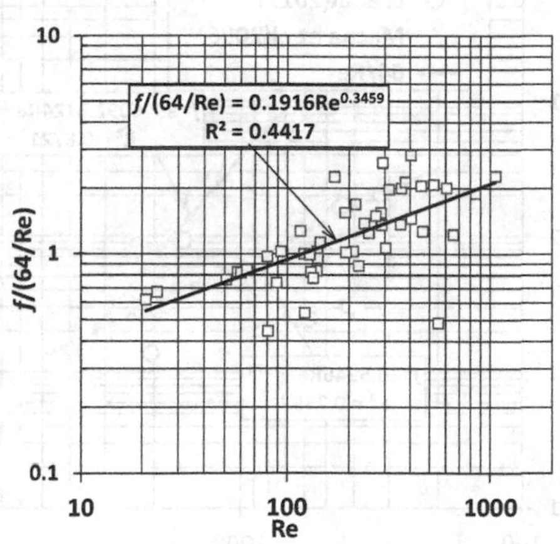

FIGURE 14.7 The ratio of the equivalent friction factor to the theoretical friction factor of the fully developed laminar flow in a circular pipe, i.e., $f/(64/Re)$ as a function of Reynolds number, the scattered data points are the same as those in Figure 14.3, the inner diameter is no longer indicated.

Recently, the flow rates in three commercial multipurpose pigtail drainage catheters at 30 mmHg pressure difference were measured *in vitro* with water by Macha, et al. (2006). The flow rates of water, three water solutions of guar gum across three pigtail catheters (two multipurpose drainage catheters and one biliary drainage catheter), were *in vitro* measured at 12 cmH_2O pressure difference under side hole unobstructed and obstructed conditions, and it was identified that the number of side holes do not affect *in vitro* biliary catheter drainage (Li et al. 2017). The catheter geometrical parameters were presented in Macha et al. (2006) and Li et al. (2017). The flow rates in Macha et al. (2006) and Li et al. (2017) for the unobstructed catheters were read and the equivalent friction factors were calculated by them with Eq. (14.2), and the results are illustrated in Figure 14.8.

Clearly, the data points in the two experiments are quite few and the Reynolds number is in the range of 300–4000, which is higher than that (20–1000) in Figure 14.3. The factors from the experimental data in Macha et al. (2006) exhibit significant variation. Even though the regression equation for them is slightly below the analytical curve of $64/Re$, its correlation coefficient is as small as 0.24.

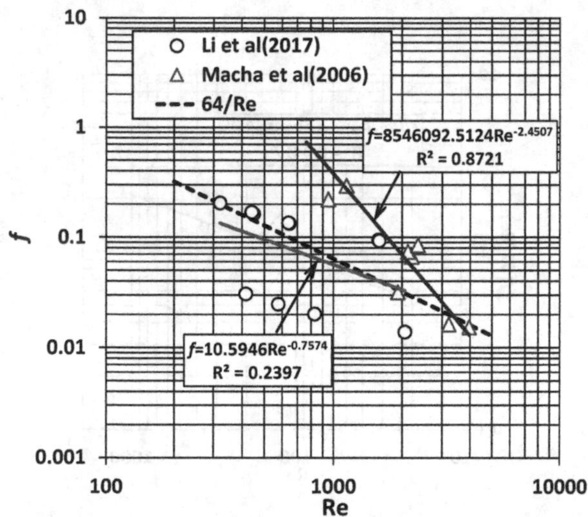

FIGURE 14.8 The experimental equivalent friction factors respectively by Macha et al. (2006) and Li et al. (2017) and fitted empirical correction of in terms of Reynolds number and its comparison with the analytical friction factor for a catheter.

The friction factors from the experimental data (Li et al. 2017) are considerably higher than the analytical curve as $Re \leq 2000$. Nonetheless further experimental confirmation is on demand.

Since there is one viscosity in the experiments and no information about the used bile rheology in Kerlan et al. (1984), the bile in fluid mechanics model is considered Newtonian. Ooi et al. (2004) measured the bile dynamic viscosity and found that the bile rheology of 20 out of 59 patients is Newtonian. Reinhart et al. (2010) found the bile of the majority samples from the CBD of 138 patients (64.5%) are Newtonian. These facts suggest that the Newtonian bile model seems reasonable. In some cases, however, the bile can be non-Newtonian (Ooi et al. 2004; Reinhart et al. 2010; Coene et al. 1994; Kuchumov et al. 2014); therefore, the correlation needs be updated in the future based on *in vitro* experimental data on non-Newtonian fluid flow through PTBD catheters.

The bile viscosity can vary significantly across patients, e.g., the dynamic viscosity of GB bile is 0.0177–0.08 Poise (Ooi et al. 2004), and even higher in GB bile of patients with cholesterol (0.05 Poise) and mixed stones (0.035 Poise) compared to hepatic bile (0.02 Poise) (Jungst et al. 2001). Therefore, more *in vitro* studies on bile flow through a catheter with a variety of viscosities need to be launched in the future.

REFERENCES

Ballard, D. H., J. S. Alexander, J. A. Weisman, et al. 2015. Number and location of drainage catheter side holes: In vitro evaluation. *Clinical Radiology* 70:974–980.

Ballard, D. H., S. T. Flanagah, H. Li, et al. 2018. In vitro evaluation of percutaneous drainage catheters: Flow related to connections and liquid characteristics. *Diagnostic and Interventional Imaging* 99:99–104.

Born, P., A. Tripttrap, E. Frimberger, et al. 1998. Long-term results of percutaneous transhepatic biliary for benign and malignant bile duct strictures. *Scandinavian Journal of Gastroenterology* 33:544–549.

Bret, P. M., M. Bretagnolle, A. Fond, et al. 1986. Use of large silicone catheters in patients in long-term percutaneous transhepatic biliary drainage. *Cardiovascular and Interventional Radiology* 9:57–58.

Carrasco, C. H., J. Zornoza, and W. J. Bechtel. 1984. Malignant biliary obstruction: Complications of percutaneous biliary drainage. *Radiology* 152:343–346.

Clark, R. A., S. E. Mitchell, D. P. Colley, et al. 1981. Percutaneous catheter biliary decompression. *American Journal of Roentgenology* 137:503–509.

Coene, P. P., A. K. Groen, P. H. Davids, et al. 1994. Bile viscosity in patients with biliary drainage. *Scandinavian Journal of Gastroenterology* 29:757–763.

Ferrucci, J. T., P. R. Mueller, and W. P. Harbin. 1980. Percutaneous transhepatic biliary drainage. *Diagnostic Radiology* 135:1–13.

Hamada, T., T. Tsujino, H. Isayama, et al. 2013. Percutaneous transhepatic biliary drainage using a ligated catheter for recurrent catheter obstruction: Antireflux technique. *Gut and Liver* 7:255–257.

Hamlin, J. A., M. Friedman, M. G. Stein, et al. 1986. Percutaneous biliary drainage: Complications of 118 consecutive catheterizations. *Radiology* 158:199–202.

Hoevels, J., A. Lunderquist, and I. Ihse. 1978. Percutaneous transhepatic intubation of bile ducts for combined internal-external drainage in preoperative and palliative treatment of obstructive jaundice. *Gastrointestinal Radiology* 3:23–31.

Joseph, P., L. S. Bizer, S. S. Sprayregen, et al. 1986. Percutaneous transhepatic biliary drainage. *JAMA* 255:2763–2767.

Jungst, D., A. Niemeyer, I. Muller, et al. 2001. Mucin and phospholipids determine viscosity of gallbladder bile in patients with gallstones. *World Journal Gastroenterology* 7:203–207.

Kerlan, R. K., G. Stimac, A. C. Pogany, et al. 1984. Bile flow through drainage catheters: An in vitro study. *American Journal of Roentgenology* 143:1085–1087.

Knap, D., N. Orlecka, R. Judka, et al. 2016. Biliary duct obstruction treatment with aid of percutaneous transhepatic biliary drainage. *Alexandria Journal of Medicine* 52:185–191.

Kuchumov, A. G., V. Gilev, V. Popov, et al. 2014. Non-Newtonian flow of pathological bile in the biliary system: Experimental investigation and CFD simulations. *Korea-Australia Rheology Journal* 26:81–90.

Lee, K. H., J. K. Han, K. G. Kim, et al. 2003. Clogging of drainage catheters: Quantitative and longitudinal assessment by monitoring intracatheter pressure in catheters and rabbits. *Radiology* 227:833–838.

Li, A. Y., D. H. Ballard, and H. B. D'Agostino. 2017. Biliary drainage catheters fluid dynamics: In vitro flow rates and patterns. *Diagnostic and Interventional Imaging* 98:355–358.

Liu, H. T., H. S. Tseng, Y. Y. Lin, et al. 2017. Percutaneous transhepatic techniques for retrieving fractured and intrahepatically dislodged percutaneous transhepatic biliary drainage catheters. *Diagnostic and Interventional Radiology* 23:461–464.

Macha, D. B., J. Thomas, and R. C. Nelson. 2006. Pigtail catheters used for percutaneous fluid drainage: Comparison of performance characteristics. *Vascular and Interventional Radiology* 238:1057–1063.

Mendez, G., E. Russsell, J. R. LePage, et al. 1984. Abandonment of endoprosthetic drainage technique in malignant biliary obstruction. *American Journal of Roentgenology* 143:617–622.

Mueller, P. R., E. van Sonnenberg, and J. Ferrucci. 1982. Percutaneous biliary drainage: Technical and catheter-related problems in 200 procedures. *American Journal of Roentgenology* 138:17–23.

Nakayama, T., A. Ikeda, and K. Okuda. 1978. Percutaneous transhepatic intubation of biliary tract. *Gastroenterology* 74:554–558.

Nghiem, D. D. 1984. Bile leakage after fracture of percutaneous transhepatic biliary drainage catheters. *JAMA* 251:892.

Ooi, R. C., X. Y. Luo, S. B. Chin, et al. 2004. The flow of bile in the human cystic duct. *Journal of Biomechanics* 37:1913–1922.

Park, J. K., F. C. Kraus, and J. R. Haaga. 1993. Fluid flow during percutaneous drainage procedures: An in vitro study of the effects of fluid viscosity, catheter size and adjunctive urokinase. *American Journal of Roentgenology* 160:165–169.

Reinhart, W. H., G. Naf, and B. Werth. 2010. Viscosity of human bile sampled from the common bile duct. *Clinical Hemorheology and Microcirculation* 44:177–182.

Ring, E. J., J. W. Husted, J. A. Oleaga, et al. 1979. A multihole catheter for maintaining longterm percutaneous antegrade biliary drainage. *Radiology* 132:752–754.

Rodkiewicz, C. M., W. J. Otto, and G. W. Scott. 1979. Empirical relationships for the flow of bile. *Journal of Biomechanics* 12:411–413.

Sirinek, K. R., and B. A. Levine. 1989. Percutaneous transhepatic cholangiography and biliary drainage. *Archives of Surgery* 124:885–888.

van Sonnenberg, E., J. T. Ferrucci, C. C. Neff, et al. 1983. Biliary pressure: Manometric and perfusion studies at percutaneous transhepatic cholangiography and percutaneous biliary drainage. *Radiology* 148:41–50.

Weber, A., J. Gaa, B. Rosca, et al. 2009. Complications of percutaneous transhepatic biliary drainage in patients with dilated and nondilated intrahepatic ducts. *European Journal of Radiology* 72:412–417.

White, F. M. 2011. *Fluid Mechanics* (7th edition). New York: McGraw-Hill Companies Inc.

White, T. T., H. Waisman, D. Hopton, et al. 1972. Radiomanometry, flow rates, and cholangiography in the evaluation of common bile duct disease. *American Journal of Surgery* 123:73–79.

Yee, A. C., and C. S. Ho. 1987. Complications of percutaneous biliary drainage: Benign vs malignant diseases. *American Journal of Roentgenology* 148:1207–1209.

APPENDIX A

Statistical Analysis of GB Pain Prediction

Table A.1 shows a 2 × 2 contingency table for the two variables A and B with a small sample size. We tend to determine the endpoints of 95% confidence interval for success rate P_A and P_B as well as to compare them. The success rates are calculated by

$$P_A = \frac{N_{AS}}{N_{AS} + N_{AF}} \tag{A.1}$$

and

$$P_B = \frac{N_{BS}}{N_{BS} + N_{BF}} \tag{A.2}$$

TABLE A.1 2 × 2 Contingency table for the variables A and B.

Variable	Success	Failure	Sample size
A	N_{AS} (P_A)	N_{AF} $(1 - P_A)$	$N_{AS} + N_{AF}$
B	N_{BS} (P_B)	N_{BF} $(1 - P_B)$	$N_{BS} + N_{BF}$

The logistic transformation is defined as for the variable A

$$\varphi_A = \log\left(\frac{P_A}{1-P_A}\right) \tag{A.3}$$

The back-transformation for P_A is

$$P_A = \frac{e^{\varphi_A}}{1+e^{\varphi_A}} \tag{A.4}$$

The endpoints of 95% confidence interval for the success rate P_A are

$$P_{AL} = \frac{e^{P_A - 1.96\sqrt{\frac{1}{P_A(N_{AS}+N_{AF})(1-P_A)}}}}{1+e^{P_A - 1.96\sqrt{\frac{1}{P_A(N_{AS}+N_{AF})(1-P_A)}}}} \tag{A.5}$$

and

$$P_{AH} = \frac{e^{P_A + 1.96\sqrt{\frac{1}{P_A(N_{AS}+N_{AF})(1-P_A)}}}}{1+e^{P_A + 1.96\sqrt{\frac{1}{P_A(N_{AS}+N_{AF})(1-P_A)}}}} \tag{A.6}$$

The endpoints of 95% confidence interval for the success rate P_B (variable B) are

$$P_{BL} = \frac{e^{P_B - 1.96\sqrt{\frac{1}{P_B(N_{BS}+N_{BF})(1-P_B)}}}}{1+e^{P_B - 1.96\sqrt{\frac{1}{P_B(N_{BS}+N_{BF})(1-P_B)}}}} \tag{A.7}$$

and

$$P_{BH} = \frac{e^{P_B + 1.96\sqrt{\frac{1}{P_B(N_{BS}+N_{BF})(1-P_B)}}}}{1+e^{P_B + 1.96\sqrt{\frac{1}{P_B(N_{BS}+N_{BF})(1-P_B)}}}} \tag{A.8}$$

The difference between the success rates of A and B can be distinguished by using the odds ratio from the two rows in Table A.1. The asymptotic standard error of the two samples is (Agresti 1996)

$$s = \log\left(\sqrt{\frac{1}{N_{\mathrm{AS}}} + \frac{1}{N_{\mathrm{AF}}} + \frac{1}{N_{\mathrm{BS}}} + \frac{1}{N_{\mathrm{BF}}}}\right) \tag{A.9}$$

The odds ratio from the two samples is (Agresti 1996)

$$\delta = \log\left[\frac{P_{\mathrm{A}}\left(1 - P_{\mathrm{B}}\right)}{\left(1 - P_{\mathrm{A}}\right)P_{\mathrm{B}}}\right] \tag{A.10}$$

The endpoints of 95% confidence interval for the odds ratio are (Agresti 1996)

$$\delta_{\mathrm{L}} = \mathrm{e}^{\delta - 1.96s} \tag{A.11}$$

and

$$\delta_{\mathrm{H}} = \mathrm{e}^{\delta + 1.96s} \tag{A.12}$$

Table A.2 illustrates the 2 × 2 contingency table for the four variables EF, R, $p_{\max}$ and $\sigma_{\max}$. The table has shown the counts of success and failure predictions compared with the clinical observations. The figures in the bracket are the success and failure rates, respectively. The success rates of

TABLE A.2 Counts, success rates of pain and no-pain predictions.

Variable	Prediction	Success	Failure	Sample size
EF	Pain (positive)	7 (0.438)	9 (0.562)	16
	No-pain (negative)	8 (0.381)	13 (0.619	21
R	Pain	5 (0.357)	9 (0.643)	14
	No-pain	8 (0.348)	15 (0.652)	23
$p_{\max}$	Pain	18 (0.581)	13 (0.219)	31
	No-pain	4 (0.667)	2 (0.333)	6
$\sigma_{\max}$	Pain	17 (0.850)	3 (0.150)	20
	No-pain	12 (0.706)	5 (0.294)	17

TABLE A.3 95% Confidence intervals of the success rates.

Variable	Success rate of pain and no-pain prediction	Confidence interval
EF	0.438	(0.225, 0.677)
	0.381	(0.203, 0.598)
R	0.357	(0.156, 0.622)
	0.348	(0.184, 0.557)
p_{max}	0.581	(0.405, 0.739)
	0.667	(0.268, 0.916)
σ_{max}	0.850	(0.407, 0.889)
	0.706	(0.416, 0.852)

prediction pain (positive) and no-pain (negative) made by using EF and R are all less than 0.5; furthermore, no significant difference exists between them. Therefore, these variables should be rejected to be as pain predictors. The success rates of prediction pain and no-pain by using p_{max} are more than 0.5 but less than 0.7, and interestingly the rate of prediction no-pain is better than the rate of pain. Consequently, p_{max} would be a no-pain prediction factor but needs to be investigated further. The success rates of prediction pain and no-pain by using σ_{max} are over 0.75, and it should be noticed that the success rate of prediction pain is so high that it is up to 0.850. σ_{max} has the best ability to identify GB pain, and it should be the variable to predict GB pain.

The 95% confidence intervals for the success rates of pain and no-pain predictions are provided in Table A.3. The sample size is so small that the logistic transformation has been applied when the intervals are decided (Agresti 1996).

The difference between the success rate of pain prediction and no-pain prediction can be distinguished by using the ratio of odds from the two rows in the 2 × 2 contingency table (see Table A.2). An inference can be made for this difference through log transform of the ratio of odds of samples with a small size (Agresti 1996). The inference for the odds ratio of pain and no-pain prediction has been summarised in Table A.4 for variables p_{max} and σ_{max}.

The 95% confidence interval for odds ratio of success rate of pain and no pain prediction with p_{max} is (0.60, 0.769). This implies the success rate of pain prediction is less 23.1% than no-pain prediction at least. Whilst the 95% confidence interval for odds ratio of success rate of pain and no pain

TABLE A.4 Inference for the odds ratio of pain and no pain predictions.

Variable	**Odds ratio of sample**	**Asymptotic standard error of sample**	**95% Confidence interval for odds ratio with normal distribution**
p_{max}	−0.386	−0.063	(0.60, 0.769)
σ_{max}	0.373	−0.009	(1.478, 1.426)

prediction with σ_{max} is (1.478, 1.426), consequently, the success rate of pain prediction is higher by 3.52% than no-pain prediction. Therefore, the peak normal stress demonstrates a slightly better ability to indicate GB pain rather than GB no-pain.

REFERENCE

Agresti, A. 1996. *An Introduction to Categorical Data Analysis*, 16–25. New York: John Wiley & Sons, Inc.

APPENDIX B

Glossary of the Biliary System

Abdomen The part of the body that lies between the chest and the thigh and encloses the ureters, intestines, liver, anus, bladder, gallbladder and reproductive system outside the breast

Acalculous Not affected with, caused by or associated with gallstones

Aetiology A study deals with the causes, e.g., of a disorder

Ampulla A dilated segment in a tubular structure

Aneurysm A localised, blood-filled dilation of a blood vessel caused by disease or weakening of the vessel wall. Aneurysms most commonly occur in arteries at the base of the brain and in the aorta. The larger an aneurysm becomes, the more likely it is to burst. Aneurysms can be treated

Antrum A general term for a cavity or chamber which may have specific meaning in reference to certain organs or sites in the body

Bile A bitter, yellow or green alkaline fluid secreted by hepatocytes from the liver of most vertebrates

Biliary	Pertaining to the bile, to the bile ducts or to the gallbladder
Bilirubin	A pigment produced when the liver processes waste products. A high bilirubin level causes yellowing of the skin
Gallbladder	A pear-shaped organ that stores about 50 ml of bile until the body needs it for digestion
Cholangiograms	An x-ray examination of the bile ducts following administration of a radiopaque contrast medium
Cholelithiasis	The presence of gallstones in the gallbladder
Cholecystogram	The radiographic record of the gallbladder obtained by cholecystography
Cholecystokinin (CCK)	A peptide hormone of the gastrointestinal system responsible for stimulating the digestion of fat and protein
Cholecystography	A procedure that helps to diagnose gallstones. In the test, a special dye, called a contrast medium, is either injected into your body, or is taken as special pills (oral cholecystography). This contrast medium shows up the structure of the gallbladder and bile duct on X-ray
Cholecystectomy	The surgical removal of the gallbladder
Choledocholithiasis	The presence of a gallstone in the common bile duct. The stone may consist of bile pigments or calcium and cholesterol salts
Cholesterol	A sterol (a combination steroid and alcohol), a lipid found in the cell membranes of all body tissues, and is transported in the blood plasma of all animals
Chyme	The food which has been acted upon by stomach juices but has not yet been passed on into the intestines
Common bile duct	The duct that carries bile from the gallbladder and liver into the duodenum (the upper part of the small intestine). The common bile duct is formed by the junction of the cystic duct that comes from the gallbladder and the common hepatic duct that comes from the liver

Cystic duct	The short duct that joins the gall bladder to the common bile duct. It usually lies next to the cystic artery. It is of variable length. It contains a 'spiral valve'
Duodenum	In anatomy of the digestive system, the duodenum is a hollow jointed tube connecting the stomach to the jejunum
Enterohepatic	Of or involving the intestine and liver
Enzyme	A protein that speeds up a chemical reaction in a living organism. An enzyme acts as catalyst for specific chemical reactions, converting a specific set of reactants into specific products. Without enzymes, life as we know it would not exist
Epigastrium	The upper central region of the abdomen. It is located between the costal margins and the subcostal plane
Epithelium	A tissue composed of a layer of cells
Fasting	Fasting is primarily the act of willingly abstaining from some or all food, drink, or both, for a period of time
Fossa	A depression or hollow in a bone or the other parts of the body
Fundus	The bottom or base of any hollow organ; as, the fundus of the gallbladder; the fundus of the eye
Gallstone	Gallstones (choleliths) are crystalline bodies formed within the body by accretion or concretion of normal or abnormal bile components
Hepatic common duct	The duct formed by the junction of the right hepatic duct (which drains bile from the right half of the liver) and the left hepatic duct. The common hepatic duct then joins the cystic duct coming from the gallbladder to form the common bile duct
Hormone	A chemical messenger from one cell (or group of cells) to another. The function of hormones is to serve as a signal to the target cells; the action of hormones is determined by the pattern of secretion and the signal transduction of the receiving tissue

Jaundice	Jaundice means the yellow appearance of the skin and whites of the eyes that occurs when the blood contains an excess of the pigment called bilirubin
Lipid	Lipids can be broadly defined as any fat-soluble (hydrophobic) naturally-occurring molecules
Lumen	The cavity or channel within a tubular structure.
Mechanoreceptor	A sensory receptor that responds to mechanical pressure or distortion. There are four main types in the glabrous skin of humans: Pacinian corpuscles, Meissner's corpuscles, Merkel's discs and Ruffini corpuscles. There are also mechanoreceptors in the hairy skin, and the hair cells in the cochlea are the most sensitive mechanoreceptors in tranducing air pressure waves into sound
Mucin	A family of large, heavily glycosylated proteins
Mucous	The adjectival form of mucus, a slippery secretion of the lining of various membranes in the body
Mucosa	The moist tissue that lines some organs and body cavities throughout the body, including your nose, mouth, lungs, and digestive tract. Glands along the mucosa release mucus (a thick fluid)
Mucus	A slippery secretion of the lining of various membranes in the body
Oesophagus	Muscular tube through which food travels from the mouth to the stomach
Oral cholecystogram	An x-ray of the gallbladder, the x-ray is taken before the gallbladder releases bile
Pancreas	A gland organ in the digestive and endocrine systems of vertebrates
Pancreatitis	An inflammation of the pancreas
Pigment	A substance that gives colour to tissue. Pigments are responsible for the colour of skin, eyes, and hair.

Physiology The study of how living organisms function, experimentally-based science. It is distinguished from other biological sciences by its emphasis on animals, how the tissues and organs interact and how the parts are integrated to make up the whole. Physiology is a core science for medicine and other biomedical disciplines

Postprandial Referring to the time after any meal

Rectum The last portion of the large intestine (colon) that communicates with the sigmoid colon above and the anus below

Rheology The study of the deformation and flow of matter under the influence of an applied stress

Secretin A peptide hormone produced in the S cells of the duodenum in the crypts of Lieberkühn. Its primary effect is to regulate the pH of the duodenal contents via the control of gastric acid secretion and buffering with bicarbonate. It was the first hormone to be discovered.

Scintigraphy A diagnostic test in which a two-dimensional picture of a body radiation source is obtained through the use of radioisotopes. For example, scintigraphy of the biliary system (cholescintigraphy) is done to diagnose obstruction of the bile ducts by a gallstone, a tumor, or another problem; disease of the gallbladder; and bile leaks

Sphincter of Oddi A sphincter muscle located at the surface of the duodenum. It controls secretions from the liver, pancreas, and gallbladder into the duodenum of the small intestine

Sonography An ultrasound-based diagnostic imaging technique used to visualise muscles and internal organs, their size, structures and possible pathologies or lesions

Symptom A symptom may loosely be said to be a physical condition which indicates a particular illness or disorder

Ultrasonography = sonography

Visceral Referring to the viscera, the internal organs of the body, specifically those within the chest (as the heart or lungs) or abdomen (as the liver, pancreas or intestines)

The definitions of the above glossary terms are adopted from those on the web sites as follows: http://en.wikipedia.org/, http://www.cancerindex.org/glossary.htm, www.nlm.nih.gov/, and so on.

Index

N

O

P

Q

R

S

T

U

V

W

Y

Z

Printed in the United States
by Baker & Taylor Publisher Services

Printed in the United States
by Baker & Taylor Publisher Services